10th ADVANCES IN RELIABILITY
TECHNOLOGY SYMPOSIUM

10th
anniversary

Proceedings of the 10th Advances in Reliability Technology Symposium held at the University of Bradford, Bradford, UK, 6–8 April 1988

Organised by

> The National Centre of Systems Reliability and the University of Bradford

In association with

> UMIST and the University of Liverpool

Support Organisations

> European Safety & Reliability Association
> Institute of Quality Assurance
> Institution of Electrical Engineers
> Safety and Reliability Society

Organising Committee

A. Z. Keller (Chairman)	University of Bradford
R. N. Allan	UMIST
G. M. Ballard	National Centre of Systems Reliability
R. F. De la Mare	University of Bradford
P. Martin	University of Liverpool
G. P. Libberton (Secretary)	National Centre of Systems Reliability
R. Campbell	National Centre of Systems Reliability

Session Chairmen

W. Vinck	Formerly of The Directorate General of Science, Research and Development, (DG XII), CEC, Belgium
B. Stevens	National Centre of Systems Reliability, UK
R. F. De la Mare	University of Bradford, UK
G. M. Ballard	National Centre of Systems Reliability, UK
A. Veevers	University of Liverpool, UK
R. N. Allan	UMIST, UK
B. K. Daniels	National Computing Centre, UK
R. Billinton	University of Saskatchewan, Canada
P. Bockholts	TNO, Netherlands
A. Z. Keller	University of Bradford, UK

10th ADVANCES IN RELIABILITY TECHNOLOGY SYMPOSIUM

Edited by

G. P. LIBBERTON

National Centre of Systems Reliability,
Culcheth, Warrington, Cheshire, UK

ELSEVIER APPLIED SCIENCE
LONDON and NEW YORK

ELSEVIER APPLIED SCIENCE PUBLISHERS LTD
Crown House, Linton Road, Barking, Essex IG11 8JU, England

Sole Distributor in the USA and Canada
ELSEVIER SCIENCE PUBLISHING CO., INC.
52 Vanderbilt Avenue, New York, NY 10017, USA

WITH 74 TABLES AND 102 ILLUSTRATIONS

© ELSEVIER APPLIED SCIENCE PUBLISHERS LTD 1988

© CENTRAL ELECTRICITY GENERATING BOARD 1988—pp. 103–113, 252–262

© CROWN COPYRIGHT 1988—pp. 223–235

© IEEE TRANSACTIONS IN RELIABILITY 1987—pp. 378–386. THIS PAPER WAS
PUBLISHED IN VOL. R-36, NO. 4 OCTOBER 1987

© UNITED KINGDOM ATOMIC ENERGY AUTHORITY 1988—pp. 189–196, 303–315
Softcover reprint of the hardcover 1st edition 1988

British Library Cataloguing in Publication Data

Advances in Reliability Technology *(10th: 1988: University of Bradford)*
    10th Advances in reliability technology symposium.
    1. Engineering equipment. Reliability.
    I. Title II. Libberton, G. P.
    620'.00452

    ISBN-13: 978-94-010-7103-1   e-ISBN-13: 978-94-009-1355-4
    DOI: 10.1007/ 978-94-009-1355-4

Library of Congress CIP data applied for

FOREWORD

On behalf of the Organising Committee of the 10th ARTS I would like to welcome all delegates, chairmen of sessions and authors of papers. I extend a particular welcome to new delegates from Europe and other overseas countries.

Little did we who were involved in the first symposia realise the degree and standard to which the symposia would develop and the worldwide and professional interest and recognition that they would gain. It is particularly gratifying to be writing this foreword for the celebratory issue of the proceedings of the 10th Symposium.

The Organising Committee particularly wish to thank Mr W Vinck who, until recently, was Chairman of the European Safety and Reliability Association (ESRA), for accepting our invitation to preside over the symposium and also to thank ESRA for supporting the symposium. Our appreciation is also expressed to Mr G Hensly of British Nuclear Fuels for being our after-dinner speaker. Mr Hensley, from the earliest days of our symposia, has always given his unstinting support.

I should particularly wish to acknowledge the very detailed and hard work (often unseen) of the members of staff of the National Centre for Systems Reliability with regard to the present and past symposia. I should also like to thank the University of Liverpool, the University of Manchester Institute of Technology and my own University for their consistent supportive attitude to the symposia.

Finally, I should like to thank all authors of papers and delegates who, in the final analysis, contribute the most important ingredients of a symposium.

Dr A Z Keller
Organising Committee Chairman

vii

C O N T E N T S

L I S T   O F   C O N T R I B U T O R S

M H Abed
Postgraduate School of Technology and Management Science, University of Bradford, Bradford, BK7 1DP, UK

R N Allan
Department of Electrical Engineering and Electronics, UMIST, Manchester, M60 1QD, UK

J I Ansel
Department of Management Systems and Science, School of Management, University of Hull, Hull, HU6 7RX, UK

J A Astley
Applied Psychology Division, Aston University, Birmingham, B4 7ET, UK

Y L Bakouros
University of Bradford Management Centre, Bradford, BD7 1DP, UK

G M Ballard
National Centre of Systems Reliability, UK Atomic Energy Authority, Safety and Reliability Directorate, Wigshaw Lane, Culcheth, Warrington, Cheshire, WA3 4NE, UK

A C Barrell
Health and Safety Executive, St Annes House, Stanley Precinct, Bootle, Merseyside, L20 3MS, UK

D M Barry
Department of Electrical Engineering, Lakehead University, Thunder Bay, Ontario, P7B 5E1, Canada

A Bendell
Department of Mathematics, Statistics and Operational Research, Trent Polytechnic, Burton Street, Nottingham, NG1 4BU, UK

R Billinton
Power System Research Group, University of Saskatchewan, Saskatoon, Saskatchewan, S7N OWO, Canada

P Bockholts
Department of Industrial Safety, Netherlands Organisation for Applied Scientific Research TNO, P O Box 342, AH 7300 Apeldoorn, The Netherlands

T B Boffey
Department of Statistics and Computational Mathematics, University of Liverpool, P O Box 147, Liverpool, L69 3BX, UK

D S Campbell
Electronic Component Technology Group, Department of Electronic and Electrical Engineering, University of Technology, Loughborough, Leicestershire, LE11 3TU, UK

R F De la Mare
University of Bradford Management Centre, Bradford, BD7 1DP, UK

J Disney
Department of Mathematics, Statistics and Operational Research, Trent
Polytechnic, Burton Street, Nottingham, NG1 4BU, UK

F A El-Sheikhi
Ras Lanuf Oil and Gas Processing Company Incorporated, P O Box 296,
Benghazi, Libya

M P Fairclough
National Centre of Systems Reliability, UKAEA, Wigshaw Lane, Culcheth,
Warrington, Cheshire, WA3 4NE, UK

A M Games
National Centre of Systems Reliability, UK Atomic Energy Authority,
Safety and Reliability Directorate, Wigshaw Lane, Culcheth, Warrington,
Cheshire, WA3 4NE, UK

J A Hayes
Electronic Component Technology Group, Department of Electronic and
Electrical Engineering, University of Technology, Loughborough,
Leicestershire, LE11 3TU, UK

R P Hughes
Central Electricity Generating Board, Berkeley Nuclear Laboratories,
Berkeley, Gloucestershire, GL13 9PB, UK

R A Humphreys
Rolls-Royce and Associates Limited, P O Box 31, Derby, DE2 8BJ, UK

A M Irving
Rolls-Royce and Associates Limited, P O Box 31, Derby, DE2 8BY, UK

Y A Jerbril
Department of Electrical Engineering and Electronics, UMIST, Manchester,
M60 1QD, UK

J C P Kam
NDE Centre, Department of Mechanical Engineering, University College
London, Torrington Place, London, WC1E 7JE, UK

C Kara-Zaitri
Postgraduate School of Industrial Technology, University of Bradford,
Bradford, BD7 1DP, UK

A Z Keller
Postgraduate School of Industrial Technology, University of Bradford,
Bradford, BD7 1DP, UK

J Knezevic
Department of Engineering Science, University of Exeter, Exeter, EX4
4QF, UK

L J B Koehorst
Department of Industrial Safety, Netherlands Organisation for Applied
Scientific Research TNO, P O Box 342, AH 7300 Apeldoorn, The Netherlands

C Leach
Software Production Engineering, Digital Equipment Corporation, Worton Grange Industrial Estate, Imperial Way, Readng, Berkshire, RG2 OTU

H Lee
Department of Electrical Engineering, Lakehead University, Thunder Bay, Ontario, P7B 5El, Canada

P T Manning
Central Electricity Generating Board, 15 Newgate Street, London, EC1A 7AX, UK

J M Marshall
Electronic Component Technology Group, Department of Electronic and Electrical Engineering, University of Technology, Loughborough, Leicestershire, LE11 3TU, UK

P Martin
Department of Mechanical Engineering, University of Liverpool, Liverpool, L69 3BX, UK

J B McDonald
Brigham Young University, 212 TMCB, Provo, Utah 84606, USA

M Meniconi
Department of Electrical Engineering, Lakehead University, Thunder Bay, Ontario, P7B 5El, Canada

N Nemat-Bakhsh
University of Bradford, Bradford, BD7 1DP, UK

M J Newby
Postgraduate School of Industrial Technology and Postgraduate School of Mathematical Sciences, University of Bradford, Bradford, BD7 1DP, UK

M Patwardhan
Power System Research Group, University of Saskatchewan, Saskatoon, Saskatchewan, S7N OWO, Canada

M J Phillips
Department of Mathematics, University of Leicester, Leicester, LE1 7RH, UK

I S Qamber
Postgraduate School of Industrial Technology, University of Bradford, Bradford, BD7 1DP, UK

D O Richards
Brigham Young University, 212 TMCB, Provo, Utah 84606, USA

J Roman
Department of Electrical Engineering and Electronics, UMIST, Manchester, M60 1QD, UK

A Saboury
Department of Electrical Engineering and Electronics, UMIST, Manchester, M60 1QD, UK

xiv

A S Sohal
Postgraduate School of Technology and Management Science, University of
Bradford, Bradford, BK7 1DP, UK

J F J Van Steen
Department of Industrial Safety, Netherlands Organisation for Applied
Scientific Research TNO, P O Box 342, 7300 AH Apeldoorn, The Netherlands

A Veevers
Department of Statistics and Computational Mathematics, University of
Liverpool, P O Box 147, Liverpool, L69 3BX, UK

J T Webb
Rex, Thompson & Partners, 'Newnhams', West Street, Farnham, Surrey, GU9
7EQ, UK

B A Wichmann
Division of Information Technology and Computing, National Physical
Laboratory, Teddington, Middlesex, TW11 0LW, UK

J C Williams
Sizewell 'B' Project Management Team, Central Electricity Generating
Board, Knutsford, Cheshire, WA16 8QG, UK

S P Whalley
Lihou Loss Prevention Services Ltd, Grays Court, 1 Nursery Road,
Edgbaston, Birmingham, B15 3JX, UK

D F Yates
Department of Computer Science, University of Liverpool, P O Box 147,
Liverpool, L69 3BX, UK

THE IMPORTANCE OF RELIABILITY IN HEALTH AND SAFETY

by A C Barrell, Director of Technology,
Health and Safety Executive, UK

Introduction

The need for reliable plant and controls for the safe operation of <u>nuclear power and the chemical industry</u> is well known and has been comprehensively covered in the literature. This paper therefore considers other industrial processes which although possibly attracting less public attention, nevertheless produce hazards which must be kept under proper control.

I have chosen examples from four different fields of interest, namely control systems, lifting equipment, machinery guarding and the use of industrial radiation equipment. In all four fields the reliability of electronic and mechanical components and systems becomes of increasing importance with the steady move away from <u>manual</u> operation of plant and processes.

Legal Requirements

Before going further it is perhaps appropriate for me to remind you of the basis for HSE's interests in any sector of workplace activity. Clearly the overall aim is to prevent accidents and promote an environment conducive to healthy working, but this has to be done with proper regard to the costs of running the enterprise. Virtually any system can be made exceedingly safe or healthy if we are prepared to spend enough time and resources to make it so. But such an approach cannot often be justified in the real world in which we live and work. This reality is embodied in the basic legislation, namely the Health and Safety at Work Act 1974, by restricting the duty on employers to that which is 'reasonably practicable'.

Specifically Section 2(2)(a) of the Act requires every employer to ensure,

> "the provision and maintenance of plant and systems of work that are, so far as is reasonably practicable, safe and without risks to health."

Clearly, this duty will encompass the concept of reliability of the plant and systems. Although the umbrella legal provision is the HSW Act, it is also set out more specifically in what are known as the 'relevant statutory provisions' - these being the Factories Acts, Explosives Acts, Mines Acts and so on.

Industrial Standards

Another HSE interest in reliability derives from its work in preparing standards and guidance at national and international level. This is particularly significant for the specialist inspectors in my own Division, who are often representing the UK whilst negotiating in Europe on CEN standards and various EC Directives. To be effective, these inspectors need to have a good awareness of the 'state of the art' in industrial plant and systems, and a confidence that the various provisions therein are based on realistic levels of reliability.

This confidence will be built up from contacts with industry, knowledge of accident and incident rates, and performance data on safety critical components. I stress the latter point because although we wish to be aware of the reliability of plant and systems in a general sense, we really do need the co-operation of industry in establishing confidence in components which are used in safety critical situations.

Perhaps I could use this opportunity to appeal for your co-operation in this area, as I hope you will see it is to our mutual advantage for my specialist inspectors to be conversant with, and confident of, various plant and systems, when they are looking after UK interests in Europe.

Control Systems

The technology of control systems is developing rapidly, particularly in the field of programmable electronics. In this respect many of you will be aware of the launch in June 1987 of our guidance note (1) dealing with the use of PES in safety related systems. This important document sets out the essential ground rules for the use of PES in applications which bear on safety, and is seen as providing a framework within which industry can develop its own more specific guidelines for various sector uses.

The range of control systems which have to be considered by HSE includes hydraulic, pneumatic, hardwired electrical as well as PES, and there are indications that the use of fibre optic systems will play an increasingly important role in future. The complexity of the systems will vary from simple interlocking arrangements on machinery guards to multiple safety linkages which are PES based, and which may form part of a high integrity shut-down system on a chemical plant.

However, despite the range and complexity of these systems, there are several common requirements, and one of these is reliability. Concern about reliability will focus on both random hardware failures and systematic failures. In the first case failures will be a function of initial quality, correct installation, working environment, maintenance and workload, and there are various well established methods (eg redundancy and voting techniques) to ensure that critical safety parameters can be met. In the case of systematic failures, we are talking about problems which may be in the original specification, the subsequent design or in the case of PES, in the software. In this area the reliability of individual elements might be first class, but the system may conceal an error, which will only manifest itself if there is a certain combination of input parameters.

So what are HSE's needs, bearing in mind its short term role of inspection, enforcement and forensic investigation and its long term role in developing standards and guidance? In the hardware area, there is a need for accurate reliability (or conversely failure) data on those particular components which play an important part in the safety protection provisions of a machine or plant. Also, there is a need for similar data on standard component sequences eg specific modules. In the software area, there is again a call for data on 'system' failures so that appropriate levels of confidence can be established in total control and protection packages.

Professionals both in industry and in HSE find that they need to analyse systems in order to establish risk levels and so put various possible hazards into context. Very often this can be done in a qualitative way to estimate the general magnitude of risk, but it may be necessary to develop this into a quantitative analysis in order to give a clearer understanding of critical areas of risk. In the latter case HSE inspectors must do this to ensure that they provide the correct advice to industry in sensitive safety areas, and that they themselves make the right decisions if HSE has licensing or approval responsibilities,. It will be recognised that equipment and system manufacturers will also want this sort of information to refine their own products.

Our need then is access to good data on system and component reliability, not for a heavy handed policing role but to ensure that the advice we give and the inputs we make into standards setting (often with regard to UK commercial interests) is sound. In this way, we can then encourage the application of new technology within a realistic framework whose aim is to achieve acceptable levels of safety.

Lifting Equipment

The range of lifting equipment inspected by HSE covers all types of cranes, automatic warehousing machinery and fork-lift trucks, and recently we have had to deal with the problem of excavators used as cranes. As you will see this is a very wide range and the complexity varies from a simple hand hoist to the sophisticated pile cap cranes used for changing fuel rods of nuclear power stations, which have a placement accuracy of $\pm$ 1mm.

In the past, designers of lifting machinery have been very conservative and this has resulted in machines which have been able to withstand a fair degree of use and abuse without failure. This meant that failures of machinery due to initial design problems were relatively few.

However, in common with other industry, designers of lifting machines are now under pressure to produce more economical products, and this together with the demand for ever larger capacities has necessitated a move away from the conservatism of the past. Within ten years we have seen the largest mobile cranes grow from 100 tonnes to over 1000 tonnes capacity and the potential for disaster grow accordingly. It is therefore of paramount importance that designers do not rest on their past laurels, but pay particular attention to reliability in the future.

Reliability is essential in the power drives of lifting machinery, especially where gearbox reduction is used. Such designs ensure a slow revving high torque output which is essential for close operational control. Braking systems if applied at the output end would be big and bulky, hence to reduce size and weight the brakes are put at the input end where the transmission is low torque high speed. Thus on all forms of lifts or cranes we may have a system where the brake is not connected directly to the load. It follows that the highest level of reliability is needed in the transmission system between the brake and the load.

New crane designs are based on a detailed prediction of the crane's pattern of use during its working life. Reliable data is therefore vital if crane components are to be designed to fulfil their role during the predicted life-cycle of the crane. Unreliable data would open up the possibility of premature failure from fatigue.

The changes in design concept have meant significant differences in the way in which the machines can be used , and even greater need for proper maintenance. However, the achievement of acceptable reliability not only places demands on the designers, but also on the users of lifting machinery. Indeed it is at the user end that problems have arisen in the past leading to the majority of lifting accidents.

There are three ways in which the designer can help the user:

(a)   By keeping the design as simple as possible. There are many examples of cranes having fallen down because there were too many unnecessary complications for the user and driver to understand.

(b)   By providing clear and unambigious operating and maintenance manuals for the machine, which clearly recognise the typical conditions of use. (Picking up test weights in the manufacturer's factory is very different from lifting a load on a congested construction site).

(c)   By providing the user with as many aids as will realistically assist in the operation of the machine. (I stress realistically because we do not want the drivers control cabin looking like the flight deck of Concorde).

Controls and instruments must also be very reliable in the environment in which they are to operate. We have recently investigated a system which is supposed to warn the driver of a tower crane when he is moving into the path of an adjacent crane. The basic system has proved most unreliable, made worse by construction plant forever damaging the communication cables between the cranes.

We have found from past experience that users decisions are crucial to the reliability of lifting machinery. All too often we find the wrong machine has been selected for a particular job (a recent example of this was when a large mobile crane collapsed over a bus station); or the job has not been correctly organised; or the equipment has not been properly repaired or maintained.

We endeavour to influence the industry in day to day matters by guidance, inspection or enforcement. Longer term we inject significant resources into the development of standards on design and operation which also cover the essential requirements of testing and examination. Our quest for reliability therefore depends on ensuring the equipment is designed and built correctly and importantly ensuring the equipment is managed and used in accordance with manufacturers recommendations.

Machinery guarding

The history of machines in the workplace contains many examples of horrific accidents where people have been caught in the mechanism in some way. For many years it was virtually taken to be a hazard of the job, and the accident statistics tell their own story of the early cotton mills or engineering machine shops. This history led to the requirement in Section 14 of the Factories Act that moving machinery must be 'securely fenced'.

However, this is an absolute standard and somewhat in conflict with the HSW Act requirement of "reasonably practicable". In practice our factory inspectors have needed to become more pragmatic in their approach to guarding, and it is likely that the forthcoming European Machinery Directive will further develop the reasonably practicable approach. This is significant for modern machines such as automatic assembly tracks, AGVs or robots where the 'secure fence' concept is somewhat difficult to interpret, indeed in some cases the machine might approach the operator rather than the reverse!

To move forward sensibly in this area of safety protection, one must balance hazards and consequences via the 'risk' approach. In other words if we do not securely fence, which may in any case be somewhat impracticable, what residual risk is one prepared to tolerate in any situation? In considering 'risk', clearly reliability forms an important input.

The current protection position is set out in BS 5304 "Safeguarding of Machinery" which was published in 1975, and which largely reflected early HSE experience with machine safety. A significant feature of this Standard relates to machine guard interlocking, where uncovenanted movements are prevented whilst the guards are open. Several years after publication of the Standard, HSE asked the UKAEA Safety and Reliability Directorate to investigate the relative reliabilities of the various interlocking methods which it covered. The results showed that the power interlock, previously hailed as the safest method, was less reliable than a cross-monitored control interlock. A revision (2) of BS 5304 will recognise the significance of the SRD work, and the section on interlocking has been re-written.

Despite these developments, interlocking probably remains a key area for further reliability study. This is not just in the narrow context of guarding, but in the wider role that interlocking plays throughout the field of machinery safety. For example, only recently, accident experience in the explosives sector has cast doubts on the overall reliability of the well known 'captive-key' method of safety protection. Photo-electric systems have been around for more than 30 years, but modern installations built to BS 6491 are far superior in terms of reliability than earlier versions. However, the question remains, are they now reliable enough?

A more recent development is in the use of PES for controlling machines, and protecting the work people,

Although there is great potential in this area, our inspectors are particularly cautious where safety protection circuits are routed through the same electronics as the machine operating controls. This is largely because validation, especially of the software, is both lengthy and difficult. There is room for significant advance in this area, particularly by using the scope of PES to improve safety knowledge via condition monitoring.

These two latter examples show how sophisticated methods of safeguarding have become, and this perhaps highlights another problem in safety reliability in the workplace. From the safety point of view, failure of a protective component which leads to a safe condition is not a problem. To the machine operator, or machine owner however, it will mean down-time and possible loss of earnings. There is always a temptation in these sort of situations to corrupt the control or protection system, and there is no doubt that this practice contributes to many accidents.

To avoid these possibilities, improved reliability can provide essential confidence. However, as with my earlier comments on control systems, improving reliability needs feedback of data on failures. This will enable the designers to build better systems, from more suitable components. Inevitably the whole process is iterative in nature, but the basic need from the point of view of the designer, the employer and the inspector is reliability data.

Control of Ionising Materials

The traditional area of radiation control falls within the nuclear sector, and as I said at the beginning, I am not covering these issues in this paper. Instead, I would like to conclude by mentioning two applications where steady improvements in reliability are enabling safety to be achieved with significant savings in installation costs.

(a)   Real time radiography - X-ray machines similar to

those used for baggage inspection at airports are being developed for non destructive testing (NDT) in industry. These real time radiography systems enable better quality assurance to be achieved in many areas of engineering inspection.

Real time radiography (RTR) replaces conventional film, with digitally enhanced images viewed on a high resolution monitor. The images are stored on a laser disc which allows compact archiving and rapid recall of information. A trained operator using the system is able to exercise almost instantaneous quality control. The all important radiation protection is achieved by the installation of engineering controls at the design stage of the equipment. However, its success, and hence safety, depend upon high component reliability and careful maintenance.

Typical applications and advantages of this system are to be found in the inspection of pressurised plant or pipeline systems. In the former case conventional NDT techniques can only give limited coverage due to the high costs associated with inspection time and radiation protection methods. In contrast RTR gives controlled and instantaneous information, thus allowing wider and more detailed monitoring (eg for corrosion) to be undertaken. In pipeline work, it has been estimated that cost savings of 30-40% can be achieved due mainly to the increased speed of weld inspection.

Our inspectors are keen to encourage these developments, and actively participate with companies in the field in ensuring that the various techniques are adequately safe. Clearly, carefully controlled enhancements of reliability in one field are enabling significant improvements of reliability (via good QA) in others.

(b)   High power industrial lasers - Laser energy as with other radiation systems requires very careful control or heavy shielding if it is not to do damage to persons and property. However, high reliability, infra red scanning systems are being developed for first line detection of errant laser radiation from the larger industrial installations, as an alternative to 'fortress' style enclosure.

My specialist inspectors are very much involved with the developments of automated machinery and flexible manufacturing systems to ensure that safety questions are adequately covered. One area of considerable potential is in applications of laser technology involving increased power and flexibility of output. For example, lasers with powers in excess of 20 kilowatts are proposed to be mounted on robots, and extended beam delivery systems are being developed which will transport the laser energy to remote workstations where it can be used to cut, drill, weld or heat treat a wide range of workpieces.

As I am sure you will agree, we are now in a different age of machinery guarding and worker protection.

A number of options exist for securing safety and improved performance at these advanced installations. We could adopt a reactionary approach and insist on massive protective enclosures with strictly controlled access, which in all but the most elementary systems would be inflexible and no doubt costly. Or, we can adopt our more usual enlightened approach(!) and seek improvement in quality control and reliability. This should be applied first in relation to the laser radiation output itself (eg beam profile, stability, alignment etc) and second, in improving the key components of the installation and the back-up safety systems.

Once performance of the system has been optimised, the release of errant radiation can be further minimised by continuous monitoring with a high reliability detection system. In this way the normal requirements for massive enclosure might be relaxed, without any diminution of safety standards, and thus expensive down-time for maintenance and repair should be drastically reduced.

Conclusion

I hope I have shown, by way of the various examples in this paper, something of the role of HSE inspectors in reliability technology and the need for better reliability information.

Certainly, many front line inspectors are engaged in day to day guidance, inspection and enforcement of existing standards and legal requirements. However other inspectors, many of them specialists in my own Division, are working in the background on new standards and technologies to ensure that the future requirements of health and safety are properly and responsibly met.

The importance of reliability in process and systems design is that it offers opportunities for improved health and safety whilst keeping costs down and providing flexible solutions to new demands.

REFERENCES

1. Health and Safety Executive publication - Programmable Electronic Systems in Safety Related Applications : General Technical Guidelines.1987, ISBN 0 11 883906 3.

2. BS 5304 : 1988, Safety of Machinery, to be published in April 1988.

EXPERT OPINION IN PROBABILISTIC SAFETY ASSESSMENT

Jacques F.J. Van Steen
Department of Industrial Safety
Netherlands Organization for Applied Scientific Research TNO
P.O. Box 342, 7300 AH Apeldoorn
Netherlands

ABSTRACT

Probabilistic Safety Assessment (PSA) is concerned with the quantification of the possible damages from hazardous materials releases and of
the corresponding probabilities. Although expert judgment is expected to
contribute substantially to improving the quality of probability assessments within PSA, it has become clear in recent years that a number of
specific problems are associated with the use of expert opinions. Thus,
the need exists for formalized procedures for expert opinion use. This
paper is based on a research project that deals with the development of
such procedures. To this end, three substantially different models for
eliciting and combining expert opinions have been formulated: (a) a
"classical" model, based on weighted averaging, (b) a Bayesian model, and
(c) a psychological scaling model. These models have been made operational and are being tested and evaluated in experiments on real-world
problems. The paper presents a discussion of expert opinion use in PSA
and of the models mentioned above, and gives a first look at the
experiments.

INTRODUCTION

Probabilistic Safety Assessment (PSA), also called risk analysis,
is concerned with the quantification of the possible damages from
hazardous materials releases and of the corresponding probabilities.
In practice, the PSA methodology consists of a number of steps:
- description of the system under consideration,
- identification of undesirable events,
- calculation of the effects of the release of hazardous materials, and
  translation of the effects into fatalities and injuries and into damage
  to buildings and installations ("damage calculations"),
- quantification of the probabilities with which the damages calculated
  can occur ("probability assessment").

A full-scale risk analysis is characterized by large uncertainties. Major uncertainties include modelling uncertainties and uncertainties in the data. Three types of modelling uncertainty are distinguished: completeness, dependent failures and human error. A prominent example of uncertainty in the data is the area of the assessment of failure rates of mechanical components, such as valves, vessels and pipes.

Current approaches to failure rate assessment are almost exclusively based on failure data as available from generic data sources, such as data bases. However, these data bases are characterized by several shortcomings (1):
- a great lack of component failure data,
- incomplete description of the failure process (failure causes and failure modes),
- incomplete description of the system in question, especially concerning operation and maintenance.
These shortcomings apply in particular in case of failure data being required for failure modes that are defined as "catastrophic failures". Pipe rupture, a safety valve remaining closed and vessel rupture are examples of this failure mode.

This situation has two consequences. Firstly, poor data cause probability assessments to be rather unreliable, thus limiting the usefulness of risk analysis results. Secondly, the analyst must resort to expert opinions for the assessment of failure rates.

This paper deals with the use of expert opinion in PSA. The organization of the paper is as follows. First some early experiences with expert opinion in risk analysis are discussed; one of the conclusions will be that the need exists for formalized procedures for expert opinion use. Then a research project that deals with the development of such procedures is briefly described. The remainder of the paper is concerned with the three major parts of this research effort: a literature study, a model development phase and an experimental program. From the literature study, two subjects are discussed: formal probabilistic tools, and recent experiences and developments with expert opinion in risk analysis. Next, a description is presented of the models that have been made operational and a first look is given at the experimental program.

EXPERT OPINION IN PSA: EARLY EXPERIENCES

The Reactor Safety Study (RSS) (2) can be considered as the first "modern" probabilistic safety analysis. There was a twofold use of expert opinion in this study (3): first, failure rates (point estimates) were collected from various experts for many different components and second, for each component the RSS analysts assessed a probability distribution of the failure rate, using the estimates provided by the experts. The initial point estimates showed an enormous spread. For example: the estimates of the failure rate per section-hour of high quality steel pipe of diameter ≥ 3 inches range from 5E-6 to 1E-10 (thirteen responses) (4).

A similar picture is reported by Okrent (5), who asked seven experts via a questionnaire to give their estimates for the probability of occurrence per year of rare earthquakes in eleven areas in the USA.

For each area substantial differences existed between the predictions of the experts.

This brief discussion of early experiences with expert opinion in PSA leads to two conclusions:
(a) Point estimates are insufficient in a situation, where apparently the experts are quite uncertain about the value of a quantity. Thus, experts should express their uncertainty as well (e.g. in terms of probability distributions) instead of just giving point estimates.
(b) The need exists for formalized procedures for expert opinion use, in order to be able to answer such questions as: how do we derive probabilistic predictions (elicitation), how do we choose experts and can we discriminate between "better" and "worse" experts (evaluation), how do we combine the opinions of different experts (combination)? These questions are addressed in an ongoing research project which is briefly described in the next section.

DESCRIPTION OF RESEARCH PROJECT

The Ministry of Housing, Physical Planning and the Environment of the Dutch Government, in recognizing the desirability of formalized procedures for expert opinion use, is sponsoring a two year research project into the use of expert opinion in probabilistic safety assessment. This project was initiated on September 1, 1986. The project coordination rests with the Safety Science Group of the Delft University of Technology, and the research is carried out jointly by the Safety Science Group, the Department of Statistics, Stochastics and Operations Research, both at the Delft University of Technology, and by the Department of Industrial Safety of the Netherlands Organization for Applied Scientific Research TNO.

The overall project goals are:
- develop operational models for the use of expert opinion in probabilistic safety assessment,
- obtain and evaluate operational experience with these models, and
- prepare protocols for the practicing risk analyst regarding the use of expert opinion in probabilistic safety assessment.

The project is divided into three major phases. The first phase of the project involved an extensive literature study into the state-of-the-art concerning the use of expert opinion. The results of this literature study are set down in a report (6). Three models for using expert opinion have been selected which appear promising and will be investigated further. The second phase concerns making the selected models fully operational; this phase also involves developing software. In the third phase of the project the models will be implemented and operational experience will be evaluated. On the basis of this evaluation, protocols for the practicing risk analyst regarding the use of expert opinion will be prepared.

The next two sections are devoted to two subjects from the literature study: formal probabilistic tools, and recent experiences and developments with expert opinion in risk analysis.

FORMAL PROBABILISTIC TOOLS

In this paper expression of uncertainty is confined to the use of subjective probabilities, or stated differently: to a probabilistic representation of degree of belief. What does the area of probabilistic predictions offer? It offers tools and techniques for:
- the assessment (encoding) of probabilities,
- the evaluation of probability assessments, and
- the combination of probabilities.
Each of these subjects is briefly discussed. More extensive treatments are presented elsewhere (6). Additional references are given where appropriate.

Encoding techniques

A variety of techniques exists for encoding subjective probabilities; several reviews have been published (7-13). Since these methods differ from each other in a number of aspects, the need exists for a taxonomy. Two major dimensions are suggested:
- response mode: direct versus indirect, and
- presence of a reward structure.
Consequently, four categories of encoding techniques are distinguished.

Direct encoding without reward structure implies that the assessor is asked to give probabilities associated with given values (P-method), values associated with given probability statements (V-method, usually called fractile, bisection or interval method) or a probability distribution: combinations of probabilities and values (PV-method). *Indirect encoding without reward structure* implies in most cases that probabilities or probability distributions are derived from comparisons which the assessor is asked to make. A good example of such a technique is the method of paired comparisons: objects are presented in pairs to one or more experts, who then are asked to indicate their preference for one of the objects concerning a particular attribute; processing of responses is done by means of methods of psychological scaling.

The presence of a certain reward structure is intended to encourage the assessor to give his/her true opinion. *Direct encoding with reward structure* is performed by so-called scoring methods. These involve the calculation of a score according to a scoring rule, which is a function of the assessed distribution and of the event which eventually occurs (7). A scoring rule is called (strictly) proper if the assessor's expected score is maximized (only) if the assessed distribution equals the assessor's true distribution. *Indirect encoding with reward structure* involves using bets (lotteries, gambles, wagers) which the assessor are offered. Probabilities are derived from indifferences between lotteries.

Evaluation of probability assessments

Several criteria exist for evaluating probability assessments. The two most important criteria are calibration and resolution. Calibration measures the degree of correspondence with reality. With discrete events, for example, an assessor is said to be perfectly calibrated if the proportions of events which occur equal the assessed probabilities over a large number of assessments. The second criterion, resolution, is a measure of the assessor's ability to move away from the average proportion correct. These and other criteria can be used to investigate the quality of probability assessments. Many experimental studies on this issue have been reported in the literature, mostly with non-experts as subjects. The general finding of these studies is that probability assessors are badly calibrated and show a significant degree of overconfidence (which is called the overconfidence bias: the uncertainty bands are too narrow) (14). A second bias that often occurs is the location bias: the estimates are shifted to higher or lower values. Much research has been done into the underlying processes leading to biased assessments. Three heuristics, as these processes have been called, are distinguished: availability, anchoring and adjustment, and representativeness (15).

Two problems have been identified concerning most of the experiments which have been performed to study the quality of probability assessments. The first problem is that many of these experiments took place in a laboratory situation which has been criticized as being artificial (16). The second problem is that in most cases the subjects were non-experts. The question then is whether experts are better probability assessors than non-experts. Although experts generally appear to be better calibrated than non-experts, this is not always the case. An example of extensively proven good calibration is found in the area of weather forecasting (17). The evidence on the quality of expert opinion in PSA is less encouraging (this is discussed in the next section).

Two more issues are the comparison of encoding techniques and the effect of training. A large number of studies have compared alternative techniques for encoding probabilities (8, 9, 11, 18-20). As different results have been reported, it appears that few general conclusions can be drawn concerning techniques. A similar picture arises from studies into the effect of training (21-24).

Combination of probabilities

Again, a variety of approaches exists for combining subjective probabilities; see (25) for an excellent recent survey. Four major categories are distinguished:
- Bayesian models,
- weighted averaging,
- psychological scaling models, and
- interaction techniques.

Bayesian models are characterized by the requirement that the analyst (or decision maker) provides a prior probability distribution which is updated with expert probability assessments by applying Bayes' theorem. *Weighted averaging* involves assigning weights to the experts and applying these weights to the assessments given by the experts. Several models have been suggested, which mainly differ in the way of defining weights. *Psychological scaling* has already been mentioned above in the subsection "encoding techniques". Several models are available, which differ in the underlying assumptions. *Interaction techniques*, of which the Delphi method is the best known example, are characterized by some degree of interaction between the experts.

EXPERT OPINION IN PSA: RECENT EXPERIENCES AND DEVELOPMENTS

This section is devoted to recent experiences and methodological developments as reported in the literature on the use of expert opinion in Probabilistic Safety Assessment.

Recent experiences

Some years after the publication of the Reactor Safety Study, sufficient operational experience with nuclear reactors became available to allow for a comparison with a number of RSS estimates for components and subsystems. This was done by Apostolakis et al. (26) by interpreting the RSS probability distributions as population variability or generic curves, reflecting plant-to-plant variations, and by using these distributions as prior distributions which were combined with statistical evidence in a Bayesian updating procedure. A similar approach was followed with failure rates obtained from another source of generic data: IEEE Standard-500 (27). As the evidence concerned plant-specific data, the posterior distributions produced could be interpreted as plant-specific distributions for the failure rates. It appeared that in two of the three cases which were considered the posterior distributions were shifted to higher failure rates when compared with the initial prior distributions. It was concluded that the initial distributions might have been biased. This assumption is supported by data on operational experience which have been collected by the Oak Ridge National Laboratory (28). For seven subsystems the failure rates as estimated on the basis of operational experience can be compared with the 90% confidence bounds as used in the Reactor Safety Study. It appears that all Oak Ridge values fall outside the RSS confidence bounds (29). Both analyses suggest that the RSS analysts are badly calibrated, which corresponds with the general conclusion of many experimental studies (mentioned above in the subsection "evaluation of probability assessments").

Since the Reactor Safety Study, expert opinions have been used in various areas of application. The existence of a substantial spread of opinion is a recurrent theme across all these applications: the analysis of nuclear risks (30, 31), the analysis of seismic risks (32), the analysis of human reliability (33), and the analysis of health risks due to air pollution (34).

What about the chemical process industries which is the main area of interest within the current research project? The PSA methodology is increasingly being applied to analyzing the risks associated with production and storage of hazardous materials in the chemical process industries. As far as Europe is concerned, the developments in this area are concentrated in England and the Netherlands (35). The best known examples from England are two studies on the chemical process industries at Canvey Island (36, 37); the best known Dutch examples are the COVO study (38) and the LPG study (39). All these analyses have contributed significantly to the development of models and methods for applying risk analysis to the chemical process industries. However, in none of these studies expert opinions have been used explicitly and in a formalized way.

Methodological developments

Various methods for the use of expert opinion have been suggested
specifically in the context of risk analysis. A strict distinction
between elicitation and combination will not be made, since these two
aspects are not treated separately in the risk analysis literature. The
proposed methods fall into three categories: Bayesian models, models for
weighted expert opinion and psychological scaling models.

Bayesian models are proposed by many authors for different areas
of application, for example component failure rates (4, 40) and the
analysis of seismic risks (41). In all these methods a distribution
representing the uncertainty is assumed beforehand (frequently the
lognormal distribution) and the parameters of the distribution are
derived from distribution characteristics (usually moments) given by the
assessor. It should be noted that the literature focuses on processing,
whereas the necessary data are assumed to be available. *Models for
weighted expert opinion* are proposed in the area of seismic risks (32).
Elicitation is done by using questionnaires, on which the experts are
asked to give percentiles of their distributions. Weights are derived
from self-ratings or colleague-ratings. *Psychological scaling models*,
in particular the method of paired comparisons, are recommended for
estimation of human error probabilities (42, 43). Related methods are
used by Human Reliability Associates in England (44).

SELECTION AND DESCRIPTION OF MODELS

A comparison of the literature on the use of expert opinion in risk
analysis with the general literature on probabilistic predictions can be
summarized as follows. In some cases the aspect of elicitation gets
little attention (Bayesian methods), whereas in general little use is
made of the possibility of incorporating a reward structure in the
encoding process. As far as processing of opinions is concerned, little
attention is paid in some cases to the possibility of incorporating an
evaluation mechanism (weighted expert opinion methods). Although the
general literature on probabilistic predictions shows some unresolved
problems, there is sufficient support for confidence in the possibilities
of improving the current practice.

Three different categories of models have been selected for further
investigation. Within each category, one model has been selected that is
hoped to be an improvement in comparison with the models suggested so far
within that category. These models are briefly described. More extensive
treatments are presented elsewhere (45). Additional references are given
where appropriate.

Weighted expert opinion model

In the weighted expert opinion models which have been suggested so
far, weights are obtained from self-ratings, colleague-ratings, ratings
by analyst or decision-maker, etc. Such weights are by no means objective
measures in such a way that the quality of the experts' assessments is
taken into account. The question then is: how should expert weights be
determined in order to meet this requirement? In the selected model for

weighted expert opinion, the weights that are used have two characteristics: (a) they are based on past performance, thus rewarding good calibration and low entropy (= high information), and (b) they are based on an extension of the traditional theory of proper scoring rules, thus rewarding honesty. The underlying theory of this model is described by Cooke (46, 47).

An implementation of the model involves the following steps: select variables whose values are known and which resemble as much as possible the variable of interest ("seed variables"), elicit distributions (in terms of fractiles) from the experts for these variables, determine the expert weights, elicit a distribution from the experts for the variable of interest, determine the weighted average for the variable of interest by applying the weights.

The main advantages of this model are that it provides a robust normative mechanism for distinguishing between "better" and "worse" experts and allows the opinions of "better" experts to weigh more heavily. Moreover, the combination algorithm is relatively simple. The main problems originate from the issue of past performance: the question is to what extent past performance is a measure of the quality of future predictions and to what extent performance on particular questions correlates with performance on the variable of interest. Furthermore, the experts are required to give probabilistic statements. Finally, it should be noted that the resulting combination of expert opinion can never be better than the opinion of the "best" expert.

Bayesian model

The Bayesian models which have been suggested so far, are characterized by simplifying and modeling assumptions and by the necessity of complicated input information (likelihood function) to be given by the decision maker. Favorable characteristics of the selected Bayesian model in comparison with other Bayesian models are that default egalitarian prior distributions are given by the model itself and expert assessments are not restricted to a given class of distributions: the experts give fractiles of their distributions. Key feature of the model is that it uses Bayes' theorem to recalibrate and combine expert assessments. The underlying theory of the model is described by Mendel and Sheridan (48).

An implementation of the model involves the following steps: select variables whose values are known and which resemble as much as possible the variable of interest ("seed variables"), elicit distributions (in terms of fractiles) from experts for these variables, elicit a distribution from the experts for the variable of interest, perform the necessary calculations (for which a computer program is required since the model is computationally complex).

The main advantages of this model are that it has a strong philosophical basis and deals with correlations in expert opinions. Furthermore, the resulting combination of expert opinion can be "better" than the opinion of the "best" expert. The main problems are similar to those of the selected weighted expert opinion model: they originate from the issue of past performance and are also related to the requirement that the experts must supply probabilistic statements. Other disadvantages are connected with the computations: these are rather complex, the possible number of experts is limited and a large number of seed variables is required.

Psychological scaling model

The method of paired comparisons has been recommended for the estimation of human error probabilities. In the models which have been suggested so far, scale values are assumed to be described by a normal distribution (Thurstone model) or by a lognormal distribution (modification of Thurstone model because of logarithmic transformation of scale values into absolute values). In the selected model, which is specifically designed for failure rate estimation, the lifetime of a component is assumed to be characterized by a negative exponential distribution. This assumption leads to a similarity with another existing psychological scaling model, the Bradley-Terry model, as far as the probability of preferring objects is concerned. This so-called NEL model is described by Stobbelaar (49).

An implementation of the model involves the following steps: select "similar" objects, elicit preferences of the experts for all pairs of objects with respect to a certain attribute, determine scale values, determine absolute values if required.

The main advantages of this model are that it requires only simple qualitative judgments from the experts (thus, the experts are not required to give probabilistic statements) and leads automatically to a group opinion. The main problems originate from the similarity of the objects and the required numbers of experts and objects (several experts are required to make judgments about several objects; for example, 15 experts and 10 objects would be a reasonable set-up). Furthermore, the experts should be selected rather carefully, since the model does not provide a formal mechanism for selecting and evaluating experts. Finally, the absolute values of one or two objects have to be known in order to allow for transformation of scale values into absolute values.

EXPERIMENTS: A FIRST LOOK

Experiments on real-world problems are currently in preparation. Purpose of these experiments is to gain operational experience with the selected models. Experiments have been identified (and will be performed) in the chemical process industries and the space industry. More specifically, the experiments concern failure rates of components, such as gaskets and flanges, and human error. Experimental results will be presented in the near future.

A preliminary experiment

The remainder of this section is devoted to giving an impression of a preliminary experiment which concerned an application of the method of paired comparisons at N.V. Nederlandse Gasunie.

The objects were seven pressure regulators. The main variables of interest were the failure rates of two different components (membrane and seal) of these regulators; the failure rates for one regulator were known. Besides, other characteristics of the regulators, such as maintainability and noise production, were defined to be of interest and were also investigated within the experiment. The experts were maintenance personnel: 21 mechanics and 6 supervisors (former mechanics). All the experts were knowledgeable with respect to the regulators; lack of probabilistic knowledge was the major motivation for using the method of paired comparisons.

Every expert was interviewed individually. Each interview consisted of two parts: an introduction, intended to motivate the expert and to explain the problem, and the actual data collection, which was done by direct interrogation (pairs of drawings of the regulators were shown while asking the questions). The subjects were not always able to make every possible paired comparison, due to lack of experience. Furthermore, the subjects were not always able to indicate a preference between two objects, although asked to do so.

It was decided beforehand to distinguish between mechanics and supervisors. Based upon an analysis of within-expert consistency, a further distinction was made within the subgroup of mechanics. Thus, five subgroups of experts could be distinguished: consistent mechanics, other mechanics, all mechanics, all supervisors and all subjects. Three models were used for processing of the responses: the NEL model and two variants of the Thurstone model (in fact, a decision as to which model is to be used does not necessarily have to be made in advance, since all these models require the same input data). The ranking of the objects was identical with the different models, but differed for different subgroups. The NEL model yielded a larger value of the largest failure rate, whereas the NEL model and Thurstone model confidence intervals showed an overlap in most cases.

The following conclusions were drawn. The method was highly appreciated by the subjects. The time required for data collection was limited: about 200 paired comparisons could easily be made within one hour. The selection of experts is important. Since different models yield similar rankings, but different absolute values, the question remains which model to use. Consequently, a comparison of paired comparison results with reality is highly needed. Finally, the possibility of indifferences and incomplete designs has to be taken into account.

CONCLUSION AND PERSPECTIVE

This paper presented a discussion of expert opinion in PSA (needs and current practice) and of the possibilities offered by the area of probabilistic predictions. It was argued that sufficient support exists for confidence in the possibilities of improving the current practice. To this end, three models were briefly described that have been made operational in order to allow for experimental evaluation. The real value of these models must of course be proven in real-world experiments, on which a first concise look was given. A preliminary experiment with one of the selected models, the method of paired comparisons, showed a hopeful picture but also left some unanswered questions for further study. More detailed experimental results will be presented in the near future.

ACKNOWLEDGMENT

This paper is based on a research project which is performed for the Ministry of Housing, Physical Planning and the Environment of the Dutch Government. The author is grateful to this Ministry for permission to publish this paper. The author is also grateful to N.V. Nederlandse Gasunie for permission to discuss the paired comparisons experiment. This experiment was performed by Mark Stobbelaar. Roger Cooke provided valuable comments on an earlier draft of this paper.

REFERENCES

1. Van der Horst, J., Technical support of safety studies. Report 84-015811, Division of Technology for Society TNO, Apeldoorn, Netherlands, 1984 (in Dutch).

2. Reactor Safety Study. NUREG 75/014 (WASH-1400), U.S. Nuclear Regulatory Commission, Washington, DC, USA, 1975.

3. Apostolakis, G., On the use of judgment in probabilistic risk analysis. Nuclear Engineering and Design, 1986, 93, 161-166.

4. Apostolakis, G., Expert judgment in probabilistic safety assessment. Paper presented at the course "Accelerated life testing and experts' opinions in reliability", International School of Physics "Enrico Fermi", S. Terenzo di Lerici (La Spezia), Italy, 1986.

5. Okrent, D., A survey of expert opinion on low probability earthquakes. Annals of Nuclear Energy, 1975, 2, 601-614.

6. Van Steen, J.F.J. and Oortman Gerlings, P.D., Expert opinion in probabilistic safety assessment, Vol. 1: Literature study. Netherlands Organization for Applied Scientific Research TNO/Delft University of Technology, Apeldoorn/Delft, Netherlands, 1988 (in preparation).

7. Staël von Holstein, C.-A.S., Measurement of subjective probability. Acta Psychologica, 1970, 34, 146-159.

8. Hampton, J.M., Moore, P.G. and Thomas, H., Subjective probability and its measurement. Journal of the Royal Statistical Society A, 1973, 136, 21-42.

9. Huber, G.P., Methods for quantifying subjective probabilities and multi-attribute utilities. Decision Sciences, 1974, 5, 430-458.

10. Hogarth, R.M., Cognitive processes and the assessment of subjective probability distributions. Journal of the American Statistical Association, 1975, 70, 271-289.

11. Morgan, M.G., Henrion, M. and Morris, S.C., Expert judgments for policy analysis. Brookhaven National Laboratory, Upton, New York, USA, 1979.

12. Stillwell, W.G., Seaver, D.A. and Schwartz, J.P., Expert estimation of human error probabilities in nuclear power plant operations: A review of probability assessment and scaling. NUREG/CR-2255, U.S. Nuclear Regulatory Commission, Washington, DC, USA, 1982.

13. Wallsten, T.S. and Budescu, D.V., Encoding subjective probabilities: a psychological and psychometric review. Management Science, 1983, 29, 151-173.

14. Lichtenstein, S., Fischhoff, B. and Phillips, L.D., Calibration of probabilities: The state of the art to 1980. In (15), pp. 306-334.

15. Kahneman, D., Slovic, P. and Tversky, A. (eds.), Judgment under Uncertainty: Heuristics and Biases, Cambridge University Press, 1982.

16. Winkler, R.L. and Murphy, A.H., Experiments in the laboratory and the real world. Organizational Behavior and Human Performance, 1973, 10, 252-270.

17. Murphy, A.H. and Winkler, R.L., Reliability of subjective probability forecasts of precipitation and temperature. Applied Statistics, 1977, 26, 41-47.

18. Spetzler, C.S. and Staël von Holstein, C.-A.S., Probability encoding in decision analysis. Management Science, 1975, 22, 340-358.

19. Ludke, R.L., Stauss, F.F. and Gustafson, D.H., Comparison of five methods for estimating subjective probability distributions. Organizational Behavior and Human Performance, 1977, 19, 162-179.

20. Seaver, D.A., Von Winterfeldt, D. and Edwards, W., Eliciting subjective probability distributions on continuous variables. Organizational Behavior and Human Performance, 1978, 21, 379-391.

21. Alpert, M. and Raiffa, H., A progress report on the training of probability assessors. In (15), pp. 294-305.

22. Schaefer, R.E. and Borcherding, K., The assessment of subjective probability distributions: A training experiment. Acta Psychologica, 1973, 37, 117-129.

23. Pickhardt, R.C. and Wallace, J.B., A study of the performance of subjective probability assessors. Decision Sciences, 1974, 5, 347-363.

24. Lichtenstein, S. and Fischhoff, B., Training for calibration. Organizational Behavior and Human Performance, 1980, 26, 149-171.

25. Genest, C. and Zidek, J.V., Combining probability distributions: A critique and an annotated bibliography. Statistical Science, 1986, 1, 114-148.

26. Apostolakis, G., Kaplan, S., Garrick, B.J. and Duphily, R.J., Data specialization for plant specific risk studies. Nuclear Engineering and Design, 1980, 56, 321-329.

27. IEEE guide to the collection and presentation of electrical, electronic and sensing component reliability data for nuclear power generating stations (IEEE Std-500). The Institute of Electrical and Electronic Engineers, 1977.

28. Minarick, J. and Kukielka, C., Precursors to potential severe core damage accidents: 1969-1979: A status report. NUREG/CR-2497, U.S. Nuclear Regulatory Commission, 1982.

29. Cooke, R.M., Subjective probability and expert opinion. Lecture notes gg2.3, Delft University of Technology, 1986.

30. Hofer, E., Javeri, V., Löffler, H. and Struwe, D.F., A survey of expert opinion and its probabilistic evaluation for specific aspects of the SNR-300 risk study. Nuclear Technology, 1985, 68, 180-225.

31. Amendola, A., Uncertainties in systems reliability modelling: Insight gained through European benchmark exercises. Nuclear Engineering and Design, 1986, 93, 215-225.

32. Cummings, G.E., The use of data and judgment in determining seismic hazard and fragilities. Nuclear Engineering and Design, 1986, 93, 275-279.

33. Brune, R.L., Weinstein, M. and Fitzwater, M.E., Peer review study of the draft Handbook for Human Reliability Analysis. SAND82-7056, Sandia National Laboratories, Albuquerque, New Mexico, USA, 1983.

34. Morgan, M.G., Morris, S.C., Henrion, M., Amaral, D.A.L. and Rish, W.R., Technical uncertainty in quantitative policy analysis - a sulphur air pollution example. Risk Analysis, 1984, 4, 201-216.

35. Cox, R.A. and Slater, D.H., State-of-the-art of risk assessment of chemical plants in Europe. In Low-Probability/High-Consequence Risk Analysis, eds. R.A. Waller and V.T. Covello, Plenum Press, 1984, pp. 257-283.

36. Health and Safety Executive, Canvey: An investigation of potential hazards from operations in the Canvey Island/Thurrock area. HM Stationery Office, London, England, 1978.

37. Health and Safety Executive, Canvey - A second report. HM Stationery Office, London, England, 1981.

38. Openbaar Lichaam Rijnmond/COVO, Risk analysis of six potentially hazardous industrial objects in the Rijnmond area, a pilot study, Reidel, 1982.

39. TNO, LPG, A Study: A comparative analysis of the risks inherent in the storage, transshipment, transport and use of LPG and motor spirit, Ministry of Housing, Physical Planning and the Environment, The Hague, Netherlands, 1983.

40. Singpurwalla, N.D. and Soyer, R., The use of expert opinion in reliability: A survey. Paper presented at the course "Accelerated life testing and experts' opinions in reliability", International school of Physics "Enrico Fermi", Italian Physical Society, S. Terenzo di Lerici (La Spezia), Italy, 1986.

41. Mosleh, A. and Apostolakis, G., The assessment of probability distributions from expert opinions with an application to seismic fragility curves. Risk Analysis, 1986, 6, 447-461.

42. Seaver, D.A. and Stillwell, W.G., Procedures for using expert judgment to estimate human error probabilities in nuclear power plant operations. NUREG/CR-2743, U.S. Nuclear Regulatory Commission, Washington, DC, USA, 1983.

43. Comer, M.K., Seaver, D.A., Stillwell, W.G. and Gaddy, C.D., Generating human reliability estimates using expert judgment. NUREG/CR-3688, U.S. Nuclear Regulatory Commission, Washington, DC, USA, 1984.

44. Embrey, D.E., Humphreys, P.C., Rosa, E.A., Kirwan, B. and Rea, K., SLIM-MAUD: An approach to assessing human error probabilities using structured expert judgment. NUREG/CR-3518, U.S. Nuclear Regulatory Commission, Washington, DC, USA, 1984.

45. Cooke, R.M., Mendel, M.B. and Van Steen, J.F.J., Expert opinion in probabilistic safety assessment, Vol. 2: Operational description of basic models. Delft University of Technology/Netherlands Organization for Applied Scientific Research TNO, Delft/Apeldoorn, Netherlands, 1988 (in preparation).

46. Cooke, R.M., Mendel, M. and Thijs, W., Calibration and information in expert resolution; a classical approach. To appear in Automatica, 1988.

47. Cooke, R.M., A theory of weights for combining expert opinion. Report 87-25, Faculty of Mathematics and Informatics/Computer Science, Delft University of Technology, Delft, Netherlands, 1987.

48. Mendel, M.B. and Sheridan, T.B., Optimal combination of information from multiple sources. Department of Mechanical Engineering, Massachusetts Institute of Technology, Cambridge, Massachusetts, USA, 1986.

49. Stobbelaar, M.F., Paired comparisons. Report 87-137, Division of Technology for Society TNO, Apeldoorn, Netherlands, 1987 (in Dutch).

A METHOD FOR IDENTIFICATION AND TRACING OF
POTENTIAL INCIDENT CAUSES IN PROCES INDUSTRY.

P. Bockholts and L.J.B. Koehorst
Department of Industrial Safety
Netherlands Organization for Applied Scientific Research TNO
P.O. Box 342, 7300 AH Apeldoorn
Netherlands.

Abstract.

Incident prevention is an area of great effort in the process industry.
Various methods have been developed for the identification and tracing of
failure sources with the aim to increase reliability and safety of process
installations. The method described in this paper is a rather general one
and is based on the analysis of many thousands of incidents. More than
with some other analyzing techniques, attention is paid on incident causes
during maintenance, repair and other than "normal operation" situations.
Especially during those situations a great many incidents occur.

INTRODUCTION

Various methods have been developed for the determination and
quantification of incident causes in the process industry. In this paper a
method is described that is derived from the analysis of many thousands of
incidents that happened in industry. The majority of those incidents were
minor ones or even near misses. Their causes however were the same as the
causes of the major accidents.
Since much more material is available about minor incidents is this
information used for the method.
The analysis learned that many incidents occur under other then the
"normal operation" conditions or circumstances. They occur due to a par
hazard presence of independent attributes or independent activities that
influence each other. Sometimes it is a minor malfunctioning in a system
that contributes to an incident under special conditions.
The method described in this paper to trace incidentcauses is based on a
matrix. In this matrix the various phases of the lifecycle of a
processplant or installation are given versus a list of all potential
incidentcauses. These causes are listed in classes to keep its length
practical.
In addition to the matrix and its use is a collection of many hundreds of
cause descriptions written in order to facilitate the use of the method.
The causes are classified according the matrix and extended with an
unitoperationcode, a unitcode and keywords.

The matrix Method for tracing potential incident causes.

The method is a rather simple one and means a step by step analysis of an
installation or unit.
The total lifecycle of the installation is divided into functional phases:

Eight different phases are distinguished for the installation. The type of
accidents also differ among these phases and by this the eight phases are
functional. The phases are listed in table I.
Fifteen different classes of incident causes have been identified among
the incidents. These classes are listed in table II.

Table I

Systemphases

- normal operation
- start up
- stop/shut down
- clean
- out of production
- modify/dismantle
- repair/maintain
- test/trial run.

Table II

Cause classes

- natural causes (lightning, earthquake, storm etc.)
- domino effects
- sabotage/vandalism
- operations by others (excavation etc.)
- operator failures
- mechanical failures
- electric failures
- system failures (clogging, waterhammer etc.)
- chemical reactions (run away reactions)
- physical phenomena (hot spots)
- incorrect environment (dirty, no ventilation etc.)
- corossion, fatique etc.
- latent failures (maintenance failures, incorrect construction etc.)
- unallowed operations (welding during production)
- risk increasing factors (abnormal practic).

Both tables are derived from classifications that are used in the FACTS*)
databank. The matrix is constructed by a combination of table I (phases)
and table II (causes). Each box in the matrix (see table III) represents a
particular phase in the lifecycle of the installation in combination with
a class of causes.
Tracing potential incidentcauses can now be carried out in a systematic
way. It means that for a given installation the situations as indicated in
each box of the matrix have to be considered.
A complete analysis means a review of all situations as represented in all
the boxes.

*) FACTS is a databank with data from incidents with hazardons materials.
   FACTS is operated by the department of Industrial Safety from TNO.

TABLE III

Cause classes	System phases	Normal operation	Start-up	Stop/shut down	Clean	Out of production	Modify/dismantle	Repair/maintain	Test/trial run
1. Natural causes									
2. Domino effects									
3. Sabotage/vandalism									
4. Operations by others									
5. Operator failures									
6. Mechanical failures									
7. Electric failures									
8. System failures									
9. Chemical reactions									
10. Physical phenomena									
11. Incorrect environment									
12. Corossion fatique etc.									
13. Latent failures									
14. Unallowed operations									
15. Risk increasing factors									

Some examples may illustrate the use of the matrix.
1. One box represents the hazards of CHEMICAL REACTION and CLEANING of the installation. It means that one should review the cleaning methods, the usual precautions, the cleaners used, the materials to be cleaned, the removing of waste and cleaner etc. Try to identify any new dangerous situation.
2. One box represents MODIFICATIONS and SYSTEM FAILURE. It means that one should consider what inpact the replacement of a component may have if this component has other specifications, c.q. a pump replaced by a larger one way need other safety values elsewhere in an installation.

These two examples may show that certain detailed considerations have to made for each box.
It could be desirable to have some kind of checklist for each box. However such checklists were not intended and will probably always be far from complete. For this reason is choosen for this rather general approach of only a simple matrix with which installationphases can be correlated with potential incidentcauses. It is expected that this rather simple approach may help to indentify potential dangers and by this contributes to efforts to avoid those.

The cause descriptions, a historical review.

Data and descriptions of many incidents that happenen in the past are stored in the FACTS* databank from TNO. A selection of many hundreds of accidents has been made among cases in processindustry that could be of interest for general preventive purposes.
All selected cases were analysed and a causedescription was made. Each description was extended with a systemsclassification, an unitoperation, a causeclassification and keywords. The causeclassification used here is the same as the one used in the matrixmethod.
This exercise resulted in a unique collection of causedescriptions that may be used for accidentprevention. The attached classifications and keywords facilitate the searches. The collection becomes available as a database as well as a handbook.
The set-up of this collection has a number of advantages. The keywordstructure is hierarchical. This means that subjects that have a functional relationship are grouped together. Tank, tankwall, tankpit but also levelindicato, high levelalarm, isolation etc. are keywords that are grouped in the same hierarchical branch. This structure facilitates the finding of causes that could be of users interest. It also allows to find causes that are not directly appointed by the initial keywords.

CONCLUSIONS.

Great effort is being spent in process industry to minimize risks and to avoid incidents. However in spite of this incidents do occurr and will occur in the future. The method for identification of potential dangers that is described in this paper has the aim to contribute to accidentprevention in processindustry. The method is derived from an analysis of accidents that happened in the past. Lessons should be learned from those events.

LITERATURE

1. P. Bockholts, L.J.B. Koehorst, march 1987
   FACTS, most comprihensive information system for industrial safety.

2. P. Bockholts
   Oorzaakanalyse
   TNO, november 1985

3. H.J. Wingerden
   Reliability Data Collection and Use in Risk and Available Assessment.
   Springer. Verlag, 1986.

THE COMPATIBILITY OF RELIABILITY

MODELS WITH THE UNDERLYING DATA

STRUCTURE

P.T. Manning
Central Electricity Generating Board
15 Newgate Street
London EC1A 7AX

ABSTRACT

The failure and repair data on the CEGB's overhead line transmission system in North Wales has been analysed to examine its data structure. The data is considerably inhomogenous, with failure and repair rates varying throughout the year; an overall time trend has also been identified.

The inhomogeneities were identified using Proportional Hazard Modelling (PHM). This paper shows that PHM is a powerful technique, but its limitations mean that the conclusions which can be drawn are strictly limited.

These limitations are discussed with respect to the CEGB's failure and repair data, and the methods used to ensure that the reliability prediction models in use are compatible with the data are briefly described.

INTRODUCTION

The CEGB operates a high-voltage transmission network whose function is to transmit electricity in bulk from power stations to demand centres. In order to ensure that the system retains a high degree of reliability without over-investment in plant it is necessary to predict its future reliability. The data for these reliability predictions is derived from historic experience, modified if necessary, and this paper discusses the interface between the historic experience and the future predictions.

The CEGB's transmission system is (in reliability terms) a classical repairable system. It is made up of components (e.g. overhead lines, underground cables, transformers, etc.) with characteristic failure and repair behaviour[1]. Both failure and repair data are known to be inhomogeneous, and hence sophisticated data-analysis techniques are required to define these inhomogeneities [2] before the data can be incorporated within reliability prediction models.

This paper concentrates on a small sub-set of reliability data, namely that for overhead lines in North Wales. Overhead lines are the most vulnerable transmission component – they are also the ones whose reliability data is least homogeneous – and North Wales is an area of particular interest. The overhead lines in that area are exposed (on occasions) to particularly severe winter weather, while the area also contains a large amount of generation – nuclear, hydro and pumped storage – including all the spinning reserve plant on the system.

Inspection of the failure and repair data for overhead lines in North Wales demonstrated that:-

. The failure rate in winter was greater than in summer.

. Failures tended to occur in "bunches" (typically bunches of 3-30 faults occurring within 12 hours) caused by adverse weather.

. Repair times of faults caused by adverse weather were shorter than average.

The data was then analysed using Proportional Hazards Analysis (PHM) [3]. This confirmed the above conclusions, and in addition that:-

. The failure rate seemed to be increasing with time (the period examined was 1968-1984).

. Repair times seemed to be decreasing with time.

Further analysis suggested that the number of short duration (1 minute) faults was increasing with time, but the number of long duration faults was constant.

Proportional Hazard Modelling not only identified the above factors but also put numberical values on them. The remainder of this paper discusses the relevance of these numerical values, and how they are used to make reliability predictions about the future transmission system performance.

PROPORTIONAL HAZARDS ANALYSIS

It is worth recapping the fundamental mathematics behind Proportional Hazard Analysis.

PHM is based on the fundamental assumption that the effect of an external influence on the hazard function can be expressed as:-

$$h(t,z) = ho(t) \exp(-\beta z) \qquad (1)$$

where: $h(t,z)$ is the hazard function;
$ho(t)$ is the "baseline hazard";
z represents external influence; and
β is a parameter of the model.

This simple expression is both the power and the drawback of the model. It is powerful because it is mathematically simple; yet its simplicity means that the assumptions are not necessarily valid. As the subject is still in its infancy, good tests and procedures to test for the validity of these assumptions are not yet fully developed.

Two aspects are particularly relevant. These are, firstly, the use of the exponential formulation; and secondly the constancy of the underlying distribution. These are now considered in more detail.

Use of Exponentiation

The proportional hazards model tests for the value of the parameter β in equation 1. If the covariate, z, is a binary function - representing, for example, the presence or absence of a design change - then the use of z=0 or z=1 will be quite adequate. However, if z is a numerical variable - e.g. time or temperature - there are a multitude of possible relationships of which the exponential is only one.

There are a large number of other possibilities; for example, if the reliability is expected to be linear in the co-variate z then the correct measure is to use log z as the co-variate. However, the use of z as a co-variate will still give a value of β which will be significantly non-zero.

A positive result from the PHM - i.e. β being significantly non-zero -therefore merely shows that the co-variate z is having a significant affect on the reliability. It cannot necessarily be used directly for reliability prediction. An example of this is given below when the "time trend" for North Wales is discussed.

Constancy of the Underlying Distribution

It is a fundamental hypothesis of the PHM formulation that the reliability distribution parameters are independant of the co-variate z. Let, for example, the underlying reliability distribution with z=0 (the baseline hazard) be exponential such that:-

$$h_o(t) = A \exp (- \lambda t)$$

If the effect of the co-variate z is simply to alter the scaling parameter A then PHM will perfectly describe the process going on. On the other hand, if the affect of the co-variate is to alter the rate parameter from λ to λ' then elementary mathematics shows that the application of PHM will give a derived value of β equal to log $((\lambda - \lambda')t/z)$. Although β is obviously not a constant it is nevertheless quite likely that application of PHM will show β to be significantly different from zero.

Proportionality plots, where cumulative hazard is plotted against time, are intended to demonstrate whether or not the above assumption is valid but in practice they are difficult to interpret. An example is given below (Figure 3).

TIME-TREND

Proportional Hazards Modelling of the data identified a time-trend. The standard form of the PHM equation (Equation 1) suggested an exponential increase in failure rate with time, of about 3% per year.

This is not to say that the failure rate increased exponentially, only that the exponential model provided a better fit than a model without a time-trend. Rather than running the PMH with different formulations it is much more appropriate to examine the raw data to ascertain the form of the time trend.

Figure 1 shows a graph of cumulative failures against time. During this period of time the total line length at risk has remained sensibly constant, and hence any effect cannot be due to population changes.

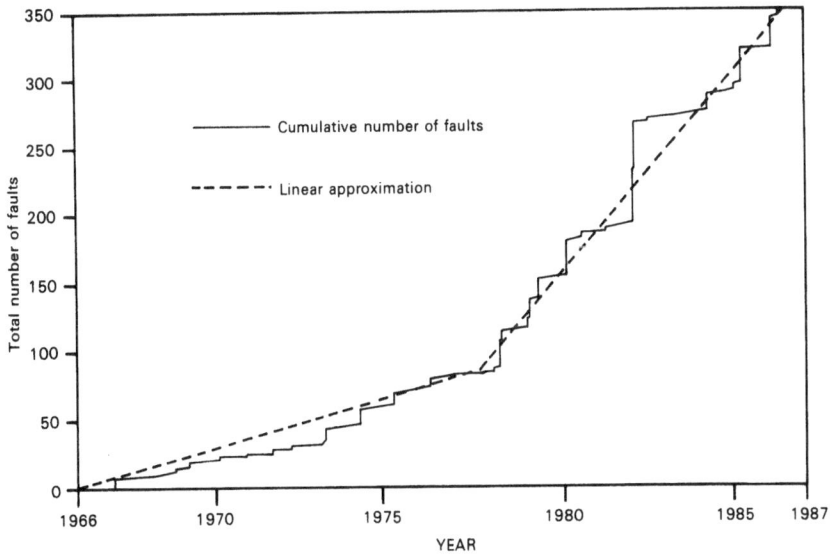

Figure 1. Cumulative number of faults on overhead
lines in North Wales.

It seems from Figure 1 that the period of time can be split into 2, pre and post 1978. Faults tend to occur in bunches in winter − giving the stepped appearance of the graph − but since 1978 the number of faults occurring in each winter seems to be greater. Although the winter of 1981/2 was particularly severe, the identification of the two time periods does not depend upon this single year.

No engineering reason has yet been identified for the apparent change in behaviour at the end of the 1970's and it is tentatively put down to a change in weather patterns, with the 1970's tending to exhibit milder winters than did the 1980's.

WEATHER

Transmission equipment failures were split into two types, depending upon whether they occurred in a bunch, or as isolated events. (A bunch was defined as a group of 2 or more faults within 6 hours and 60 km of each other, at least one of the faults having been ascribed to an environmental cause). The bunch faults were ascribed to "adverse" weather, and the non-bunch faults to "normal" weather. PHM confirmed that the two distinction between "bunch" and "non-bunch" faults was a valid distinction, although the element of circularity in the argument was succinctly described in [3].

Although PHM confirmed the presence of two distinct types of faults, the results are somewhat misleading. This is because the "time between failure" was calculated as the time from the previous failure, and hence is a mixture of intra-bunch times and inter-bunch times, depending on whether or not the previous failure was part of the same bunch.

Figure 2 shows a hazard plot for the "baseline hazard" for the times-between-failures, on Weibull axes. The plot was interpreted [3] as showing the effect of the difference between "intra-bunch" and "inter-bunch" times, each portion individually displaying approximately exponential distribution but the whole being significantly non-standard and difficult to parameterise.

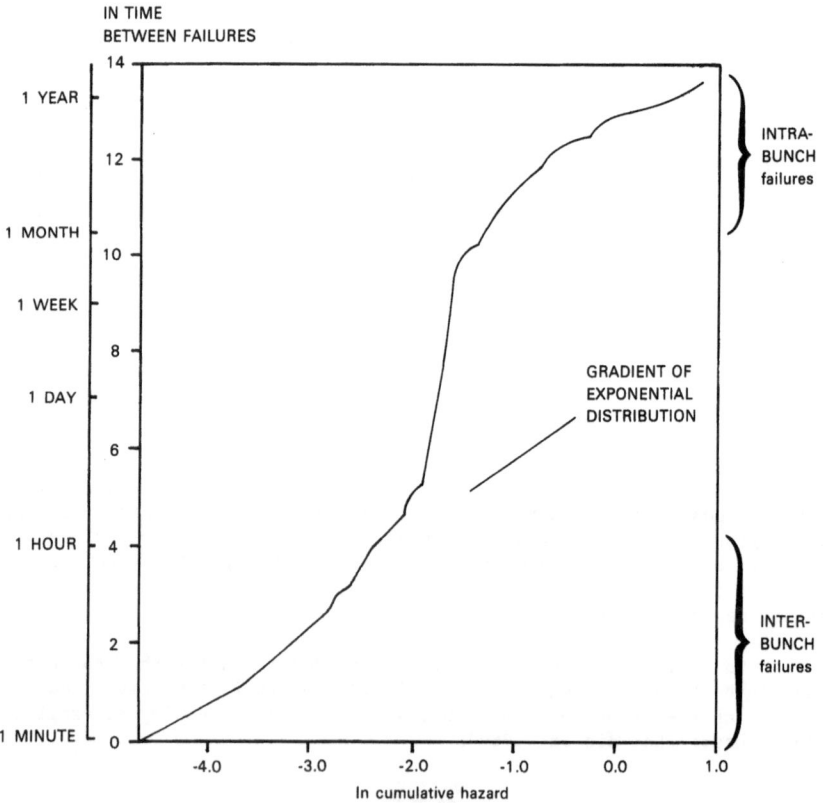

Figure 2. Base-line hazard plot for overhead line failure in
North Wales (Weibull axes).

The proportionality plot shown in Figure 3 further emphasises the shortcomings of PHM in this particular case. If the assumptions within PHM are valid, the two lines (which represent data for the two cases $z=0$ and $z=1$) will have a constant vertical separation. In Figure 3 the two lines cross, the vertical separation being equivalent to $\pm$ half an order of magnitude (note that the vertical axis is logarithmic). It is clear that the basic assumptions of PHM are not valid in this case.

It should be emphasised that this does <u>not</u> invalidate the conclusion that "weather" is affecting the reliability data. The hypothesis being tested by PHM is that weather does <u>not</u> have an influence, and PHM has clearly shown this hypothesis to be invalid. However, although PHM provides a better explanation than no explanation at all, it has a number of shortcomings and a better model still should be sought for.

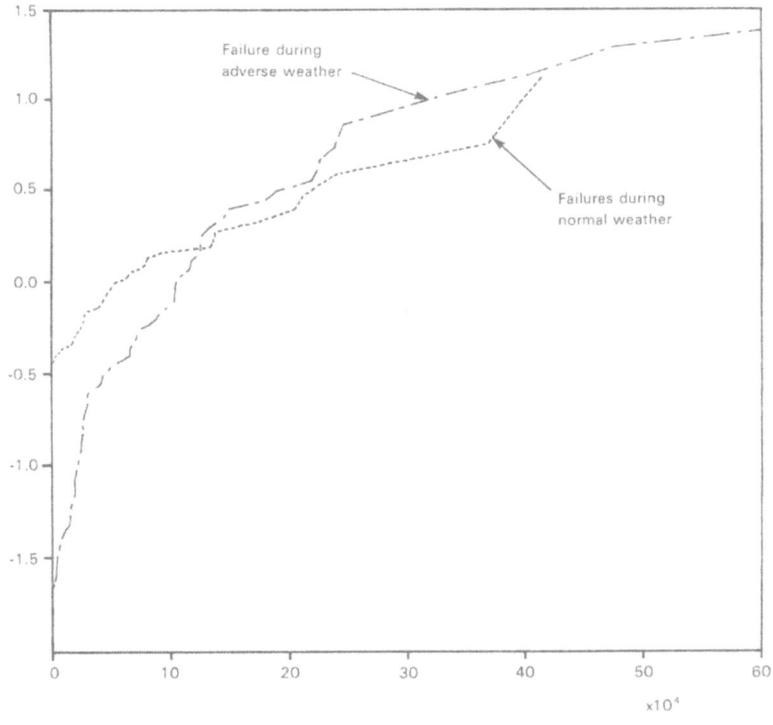

Figure 3. Proportionality Plot for overhead
line failure data.

RELIABILITY PREDICTION

It is possible to construct reliability models which faithfully mirror the PHM results. For example, the CEGB has a Monte-Carlo Simulation Program TRANSIM[4] which utilises reliability data sampled from various reliability distributions. It would be conceptually simple to utilise the base-line hazard as the fundamental distribution, modified by the values of the co-variates appropriate at that time.

Such an approach is not adopted, because of the doubt that such a model adequately describes reality. The use of the PHM formulation does not help in understanding the phenomena at work, and doubts have been expressed above that the formulation is valid. Instead the following methodology has been adopted.

The results of the "time-trend" parameter are assumed to represent two different time periods. Rather than a steady deterioration at 3% per year, the data is divided into two portions, namely pre and post 1978. The post 1978 data is assumed to apply to the future, and the possible effects of an improvement (i.e. back to pre-1978) or a further worsening are examined using sensitivity studies.

The other co-variates of relevance are season, weather and cause. "Weather" refers to faults being bunch or non-bunch, and is significant for failure data; "cause" is whether or not the fault was recorded as being due to an environmental or non-environmental reason, and is significant only for repair data. "Weather" and "cause" are closely correlated, and the effect of one is likely to hide the effect of the other three [3].

Earlier analysis by CEGB has concentrated on two-state weather state models, where both failure and repair data is influenced by the weather (itself a stochastic variable) and season, and software to model this has been developed [4]. As the results of the PHM confirm that this is a realistic way of modelling the system, the fault data is therefore divided into four data-sets (two weather states and two seasons) and each data-set is analysed individually before being utilised in the reliability prediction software.

CONCLUSIONS

Proportional Hazards Analysis is a powerful tool for examining the data for significant external influences. It has not only confirmed results which had been inferred by inspection, but also identified other external influences on the reliability data.

It is, however, difficult to confirm that a proportional hazard is the correct representation of the reliability data. Although a number of diagnostic tools are available, their use and interpretation currently presents difficulties and a more detailed theoretical examination of the statistics will prove necessary.

Proportional Hazards Modelling should be used to disprove the null hypothesis, i.e. to demonstrate the falsity of the assumption that an external influence (co-variate) does not affect the reliability data. Once this has been achieved, until the statistical theory behind PHM is much further advanced, the blind use of PHM results in reliability predictions is to be deprecated.

ACKNOWLEDGEMENT

This paper is published by permission of Central Electricity Generating Board. The Proportional Hazard Modelling work was carried out by Professor Bendell and Dr Baxter of Trent Polytechnic.

REFERENCES

1. Manning P.T., Evaluating the Reliability of the CEGB transmission system. 21st Universities Power Engineering Conference, London, 1986.

2. Argent S.J., Jarrett K.R. and Pearson A.R., Modelling the Effect of Weather on Electrical Power Transmission Systems. 7th Advances in Reliability Technology Symposium, Bradford, 1982.

3. Baxter M., Bendell A., Manning P.T. and Ryan S.G., Proportional Hazards Modelling of Transmission Equipment Failures. Reliability '87, Birmingham, 1987.

4. Coxson B.A. and Gilbert J., Advances in Transmission Reliability Assessment : the Simulation method using TRANSIM. 12th Inter-RAM Conference for the Electric Power Industry, Baltimore, 1985.

AN ELECTRONIC COMPONENT RELIABILITY DATA BASE

J. M. Marshall, J. A. Hayes & D.S. Campbell
Electronic Component Technology Group,
Department of Electronic & Electrical Engineering,
University of Technology,
Loughborough,
Leicestershire LE11 3TU, U.K.

and

A. Bendell
Mathematics, Statistics & Operational Research,
Trent Polytechnic,
Burton Street,
Nottingham NG1 4BU, U.K.

ABSTRACT

The Paper discusses the problems arising in the establishment of a computerised database for the field failures of electronic components. The purpose and aims of the program are described, together with the difficulty in establishing a satisfactory data format of sufficient generality and the identification of a suitable database structure. The practical need for data validation is discussed, together with the nature of validation applied.

Analysis of the data in the base is in the early stages but an overview is provided of early results together with implications for the future.

INTRODUCTION

The majority of reliability studies on electronic components have been concerned with the determination of failure rates under 'artificial' conditions such as simulation or accelerated life tests. Failure rates determined from such tests are then extrapolated to what might be expected under field conditions. Unfortunately, these extrapolations are often subject to doubt if only because multiple failure mechanisms may well be involved. Even in the most simple devices bimodal failure distributions can be found, and the level of modality may rise as the complexity of the device increases [1].

The availability to equipment manufacturers of failure rate data for electronic components is essential for system reliability prediction. For this purpose the failure rate models published in MIL-HDBK.217(E) [2] are those predominantly used by the industry. These models have been derived in the main from test bed and accelerated life studies. Manufacturers have commented that the models can be wildly pessimistic when compared with observed field performance, particularly in the case of modern microelectronic devices. The complexity of some of the models and the difficulty in obtaining satisfactory values for all of the terms illustrates the problem in predicting failure rates under field conditions. This is particularly demonstrated by the failure rate model for microelectronic devices:-

$$\lambda_p = \pi_Q[C_1\pi_T\pi_V + (C_2 + C_3)\pi_E]\pi_L$$

where λ_p is the failure rate expressed as the number of failures per 10^6 hours, π_Q is the quality factor relating to the quality level of the production, π_T is the temperature acceleration factor based on technology, π_V is the voltage derating stress factor, π_E is the application environment factor, C_1 and C_2 are the circuit complexity failure rates based on gate count, C_3 is the package complexity failure rate and π_L is the device learning factor.

A study centred at Loughborough University of Technology (LUT) and sponsored by the MOD (DCVD) commenced in September 1984 to investigate the failure behaviour of electronic components when used in equipments in the field. The aim of the study is to identify component types which are a major cause of field failure in equipment, the conditions under which failure occurs and the nature of the failure mechanisms. The intention is to build up a structured component failure database over a period of time.

By its very nature such a study involves the co-operation of several electronic equipment manufacturers. LUT held discussions with many systems manufacturers, and although a number of companies expressed a willingness to become involved, LUT finally entered into detailed discussions with the research centres of three major British companies, viz:

Plessey Research (Caswell) Ltd.
Marconi Research Centre, GEC Research, Great Baddow
STC Components, Harlow

It was recognised that the research centres were in the best position to identify suitable data sources within their groups. Given the importance of the project, separate contracts were let by the MOD to the companies. Subsequently a contract was also let to Trent Polytechnic to provide a statistical input to the project.

Earlier, a pilot study involving two Danish companies, co-ordinated by the Danish Engineering Academy (DIA) under sub-contract to LUT had been initiated. Figure 1 shows the organisational structure of the project team. The number of data sources varies between companies and is growing with time.

The project is unique in that three major companies in competition in the market place are jointly involved in a field study of this type. The success of the project, under such circumstances, is partly due to the confidentiality guaranteed by LUT to each of the individual companies. Individual data sets are desensitised by coding and lose their identity when the data is pooled.

DATA SOURCES

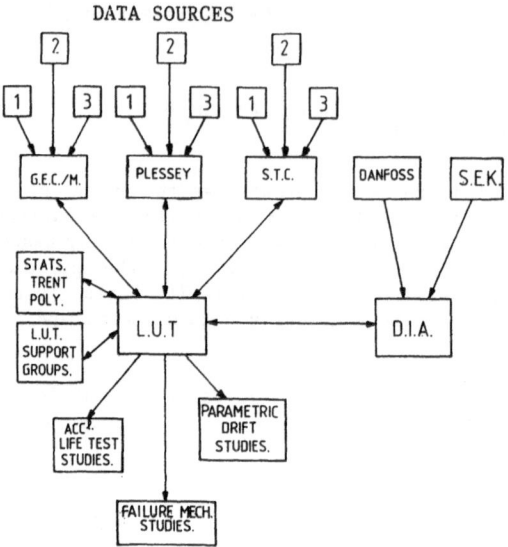

Figure 1   Organisation of Information Flow

Within the framework of the organisation two levels of group meetings take place. Co-ordination meetings are held twice yearly. These meetings are a forum which enables senior members of the consortium to meet and discuss progress and future policy.   However, it was realised in the early stages that a Working Party, including those involved in the acquisition, transfer and processing of data was needed to undertake the detailed work of the project.

THE FORMULATION OF A DATABASE

The main task initially confronting the working group was to establish a data input format. This format requires information at the system level (equipment) and subsystem level (printed circuit board (pcb)) in order to obtain the necessary data at the component level. Due to the hierarchical structure of the data, trackability at the system and subsystem level is essential in order to obtain dependable data at the component level. Such trackability, for example, would require a detailed knowledge of the repair loops of pcb's, including knowledge of their previous use and storage.

To achieve meaningful statistical analysis, information is required, not only on component failures, but on those components which have not failed and are at risk in the field during the period of observation.

The final data format was arrived at after considerations of the minimum information necessary for analysis. Typically, this basic information would be of the form:

(i)   equipment identification (type, serial-no)
(ii)   operational environment (ground benign etc.)

(iii)   installation, failure, first and subsequent repair dates etc.
(iv)   failure mode of equipment
(v)   equipment condition when failure occurred
        - mode of operation immediately after repair
(vi)   identification of failed component
        - component type
        failure mode/position or circuit reference [3]

Discussions of the practicalities of obtaining this data resulted in a final data format containing additional information and was more detailed in relation to the reporting of the minimum information required.

After the establishment of the data input format, discussions regarding the selection of types of equipments to be followed in the field needed to be made. The criterion for equipment selection was based on the dependability of the data only.

In order that the companies could establish suitable sources within their organisations, a check list based on the requirements of the data format was devised. The final equipment selections were based on these check lists which were discussed in detail by the Working Party. Not all approached sources could meet the data minimum requirements. In some cases it was possible to make minor modifications to the existing reporting systems to satisfy the minimum requirements. However, major modifications of reporting systems were impractical. This in practice meant that the company co-ordinators had to assess several sources within their organi- sation before final selection.

After the initial equipment selection had been finalised, the working group addressed the practical problems associated with the detailed coding of the data input.

The type and manner of data recording, not surprisingly, differs between data sources. Since the data to be transferred to LUT was required in a fixed detailed format, varying degrees of translation from either existing databases or paper records had to be performed by the companies. This information, however, is not always recorded in a manner that allows it to be directly extracted from the data records. Of particular importance, requiring special attention, are the generic coding descriptions of the components, and the differences in the recording of the time metric between the different sources.

Descriptions of the same component type can vary between data sources; some descriptions are based on use and others, for example, by power rating. No suitable existing listings of component descriptions, e.g. CODUS, MIL-217, were compatible across all data sources. Indeed, these descriptions were inconsistent with the wider aim of the project, which in addition to the statistical output requires failure mechanism analysis on the problem components to be carried out at LUT. With this in mind, a component generic type listing was devised at LUT, based on material and construction to be used by all companies.

The detailed generic coding requirement means, in practice, that the company co-ordinators are involved in extensive translation from parts lists to the LUT Format. Although slowing the rate of data transfer somewhat, the final result is a database which will allow component field failure behaviour to be linked to specific failure mechanisms. This is an important relationship to determine, in any component reliability study.

The need for component times to failure requires, because of the hierarchical nature of the data, knowledge of the operating time of equipment, downtime, circuit board replacements etc. In some instances equipments are fitted with modern electronic elapsed time indicators giving use to accurate time metrics. This is not, however, always the case and the required times can only be arrived at from knowledge of delivery dates, commissioning dates and up and running dates etc. The method of computation of the various times required in the data format is indicated by a suffix letter. This letter is used to highlight the relative accuracy of the data. The use of such letter indicators ensures the integrity of the data used in any statistical analysis.

Although it is not possible in this paper to discuss in detail every coding consideration of all the elements in the data format, the following examples illustrate some of the topics which warranted detailed Working Party discussions regarding methods of reporting and/or obtaining the information:-

(i)   Environment codings

(ii)   Component encapsulation and mounting technology
        - this detailed information is not normally given in parts lists

(iii)   Screened level of components

(iv)   Coping with either simultaneous failures or replacements

(v)   Circuit reference position of components etc.

It is seen that the construction of the finalised data format has been arrived at only by a team effort requiring both engineering and statistical expertise. The identification of the required data, associated codings and the protocols of data transfer have been addressed, and resulting in a practical working document. This working document is in ring binder form facilitating updates and replacements.

Although the data input format has been finalised and a final working document compiled, supplementary information on the data is required by LUT. A soft information document has been compiled in order to highlight the details on, for example, the repair loop of an equipment, the computation methods, the reasons for using particular indicators etc. This soft information provides additional knowledge about the data.

Initially it was thought, based on pilot studies in Denmark, that the database could be held on an IBM PC AT. It soon became evident, however, that this would be limiting with regard to the storage facilities required for the data and the necessary statistical software packages. It was, therefore, decided to use the Honeywell Multics mainframe based in the Computer Centre at LUT, for data storage and analysis. The initial concept of storing the data comprised of 4 data files, namely:-

                (i)   Failure information
                (ii)   Population information

(iii)   Component parts lists
(iv)   Failure analysis

However, since this required extensive programming in order to explore and retrieve the data, a structured database package was considered. Already in existence on the mainframe was a relational database (Multics Relational Data Store), which with an efficient database design, would fulfill the needs for storage, retrieval, updating and exploration of the data. Designing a relational database [4] for this hierarchical data structure posed a few problems. Some of the problems considered include:-

(i)   The need for updating the database. This is a basic consideration since the collection of data on equipments in the field is an ongoing situation.

(ii)   The hierarchical nature of the data does not naturally conform to the relational database structure.

(iii)   Minimising duplication in the data not only to improve the efficiency of storage and retrieval, but to reduce some of the workload of companies.

(iv)   The realisation of a database, which is not an accounting system performing the same routines periodically, but a database in which retrieval of data is flexible and exploration possible.

Together with advice from the Computer Centre at LUT, a final design was achieved.

As illustrated in Figure 2, the database consists of 4 relations. Relations 1 and 2 contain the parts list of a piece of equipment, i.e. the types and totals of subassemblies in a piece of equipment, with the corresponding component types and totals in each subassembly.

1. BOARD STRUCTURE

Co	Board Design	Component Type	No of Components

2. EQUIPMENT STRUCTURE

Co	Equip Type	Board Design	No of Boards

3. EQUIPMENT IDENTIFICATION

Co	Equip Type	Equip Serial No	Up and Running Date	End Date

4. FAILURE

Co	Equip Type	Equip Serial No	Board Design	Board Serial No	Up and Running Date of Board	Component Type	Time to Failure

Figure 2   Database Design

The third relation gives the details on the information concerning which equipments are in the field and over which periods they are operating. Relation 4 includes all the data on a component failure. For example:-

    (i)   which equipment it was in
    (ii)   which board it was part of
    (iii) when it failed
    (iv)   which position it failed in
    (v)   why it failed etc.

This design is now up and running and has proved to be working efficienctly. The advantages of this database design include:-

(i)   Updating - population updating need only occur in relation 3, the period of observation would change unless there has been a redesign of a piece of equipment or printed circuit board.

(ii)   Obviously, defining the structure of a piece of equipment (from relations 1 & 2) indicates that the company need only code and transfer this data once, thus reducing some of the work load of the companies.

(iii) Exploration of the data can take place by writing the necessary software. This exploration could take the form of, for example, differences in the reliability of different screened components, different encapsulation methods, different mounting technology etc.

These examples highlight the advantages of a database which has been designed for research purposes.

Apart from the relations discussed above, a fifth data file exists. This file contains the results of any failure analysis carried out on a particular failed component, and therefore provides additional information regarding this failed component. Included in one of the fields in the Failure Relation (relation 4) is a reference number which directly points to a failure analysis record in the fifth file, thus linking some of the component failures with their corresponding failure mechanisms and helps to highlight some of the differences between sources.

Since the collection and storage of data have been finalised, together with the compilation of a full working document, the protocols of data transfer and data validation were considered.

TRANSFER AND VALIDATION OF DATA

The data transfer procedure to LUT is displayed in Figure 3. The communication medium for three of the companies is magnetic tape, while the fourth, the DIA, communicates via IBM compatible floppy disks. Reading the magnetic tapes has not posed many problems. The data on floppy disks is transferred from the IBM PC AT using Kermit Protocols, and in general, is fairly successful. However, one recurring problem happens when non-ASCII characters are contained in one of the files. The file cannot be transferred to the mainframe until they are detected and deleted.

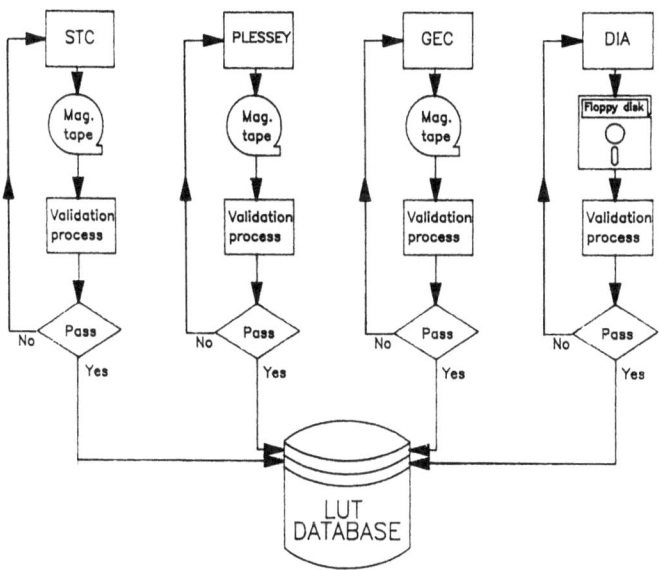

Figure 3   Data Transfer Procedure

After successfully reading the data into the system it is rigorously examined. This is known as data validation [5].

The data processing system must obviously guard against the possi- bility of erroneous data being admitted to the database. The probability of such errors occurring is very high, and the effect of not discovering errors is detrimental to the quality and integrity of the data within the database. Therefore, all input files are subjected to examination by a data validation program which exhaustively inspects all file records for possible errors. The detection of most errors is at rapid computer speed, but correction tends to be a human process which involves collecting background information, correcting the data and finally proceeding through the data preparation and transcription stages again.

The data validation procedure used in this study has three levels of checking.

Company Validation

Each of the companies has its own database system which stores and transforms the information into the format required by LUT. They each employ their own data processing systems, including data validation pro- cedures, before writing the data onto a magnetic tape.

Checks at Data Entry

After successfully reading and storing the data, the validation software is run. If the data passes, it is automatically entered into a database package before being pooled in the main component database. However, if the data fails the checks, the program flags up the erroneous fields and corresponding locations. Having detected the errors, amendments then have to be carried out (Fig. 4). The procedure for correcting erroneous data involves discussions with the relevant company contact. If the problems are minor, then amendments are made almost immediately. However, with more complex problems the complete data set would be investigated and amended by the company before being resubmitted to LUT.

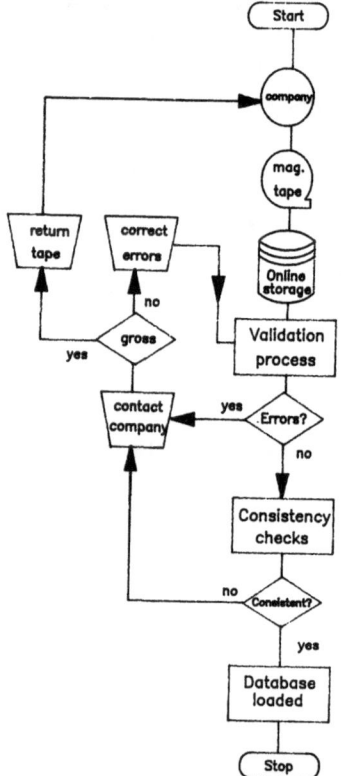

Figure 4   LUT Data Validation Procedure

The validation programs detect the following types of errors:-

(i)   Fields containing illegal characters. For example, numeric fields containing I instead of 1.

(ii)   Duplicate records.

(iii)   Wrong Codings – the component generic description code contains, in general, three characters. Errors have occurred when, at most, only two of these characters have been present in the field. A list of all valid component types exists. Therefore, if anything other than these occur, an error is flagged. An example of this would be TIA instead of T1A. Obviously, this is a keying error.

(iv)   Out of Range Values – an error will occur if the equipment operating time is less than the circuit board operating time. Similarly, if circuit board up and running date is less than manufacturing date.

(v)   Missing Information – contained in the working document is a list of essential fields. This means that these fields must be fully completed in the data format. If any of this essential data is missing, then an error will occur. For example, missing screened level of a component, missing mounting, substrate or encapsulation of a component type.

(vi)   Numeric Fields – previous use is recorded in days, thus if either a letter or a space occurs in this field an error message is flagged.

(v)   Valid Dates – for example, if the month number is zero or greater than 12, then an error has occurred.

(vi)   Length of Fields or Records

These are practical examples of the types of checks employed. The errors mentioned above are actual problems encountered during validation.

Consistency Checks Using Database Manipulation

These routines use the database manipulation procedures which check the consistency across the four relations. Similarly, if the files being validated are updates, the routines check the consistency between those records previously stored and validated for this data set. For example, the type of questions asked include:-

(i)   Are the failed component types in the population?

(ii)   Are all the failed equipments and pcb's described in the population?

(iii) Is the operating time of a piece of equipment less than its total observation period.

If any errors are flagged while running these procedures, then discussions are held with the relevant company to obtain the best possible solution to the problem.

On average, this validation loop including company interchange has a cycle of two days. However, this cycle is dependent on the complexity of the errors identified.

Apart from computer validation procedures, manual exploration of the data is carried out. This involves scanning across each field and down each record, searching for any data which may appear odd. For example, in one data set, it was found that the serial number of a circuit board was the same as the serial number of the equipment, high-lighting the need for further investigation of the data. Identifying this occurrence aided in the comprehension of the repair loop for the particular equipment under discussion.

During discussions concerning the data collection, a debate concerning the need for validation checks occurred within the Working Party. It was felt, by some members, that the validation procedures carried out by the companies would be sufficient to detect the majority of errors. However, this has been shown to be untrue. Indeed, the validation process described above has proved to be successful in identifying erroneous data and thus ensuring the quality and integrity of the data in the pooled component database.

OVERVIEW OF EARLY RESULTS

Much of the work to date at LUT has been concerned with the setting up of the database, the coding format and the establishment of dependable sources of data. Therefore analysis of the data is in the preliminary stages.

The database currently contains information on 70 out of 125 different component generic descriptions with an approximate total of 3.9 x 10^{10} component hours. Figure 5 illustrates the percentage number of components in the data base per component grouping.

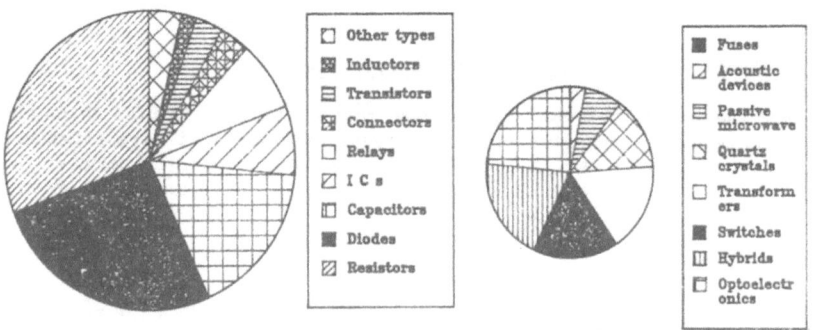

Percentage breakdown of other types.

Figure 5   Percentage Number of Components in the Database broken down by Component Type

Failure rate calculations have been carried out together with their associated 95% confidence limits, as shown in Figure 6. The overlapping of confidence limits around the failure rates of different component types, indicates that the confidence surrounding this hierarchy of problem components is low. One slight change in any of the failure rates would most certainly result in an alteration of the hierarchy. Similarly, the confidence surrounding each individual failure rate is low, in particular those components showing the highest failure rates. Figure 6 highlights those component types for which more data is needed. However, as more data arrives, the confidence limits will become narrower and eventually the hierarchy of problem components will reach a steady state. The legend describing the component types in Figure 6 is shown in Table 1.

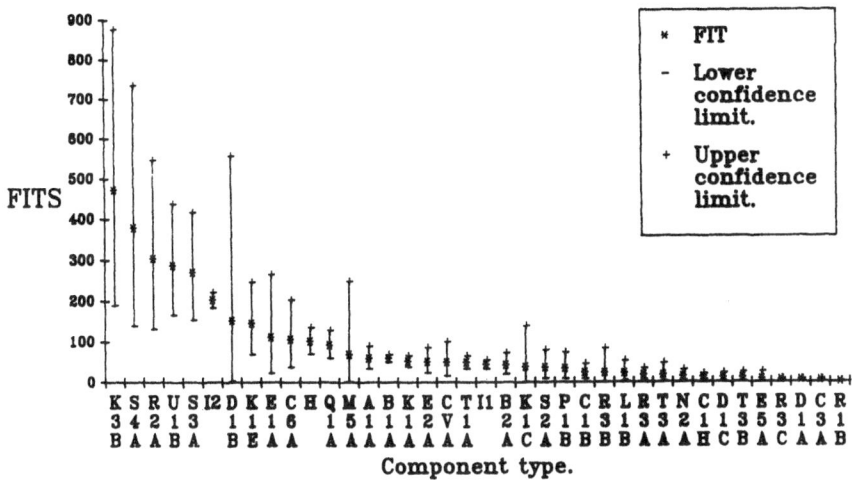

Figure 6   95% Confidence Limits for Failure Rates per Component Type

TABLE 1

LEGEND FOR FAILURE RATE GRAPH (Figure 6)

COMPONENT TYPE	COMPONENT CODINGS	DESCRIPTION
Integrated Circuits	I1	Bipolar
and Hybrids	I2	MOS
	H	Hybrid
Optical Devices	E1A	Photosensors
	E2A	LEDs
	E5A	Optocouplers
Transistors	T1A	Bipolar
	T3A, T3B	MOS
Diodes	D1A	PN-Junction
	D1B	Varactor
	D1C	Avalanche
Switches	S2A	Rocker
	S3A	Keyboard
	S4A	Rotary
Relays	B1A	Coil Activated
	B2A	Reed
Connectors	K1A	Rectangular
	K1C	Cylindrical
	K1E	Coaxial
	K3B	Feed Through Filter
Transformers	A1A	General Types
Fixed Resistors	R1B	Carbon Film
	R2A	Metal Oxide
	R3A	Wirewound
	R3B	Metal Foil
	R3C	Metal Film
Variable Resistors	P1B	Carbon Film
Non Linear Resistors	N2A	Varistor
Capacitors	C1B	Aluminium, foil, solid electrolyte
	C1H	Tantalum, sintered, solid electrolyte
	C3A	Ceramic Multilayer
	C6A	Polystyrene Foil
	CVA	Tuning/Trimmer Variable
Acoustic Devices	U1B	Loudspeaker
Quartz Crystal	Q1A	Crystal Oscillator
Inductor	L1B	Loaded Inductor
Passive Microwave Devices	M5A	Filter

IMPLICATION FOR THE FUTURE

Despite major progress, it is of course early days in the development and use of the database. Initial results are promising, but it is only as the data content builds up that meaningful comparisons and assessment, and the physical investigation of major failure mechanisms, will become possible.

As expressed above, in carrying out such comparisons and analysis of electronic component reliability, an exploratory approach will be taken. Exploratory Data Analysis (EDA) is now an established methodology in statistical analysis [6], and is beginning to be widely used in Reliability work [7]. The emphasis is on searching the data for unexpected structure that is then employed in its analysis, rather than on the confirmation of a priori assumptions.

One particular statistical tool to be used in conjuntion with this exploratory approach is Proportional Hazards Modelling (PHM) [8 & 9]. PHM can in a sense be regarded as a generalisation of the MIL-HDBK-217 approach [10]. In contrast to MIL-217, however, PHM does not assume constant failure rates, nor a priori determined π-factor. Indeed, the approach makes no distributional assumption nor assumptions as to what factors effect the failure rates of components, nor how or by how much. Instead, these are determined by the data.

The fundamental equation on which PHM is based is the assumed decomposition of the hazard function for a component into the product of a generic or base-line hazard-rate (ho(t)) and a term incorporating the effects of possible explanatory factors, $(Z_1, \ldots, Z_k)$ corresponding to component characteristics, use and environment:

$$h(t;z_1,\ldots,z_k) = h_o(t)e \exp (\beta_1 z_1 + \ldots + \beta_k z_k)$$

The β_i's are the parameters, like π-factor, that can be estimated from the data by the method of Partial Likelihood and tested for statistical significance. The base line hazard can then be studied using hazard plots, graphical and other methods.

In the current context, like most engineering applications, it is likely to be necessary to try several versions of the model exploratorily for a particular component type, with varying time metrics, and methods of defining factors. Any attempt to predict the nature of the results likely to be obtained is accordingly premature.

The work of the project has, however, already established many important precedents with implications for the future. Not least of these is the establishment of an active and co-operating working group with participation from competing companies, academic institutions and MOD. This achievement, together with the construction of an efficient data acquisition system and a feedback loop on field problems, bodes well for the future of the project and is very much in line with the principles of R & M 2,000 and the move from prediction to achievement (IEEE 1987). It is to be hoped that in the fullness of time, by co-operation throughout Europe, the base will take on, or become part of, a wider European perspective.

ACKNOWLEDGEMENTS

The work is being carried out with the support of the Procurement Executive, Ministry of Defence, and we are grateful to them for permission to publish this paper.

REFERENCES

1. Campbell, D. S., Hayes, J. A. and Hetherington, D. R., 'The organisation of a study of the field failure of electronic components', Quality and Reliability Engineering International, 3, pp 251258, 1987.

2. 'Reliability prediction of electronic equipment', Military Handbook, October 1986.

3. Blanks, H. S., 'The generation and use of component failure rate data', Quality Assurance, 3, No. 3, pp 85-95, 1977.

4. Date, L. J., 'An Introduction to Database Systems', Addison-Wesley, 1976.

5. Elder, J., 'Construction of Data Processing Software', Prentice-Hall International Inc., pp 146-168, 1984.

6. Tukey, J. W., 'Exploratory Data Analysis', Addison-Wesley, 1977.

7. Bendell, A. and Walls, L. A., 'Exploring reliability data', Quality and Reliability Engineering International, 1, pp 37-51. 1985.

8. Cox, D. R., 'Regression models and life-tables', J. Roy. Statist. Soc. Sci. B., 34, pp 187-220, 1972.

9. Wightman, D. W. and Bendell, A., 'The practical application of proportional hazards modelling', Reliability Engineering, 15, pp 29-53, 1986.

10. Landers, T. and Kolarik, W., 'Proportional hazards models and MIL-HDBK-217', Microelectronic Reliability, 26, No. 4, pp 763-771, 1986.

THE RELIABILITY ASSESSMENT OF OFFSHORE STRUCTURES UNDER THE INFLUENCE OF FATIGUE CRACK GROWTH

J.C.P. Kam
NDE Centre
Department of Mechanical Engineering
University College London
Torrington Place
London WC1E 7JE

ABSTRACT

Reliability concept has been applied to the offshore steel structural design for some years. The design and analysis, however, was concentrated on the strength aspects of welded joints. The applications in fatigue crack growth assessment have only been started very recently. It is partly because the fatigue behaviour of the special welded joints was not well known; and also because the reliability of the crack inspection / monitoring techniques was not well established. With the recent progress in fatigue fracture mechanics methodology and findings in the inspection reliability evaluation, it is now possible to apply the reliability concept in assessing the effects of fatigue cracking upon the integrity of offshore steel structures.

This paper briefly reviews the literature, highlighting the advances of the applications of reliability techniques in offshore fatigue analysis.

Furthermore, a newly developed probabilistic fracture mechanics crack growth model is presented, and the applications of such a model in reliability assessment and inspection planning are discussed and illustrated with practical examples.

INTRODUCTION

The increased offshore activities worldwide and particularly in the North Sea during the late sixties and seventies, have promoted the research in many areas of advanced structural engineering. The general applications of reliability methods, however, did not emerge until very recently. This was partly because the physical understanding of much offshore structural behaviour was not well established and, partly because reliability methods had not been widely used in civil and structural engineering.

The introduction of reliability methodologies into structural engineering was to assist in developing limit state design codes [1-4] such that all parts of a structure can be designed to have similar reliability (or probability of failure) under different modes of failure. The methodologies employed were essentially the analyses of component reliability (as opposed to system reliability), and these remain the most important areas of reliability research for offshore structures and is the first area to be reported in this section.

Component reliability concerns mainly with the strength (S) and load (L) (or supply and demand) relationship. A state function which describes the component behaviour is expressed as,

$$g = S - L \qquad (1)$$

where g, S, L are all random variables (Figure 1). When $g \geq 0$, the component is "safe". When g < 0, the component fails.

If g is normally distributed, $N(\mu_g, \sigma_g)$ (Figure 2), then $(g - \mu_g) / \sigma_g$ is standard normal N (0,1). μ_g and σ_g are the mean and the standard deviation of g respectively. The probability of failure (POF) becomes,

$$
\begin{aligned}
POF = CD_g(0) \\
= \Phi(-\mu_g / \sigma_g) \\
= 1 - \Phi(\mu_g / \sigma_g)
\end{aligned}
\qquad (2)
$$

where $CD_g(x)$ is the cumulative distribution of g at g = x. Φ is the standard normal cumulative distribution function.

The safety index (or reliability index) β, is defined so that, even for the case when g is not normal, a simple index can be used to indicate the relative level of reliability. In the case of normal distribution,

$$\beta = \mu_g / \sigma_g \qquad (3)$$

which is a ratio relating the mean of g and the dispersion in g. This index, in general reliability literature, is also known as the safety margin. This second name, however, is used in structural engineering to refer to "g".

The component reliability analyses have found many applications and have been used in almost every modes of strength (failure) assessment including yielding (axial or bending), buckling and instability, shearing and punching, slab strength, fracture and fatigue [such as 5-10]. However, the majority of investigations have been concentrated on the design aspects of offshore structures. During operation, the main cause for the degradation of structural reliability has been identified as fatigue [11]. Although there are a large number of platforms requiring maintenance due to past and continuing offshore activities, it was not until very recently there appeared reliability analysis which bring together fatigue crack growth and inspections for the maintenance of offshore structural integrity and reliability [12, 13]. The safety index (β) monitoring strategy was pioneered by AS Veritas of Norway and has the potential of optimizing the inspection intervals to ensure integrity.

The second area to be reported is the applications of system reliability. These methodologies are probably better known in general reliability research. The main applications have been to estimate the overall risk (such as the risk of collapse of offshore structures) [14, 15]. There were also applications in assessing the multi - modes of failure of a structural component.

The latest development in the European BRITE projects [16], extends the methodology into design optimization of offshore structures. The optimized design is reached under a set of constraints formulated in terms of reliability requirements. Multi-failure modes are automatically taken into account. Theoretically, these techniques can be extended to included inspection to formulate optimal maintenance policies for the overall structure.

However, before optimal maintenance policies can be formulated, the reliability of inspection must also be known. This is the third area of applications of reliability methods to be reported in this paper.

NON DESTRUCTIVE INSPECTION (NDI) RELIABILITY

The methods employed in this area are different from those used in the reliability research reported above. Other industries such as the aerospace and the nuclear industries have carried out large trial programmes to study the reliability of different inspection methods. The trials relevant to offshore industry have only been started very lately [17]. The major lessons learnt in all these trial programmes can be summarised as the following:

(a)   NDI reliability in realistic (in situ) situation differs significantly from ideal (laboratory) situation. Therefore, trials must be carried out under realistic simulation of the working environment.

(b)   The performance of sizing accuracy is significantly inferior to the performance of general detection. The best solution is to link the definition of detection with sizing performance.

(c)   The binomial statistics are generally acceptable for analysing the trial results.

The new offshore NDI trial programme has taken all these factors into account. A new sizing performance linked detection statistics has been defined and is called the overlap accuracy (Figure 3). This accuracy criterion can be applied to defect length or area measurements. In the case of measuring the depth of surface breaking cracks (these are typical in fatigue situation), the accuracy criterion is slightly simplified (Figure 4). By defining a required level of accuracy, a probability of detection (POD) curve can be constructed from the inspection trials. An ideal POD curve is shown in Figure 5. An important feature concerning the POD curves is that the critical crack size (a $_a$) varies according to the specified accuracy and confidence levels. There are many other features concerning this type of POD curves and are currently being explored extensively. This paper will restrict itself to only one of the many applications of such a curve. Figure 6 shows a typical POD curve for a good, practical NDI system. Normally, the POD curve does not reach the 100% level until the crack is very large or unless the accuracy requirement is greatly reduced.

FATIGUE CRACK GROWTH BEHAVIOUR AND MODELLING

The main feature in fatigue crack growth in offshore welded tubular joints is that, more than half of the fatigue life is usually spent on crack propagation. In other words, there will be relatively large cracks existing in the joints even within the "safe" design life.

Fatigue crack growth is described by three fracture mechanics equations. First of all, the Paris crack growth law,

$$da/dN = C(\Delta K)^m \tag{4}$$

where da / dN is the instantaneous crack growth rate (in m/cyc). C and m are materials properties, the values of which depend on the materials, the loading conditions and the environment. ΔK is the stress intensity range and can be expressed in the second equation,

$$\Delta K = Y S_c S_h \sqrt{\pi a} \tag{5}$$

where a is the instantaneous crack size. S_h is the equivalent applied stress range and S_c is the so called "hot spot" stress concentration factor. The stress information normally comes from load and stress analyses of the structure. Y is the stress intensity modification which is related to the crack shape, dimensions and stress distribution over the cracked region.

Early investigations have shown that experimental data can be reasonably described [18] by the third equation,

$$Y = F(T/a)^l \tag{6}$$

where T is the chord wall thickness of the joints. The values of F and l depend on the fatigue loading and joint geometries.

The offshore industry has been attempting to predict fatigue crack growth by using many different numerical and theoretical models to predict the Y functions. In fact, the predictions made by any reasonable crack growth models could be expressed in the form of equation (6). Therefore, the following discussion is general and is not restricted to the particular model used in this paper.

The crack size grown from an initial size distribution a_i after N_L stress cycles (or after time $t_L = N_L / N_y$ where N_y is the cyclic frequency of stress cycles) can be evaluated by integration, using equations (4), (5) and (6). The final crack size distribution is therefore,

$$a = \left\{ vCS_h^m S_c^m \pi^{\frac{m}{2}} F^m T^{ml} N_y t_L + a_i^v \right\}^{\frac{1}{v}} \tag{7}$$

where $v = 1 + ml - m/2$

All variables in equation (7) can be random variables having different types of probability density distribution. The effect can be analysed by the standard level II structural reliability method.

LEVEL II RELIABILITY METHOD

The state function of the fatigue reliability can be expressed in terms of the crack size criterion,

$$g = a_f - a \tag{8}$$

where a_f is the failure crack size and can also be a random variable and a is given in equation (7). The probability of failure is therefore the convolution,

$$POF = \int_{g<0} f_{g(X)}(X) dX \tag{9}$$

where X is the vector containing all the variables (x_i) in equation (7) and (8), and $f_{g(X)}(X)$ is the probability density function of g expressed in terms of the variables.

The level II reliability method evaluates the safety index β by making use of the fact that β is the minimum distance from the origin to the tangent hyperplane to the failure surface ($g = 0$), in the transformed space. The transformed space consists of multiple dimensions. Each dimension represents one normalised state variable,

$$\frac{x_i - \mu_i}{\sigma_i} \tag{10}$$

The idea can been seen in Figure 7.

If x_i is not normally distributed, μ_i and σ_i (the mean and standard deviation) will be assigned some special values so that the cumulative distribution function is the same as the normal distribution at the failure point. For a more detailed description of the method and for solution algorithms, reference should be made to texts such as [2] and [20].

APPLICATIONS

Having understood the fatigue crack growth behaviour and NDI reliability, the maintenance of offshore structures can be studied from a new perspective. The following examples will demonstrate the power of the analysis technique outlined above.

(a)   Example A: Normal Operation

A tubular welded T joint under axial loading has been designed with the following parameters,

Chord: Wall thickness = 45 mm
            Diameter = 927 mm
Brace: Wall thickness = 45 mm
            Diameter = 658 mm

N_y = 10 Million cycles / year, Design life = 20 years.

Table 1 also contains the random variables used in the analyses. Obviously, all variables in Equations (7) and (8) could be expressed as random variables. Normally, the uncertainties in crack growth and modelling are expresses in C and F respectively and therefore the values m and l can be assumed constant.

The C statistics (and m values) are obtained from materials testing. F and l are determined by an engineering crack growth model, TPM [21]. The statistics of F were obtained by comparing the model prediction with experimental data. As discussed above, other values of F and l obtained by any reasonable prediction model could also be used. The statistics of a_i were obtained by Wolfram [11] from back extrapolating experimental fatigue crack growth data. The statistics for S_h and S_c are typical for offshore applications. Finally, the failure crack size is taken as the chord wall thickness as experience shows that this is a realistic definition of failure. In the examples, N_y is assumed to be constant. The NDI to be used has $a_\alpha = 6$ mm at a probability of detection at 90% and the accuracy criterion is 80%.

The safety index (β) development curves are shown in Figure 8. Usually inspections are carried out at 5 years intervals (as suggested by the Department of Energy [22]). Taking into account the unreliability of the NDI, β is restored after an inspection, but to a lower level. The level of β has also become lower at the time of the second and the third inspections. For a consistent reliability requirement at the time of all inspections, it can be suggested that either the first inspection could be delayed to the 6.8 th year or the inspection intervals except the first one should be shortened to 4 years.

(b)   Example b: Crack Found In-Service

If a crack of 10 mm deep is found in the joint after 10 years in service, what is the effect on structural integrity?

The analysis in Example A really only applies to situations where no crack is found. The current offshore practice is to repair all defects identified in service. Due to the pressure of low oil price, high maintenance cost and as the platforms are aging (and therefore more likely to have cracks), this "repair all" practice may no longer be the most effective policy. An alternative practice is to assess the effect of the crack on the structural integrity and then make a decision on repair using that information. The matter is further complicated by inspection unreliability.

A new approach which is called the Worst Case Growth Curve (WCGC) can be used to formulate an objective assessment and repair criterion. The approach requires the construction of a worse case growth curve (WCGC) so that the reliability (safety index) on the curve is the same at all crack depths. The common index can be specified by legislation or chosen to be the same as the original β at the end of the design life. The latter is chosen in this example and β is 1.32. If the crack grows faster than the WCGC, there is a case for repair, otherwise the crack is considered to be growing within the design.

The NDI reliability data is incorporated into the WCGC by constructing two extra lines, the alert line (AL) and the significant line (SL) (Figure 9). The two lines are determined in the following way. If 90% of the trial data fall within the 80% accuracy limit (80% to 125% of the true size), the coefficient of variance in sizes will be 13.7%. If a 95% confidence is required from the in-situ inspection sizes, the AL and SL should therefore be set at 0.67a and 1.38a for all a > a_α respectively.

If the measured crack size falls below AL, the crack can be left alone with confidence that it will not propagate faster than the expected rate (at $\beta = 1.32$), until the next inspection. If the measured size is above the SL, remedial actions are inevitable. Only when the size falls between AL and SL, there will be needs for reinspection and possibly re-analysis. Therefore, this approach provides a rational and fast assessment of cracks found in service. In the example, the crack having a measured size of 10 mm in the 10th year can be left until the next inspection.

The β development curves will also be changed (Figure 10). If inspection is required at the same level of reliability as the 10th year, the next inspection should be carried out in the 14.4th year.

(c) Example C: Extended Life

At the end of the 15th year, the life of the structure is required to be extended for another 5 years (possibly due to breakthrough in field exploitation technology or new field discovered nearby). An inspection carried out at that time showed that the crack is still at 10 mm. The new WCGC, AL and SL based on this new information and the extended life requirement can be constructed (Figure 11). On this occasion, β is chosen to be 1.32 again. If the crack information is not used to update the β development curve, Figure 10 can be used and inspections can be repeated until the end of the 25th year.

(d) Example D: Increased Loading

Another common problem is caused by increased loading. After 18.8 years of service, due to better information on wave loading and the necessary upgrading of the deck facilities, the mean nominal stress range is increased to 1.75 MPa, but the coefficient of variation is reduced to 0.25.

It can be observed from the WCGC curve (Figure 12) that the crack growth rate has increased, and this is due to the increase in the mean value of the loading. Moreover, the increased reliability of the loading information extends the inspection interval to 4.6 years (Figure 13). However, the rate at which β decreases will be higher at the later stage of the fatigue life.

DISCUSSION

This paper has chosen one of the most ideal cases for probability of detection. In reality, the POD tends to increase steadily with crack sizes. To take this into account, alternative AL and SL can be constructed with variable intervals at different crack depths. The uncertainties in size distribution after undetection will also be increased. And for example, in Figure 9, the bounds contained by AL and SL when the WCGC is below 6 mm should be larger than shown.

The β development curves could be expressed in the equivalent probability of failure (Figure 14). The POF information can be used for the evaluation of the expected inspection and failure costs. Coupled with the crack growth and the monitoring curves, the expected costs for different repair criteria can also be found. The "repair all" approach is the same as drawing a line parallel to the time axis and passing through the 6 mm mark. Any crack size above the line will be repaired and those below will be left alone. Obviously, a more flexible crack acceptance criterion such as using the AL, can be more cost effective and at the same time, the safety requirement is not being infringed.

The WCGC and β curves will therefore provide essential information in order to evaluate different inspection, repair and maintenance (IRM) policies for maintaining offshore structural reliability. Further development involves the applications of these analyses in formulating optimum IRM policies.

CONCLUSIONS

The applications of the reliability methodologies in offshore structural engineering have been reviewed briefly. It appears that the physical understanding of the fatigue crack growth, which causes the major degradation of structural integrity during operations, has progressed to the stage where reliability methods can be used in assessing structural integrity. Coupled with the NDI reliability information, methods are developed in order to assist decision making concerning inspection, repair and maintenance (IRM).

Simple examples have been used to demonstrate the potential of the β monitoring and WCGC approaches in planning and assessing IRM policies under different circumstances.

Table 1 Statistical Input For The Examples

	μ_i	σ_i / μ_i	Type
Sh	1.5 MPa	0.4	Lognormal
Sc	10.0	0.15	Normal
a_i	1.4 mm	0.48	Normal
F[1]	0.5152	0.028	Lognormal
C[2]	4.5e-12	0.03	Lognormal

1 : The corresponding l = 0.3566
2 : The corresponding m = 3.3 and da / dN is in m/cyc

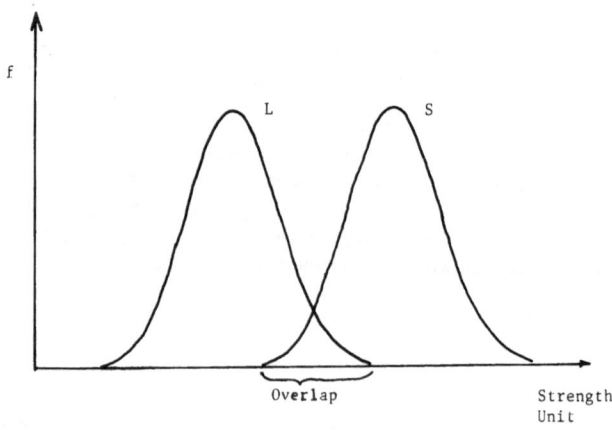

FIGURE 1   DIAGRAM SHOWING THE DISTRIBUTIONS OF
STRENGTH AND LOADING

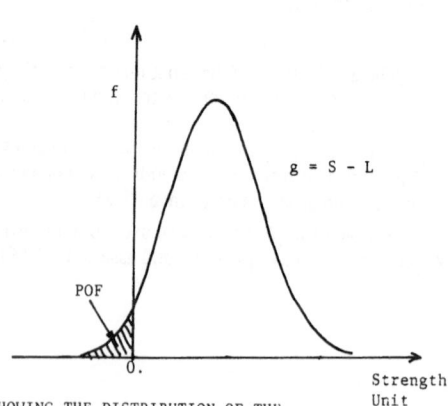

FIGURE 2   DIAGRAM SHOWING THE DISTRIBUTION OF THE
SAFETY MARGIN (g)

OVERLAP = MIN (A/L, A/D)

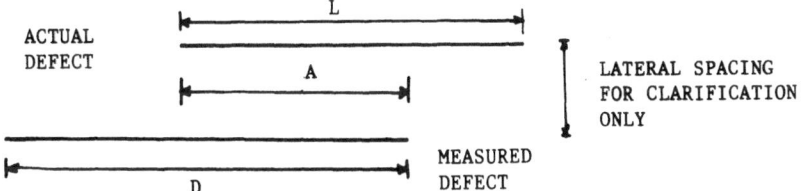

ACTUAL
DEFECT

LATERAL SPACING
FOR CLARIFICATION
ONLY

MEASURED
DEFECT

FIGURE 3   COMPARISON OF MEASURED AND
ACTUAL DEFECTS USING AN OVERLAP CRITERION

OVERLAP = MIN (D/L, L/D)

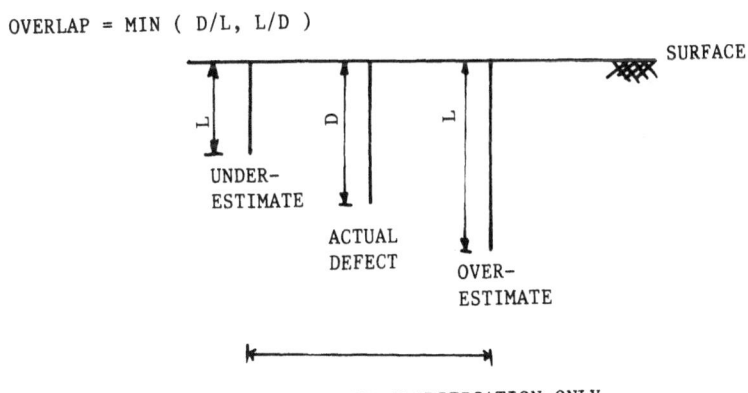

SURFACE

UNDER-
ESTIMATE

ACTUAL
DEFECT

OVER-
ESTIMATE

SPACING FOR CLARIFICATION ONLY

FIGURE 4   OVERLAP CRITERION USED IN
DEPTH MEASUREMENTS

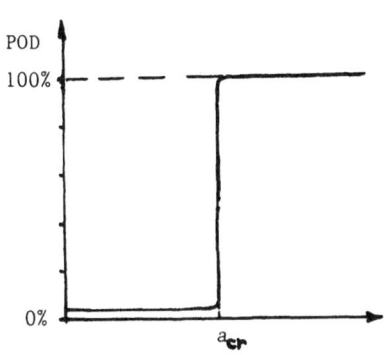

POD
100%

0%

a_{cr}

FIGURE 5   THE POD CURVE FOR AN   IDEAL
INSPECTION   SYSTEM

62

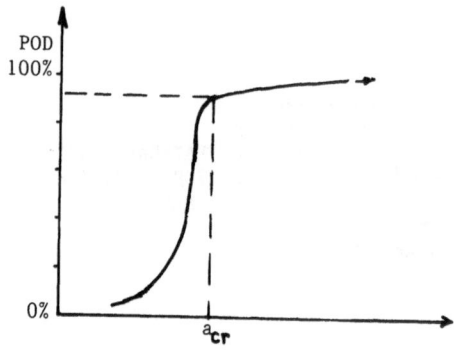

FIGURE 6   THE POD CURVE FOR A GOOD,
            AND PRACTICAL INSPECTION SYSTEM

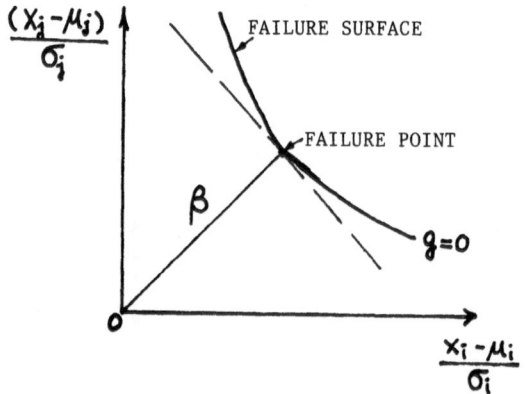

FIGURE 7   LEVEL II RELIABILITY
            SHOWING β IN THE TRANSFORMED
            TWO DIMENSIONAL SPACE

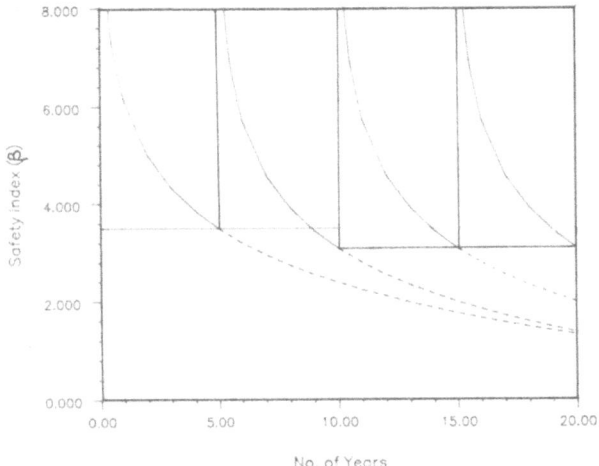

FIGURE 8   Time History of The Safety Index (β)
With Imperfect Inspection
At Fixed Inspection Intervals

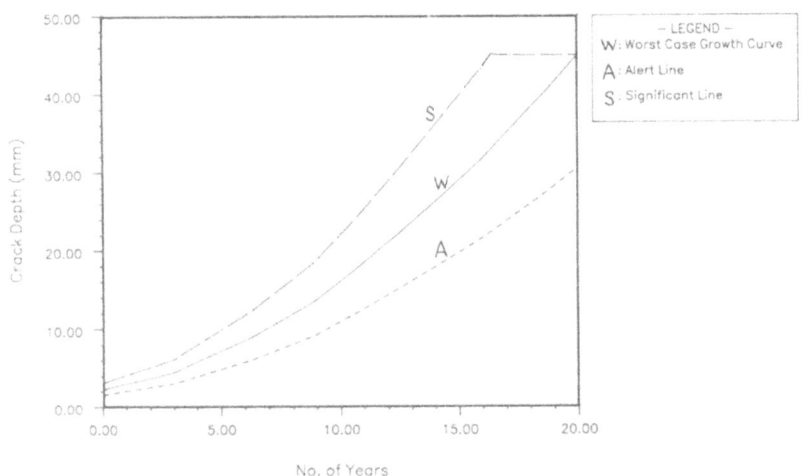

FIGURE 9   The Worst Case Growth and Monitoring Curves at $\beta = 1.32$

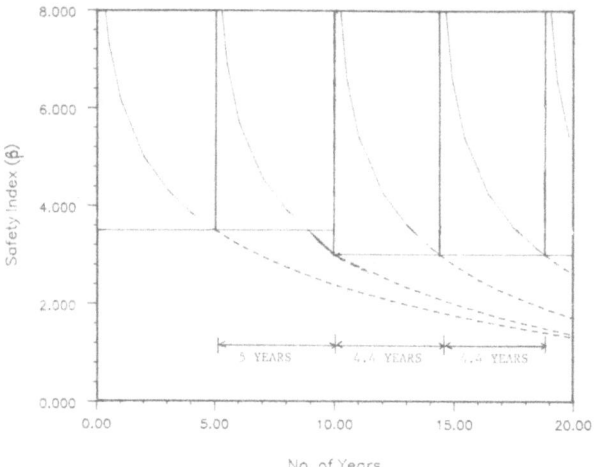

FIGURE 10 Time History of The Safety Index (β)
With Imperfect Inspection
And Crack Found In Service

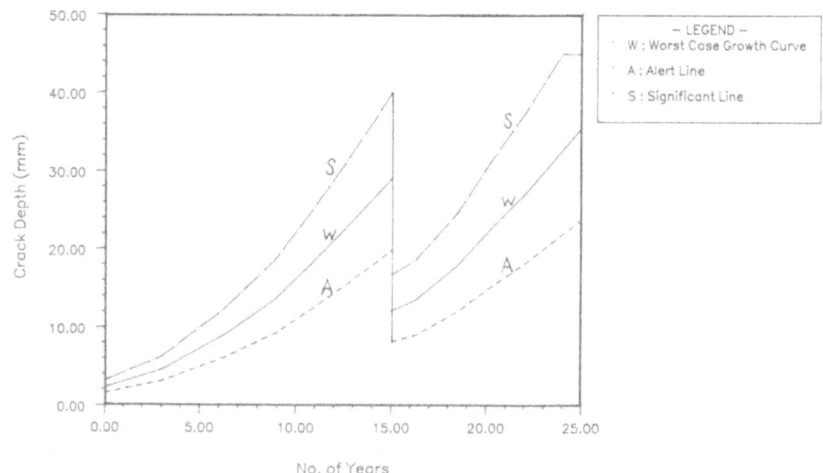

The Worst Case Growth and Monitoring Curves at $\beta = 1.32$
for extended life

FIGURE 11

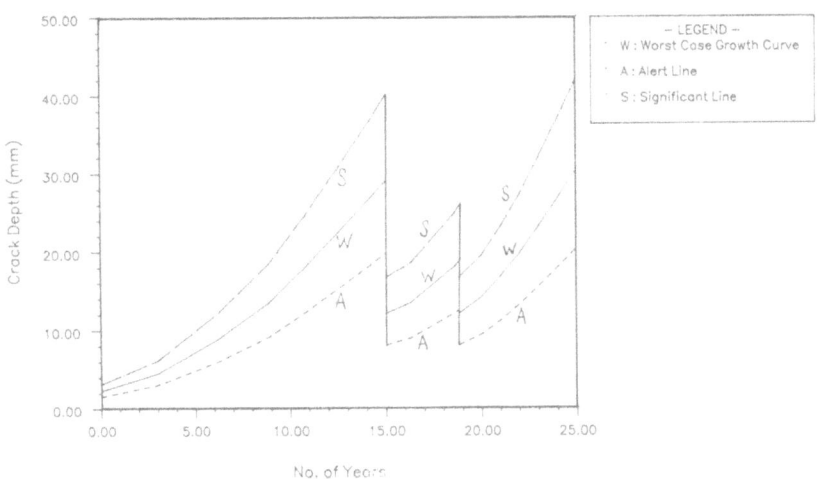

The Worst Case Growth And Monitoring Curves At $\beta = 1.32$
For Extended Life And Increased Loading

FIGURE 12

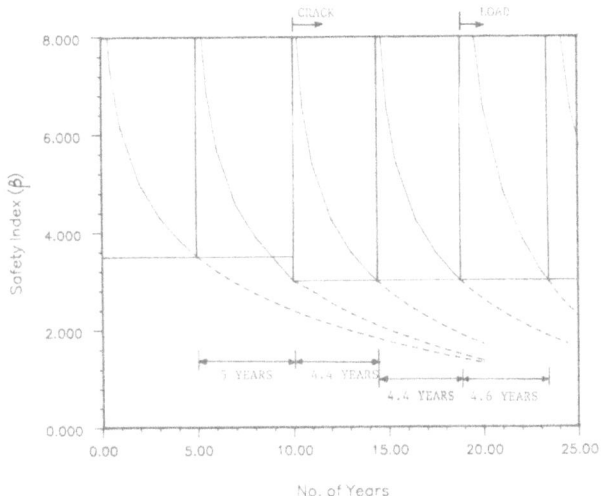

FIGURE 13   Time History of The Safety Index (β)
With Imperfect Inspection
And Increased Service Loading

66

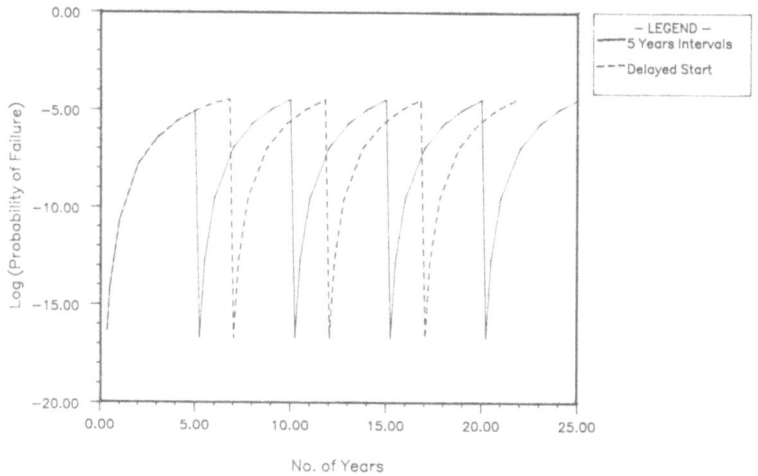

FIGURE 14   Time Histories of The Probability Of Failure
Under Two Inspection Policies

REFERENCES

1.   BSI: CP110: The Structural Use of Concrete. British Standards Institution, London 1972.

2.   CIRIA: Rationalisation of Safety and Serviceability Factors in Structural Codes. Construction Industry Research and Information Association, London, CIRIA Report 63, 1977.

3.   BSI: BS6235: Fixed Offshore Structures. British Standards Institution, London, 1979.

4.   BSI: BS5400: Steel, Concrete and Composite Bridges. British Standards Institution, London 1982.

5.   Faulkner, D., et al, Limit State Design Criteria for Stiffened Cylinders of Offshore Structures. Proc. ASME 4th National Congress of Pressure Vessels and Piping Technology, Portland, OR, June, 1983.

6.   Crohas, H., et al, Reliability of Offshore Structures Under Extreme Environmental Loading. Proc. Offshore Technology Conf., Houston, Paper OTC4826.

7.   Wirsching, P.H., Probability Based FAtigue Design Criteria for Offshore Structures. Final Report API PRAC 15. American Petroleum Institute, January 1985.

8.   Frieze, P.A., Plane C.A., Partial Safety Factor Evaluation for Some Aspects of Buckling of Offshore Structures. A Pilot Study Based on BS5400 - Part 3. Prepared by the University of Glasgow for the U.K. Department of Energy. To be published by HMSO, London.

9.   Madsen, H.O., Moghtaderi-Sadeh, M., Reliability of Plates Under Combined Loading. Proc. of Maine Structural Reliability Sym., Arlington, VA, 1987.

10. Plane, C.A., The Determination of Safety Factors for Defect Assessment Using Reliability Analysis Methods. Proc. of 3rd Int. Conf. on Integrity of Offshore Structures, Glasgow, 1987.

11. Wolfram, J., The Effects of Fatigue Cracking upon the Reliability of Tubular Members of Offshore Steel Structures. Proc of 5th Int. Offshore Mechanics and Arctic Engineering Sym., Tokyo, 1986.

12. Madsen, H.O. et al, Experience on Probabilistic Fatigue Analysis of Offshore Structures. Proc of 5th Offshore Mechanics and Arctic Engineering Sym, Tokyo, Japan 1986.

13. Madsen, H.O. et al, Probabilistic Fatigue Crack Growth Analysis of Offshore Structures with Reliability Updating Through Inspection. Proc. of Maine Structural Reliability Sym., Arlington, VA, 1987.

14. Guenard, Y.F., Application of System Reliability Analysis to Offshore Structures. John A. Blume, Earthquake Eng. Centre, Standford Univ., Report No. 71, Nov., 1984.

15. Bjerager, P., System Reliability of Offshore Jacket Structures by Ideal Plastic Analysis. Proc. of Maine Structural Reliabilty Sym., Arlington, VA, 1987.

16. Thoft-Christensen, P., Private Communication.

17. Underwater NDE Centre Project on Inspection Reliability, NDE Brochure, Doc No. NDE/86/108.

18. Dover, W.D., Dharmavasan, S., Fatigue Fracture Mechanics Analysis of T and Y Joints, Paper OTC4404, Offshore Technology Conf., Houston 1982.

19. Hasofer, A.M., Lind, N., An Exact and Invariant First Order Reliability Format. Jnl. of Eng. Mechanics, ASCE, Vol 100, No. EM1, Feb., 1974, pp. 111-121.

20. Thoft-christensen, P., Baker, M.J., Structural Reliability Theory and Its Applications, Spring Verlag Publishers, 1982.

21. Kam, J.C.P., Dover, W.D., Structural Integrity of Welded Tubular Joints in Random Load Fatigue Combined with Size Effect. Proc. of the 3rd Int. Conf. on the Integrity of Offshore Structures, Glasgow, UK, 1987.

22. D.En, Offshore Installations: Guidance on Design and Construction. HMSO, London, 1984.

STATISTICAL APPROACHES TO FATIGUE
CRACK GROWTH MODELS

Martin Newby
Postgraduate School of Industrial Technology

and

Postgraduate School of Mathematical Sciences,
University of Bradford,
Bradford,
BD7 1DP,
UK

ABSTRACT

A critical review shows that the models widely used for fatigue crack growth are equivalent to a general Markov diffusion process. Parameter estimates are obtained and a method for grouping data is shown. The results are illustrated by the analysis of some crack growth data.

DISCRETE MARKOV MODELS

In characterising crack growth it is natural to assume that the state variable is continuous. However, a number of effective models are constructed on the assumption of a discrete state space. Since observations are most commonly a sequence (x_n, c_n) where x_n is a measure of crack length and c_n is the number of cycles. The data is thus essentially discrete and so the assumption is that if the underlying process is Markov then the observations x_n form a Markov process. Indeed by using embedding techniques the observation process may be Markov even if the underlying process is not. Substitution of a surrogate for x_n may also render the process Markov as is done in a series of papers by Bogdanoff and his co-workers (1-5).

The simplest assumption is to assume that the crack develops as a simple random walk with probability p of an increase and q of no change. The probability that the state variable X_n takes the value k is then the binomial

$$P(X_n = k) = \binom{n}{k} p^k q^{n-k}.$$

By proceeding to the limit (6) we find that x is normally distributed with density

$$p(x,t) = \frac{1}{\sigma\sqrt{2\pi t}} \exp\left\{-\frac{1}{2}\left[\frac{x - \mu t}{\sigma\sqrt{t}}\right]^2\right\}.$$

Krausz (7) offers an extension of the simple random walk by assuming that p and q are determined by the activation energies for the breaking and healing of bonds. More complex models may be constructed (8) but in the continuous limit all lead to a model of the same form (8).

In his paper (1) Bogdanoff and in the review (5) the discrete models are exploited by making use of the Markov property and the semi-group properties of the transition probability matrices, that is that if the crack stages (not necessarily length) are labelled by the integers then

$$P_{m+n} = P_m P_n$$

and it is easy to handle computationally. More illuminating is the representation (1) as a cumulative damage model

$$x_n = x_{n-1} + d_{n-1}$$

where d_n is the damage at stage n $-$ 1 which has a discrete distribution. In the simple random walk $d_n = 1$ or $d_n = 0$ with probabilities p and q.

DIFFUSION PROCESS MODELS

In the discrete models discussed above a passage to the limit (9) leads to a version of the diffusion equation

$$\frac{1}{2} \sigma^2 \frac{\partial^2 p}{\partial x^2} - \mu \frac{\partial p}{\partial x} = \frac{\partial p}{\partial t}$$

for $p(x,t)$, the density function of the length x at time t, where μ is the drift and σ^2 the dispersion parameter.

A more complex discrete model may be expressed through the Chapman-Kolmogorov equation for $P_{jk}^{(n)}$, the probability of going from stage j to stage k in n steps. If φ_i and θ_i are the probability of a positive unit step and a negative unit step at stage i and $1 - \varphi_i - \theta_i$ is the probability of no step, then

$$P_{jk}^{(n)} = P_{jk-1}^{(n-1)} \varphi_{k-1} + P_{jk}^{(n-1)} (1 - \varphi_k - \theta_k) + P_{jk+1}^{(n-1)} \theta_{k+1}.$$

The probabilities φ and θ may be state dependent. If a passage to the limit in this equation is taken then it can be shown (6) that

$$\frac{1}{2} \frac{\partial^2}{\partial x^2} \left[\sigma^2(x)p(x,t) \right] - \frac{\partial}{\partial x} \left[\mu(x)p(x,t) \right] = \frac{\partial p}{\partial t},$$

where $p(x,t)$ is the probability density of x at time t.

The functions μ and σ are the infinitesimal mean and infinitesimal variance defined as

$$\mu(x) = \lim_{\Delta t \to 0} \frac{E\left[X(t + \Delta t) - X(t) \big| X(t) = x \right]}{\Delta t}$$

$$\sigma^2(x) = \lim_{\Delta t \to 0} \frac{E\left[(X(t + \Delta t) - X(t))^2 \big| X(t) = x \right]}{\Delta t}.$$

The fact that μ and σ^2 are functions of x alone define the process as a homogeneous Markov process.

Under fairly mild conditions a more general continuous Markov process may be described by its transition probability $p(x,t)$ which satisfies the Fokker-Plank equation

$$\frac{1}{2} \frac{\partial^2}{\partial x^2} \left[\sigma^2(x,t)p(x,t) \right] - \frac{\partial}{\partial x} \left[\mu(x,t)p(x,t) \right] = \frac{\partial p}{\partial t}.$$

WHITE NOISE MODELS

Rather than starting from a model in which the stochastic nature of the problem is an intrinsic element other authors (9) have looked at ways of introducing randomness into the deterministic model. The natural way to do this is to look at random initial or boundary conditions and to assume that noise in measurement and other errors are superimposed on the deterministic model. This approach is taken by Sobczyk (9) who starts from the Paris-Erdogan law and looks at random effects. This model assumes that the deterministic law expressed in terms of length, x,

$$\frac{dx}{dt} = m(x,t)$$

is modified by a white noise so that

$$\frac{dx}{dt} = m(x,t)\left\{ 1 + \gamma(t)S(t) \right\}$$

where γ^2 is a variance and S represents white noise.

STOCHASTIC DIFFERENTIAL EQUATIONS

An alternative formalism (10) may be used to represent the models described above. The direct construction of stochastic models which lead us to the Fokker-Plank equation also allows the growth of the random variable X_t, crack length at time t to be described by

$$dX_t = \mu(X_t,t)dt + \sigma(X_t,t)dW_t$$

where W_t denotes a Wiener process with $E(W_t) = 0$, $var(W_t) = \alpha^2 t$. μ and σ are defined exactly as in the Fokker-Plank equation.

The white noise equation is also reducible to a stochastic differential equation, an application of the Ito (10) stochastic calculus shows that the white noise equation is equivalent to

$$dX_t = \left[m(X_t,t) + \frac{1}{2} \gamma^2 m(X_t,t) \frac{dm(X_t,t)}{dx} \right] dt + m(X,t)\gamma dw_t$$

$$= \mu_1(X_t,t) dt + \sigma_1(X_t,t) dW_t.$$

so that both models reduce to the same form.

COMMENTS

The survey and discussion above show that models which assume that crack growth is intrinsically random and those that assume that the process is deterministic but that the observations superimpose randomness both lead to models of the same form, that is as a stochastic differential equation

$$dX_t = \mu(X_t,t) dt + \sigma(X_t,t) dt$$

or with transition density $p(x,t)$ given $x = x_0$ when $t = t_0$ as a solution of

$$\frac{\partial}{\partial x} \left[\frac{1}{2} \sigma^2(x,t) \frac{\partial p}{\partial x} \right] - \frac{\partial}{\partial x} \left[\mu(x,t)p \right] = \frac{\partial p}{\partial t}.$$

RESULTS

Using the Paris-Erdogan equation in the form

$$\frac{dx}{dt} = kx^{2m}$$

Sobczyk (9) shows that x by imposing white noise satisfies the stochastic differential equation

$$dX_t = \left[\alpha X_t^m + \frac{1}{2} \beta^2 m X_t^{2m-1} \right] dt + \beta X_t^m dW_t$$

and that

$$Y_t = X_t^{1-m}$$

satisfies

$$dY_t = -(m - 1)\alpha dt - (m - 1)\beta dW_t.$$

Given the initial value $Y_0 = X_0^{1-m}$ then $Y_0 - Y_t$ is normal with mean $\alpha(p - 1)(t - t_0)$ and variance $\beta^2(p - 1)^2(t - t_0)$.

It is then easy to find

$$p(x,t) = \frac{x^{-m}}{\sqrt{2\pi} \, \beta \, \sqrt{t - t_0}} \exp\left\{ -\frac{1}{2} \frac{[x_0^{1-m} - x^{1-m} - \alpha(m - 1)(t - t_0)]^2}{\beta^2(m - 1)^2(t - t_0)} \right\}$$

Since $p(x,t)$ is the transition probability density for the change from (x_0,t_0) to (x,t) we may, on writing

$$q(x,t|x_0,t_0) = p(x,t),$$

define the path likelihood for a crack observed as a sequence $\{(x_i,t_i)\}_{i=1}^{n}$ as

$$L = \prod_{i=1}^{n} q(x_i,t_i|x_{i-1},t_{i-1})$$

and obtain maximum likelihood estimators for α, β and m.

When the data comes from a number of cracks where the j-th crack is a sequence $\{(x_{ji},t_{ji})\}_{i=1}^{n}$ we may pool the observations into a likelihood for the experiment since

$$L = \prod_{j=1}^{m} L_j$$

where

$$L_j = \prod_{i=1}^{n} q(x_{ji},t_{ji}|x_{ji-1},t_{ji-1}).$$

This process of pooling data allows for the construction and estimation of models of crack behaviour through the effects of test/environment on α, β and m. Another advantage is that the data can be pooled even when experimental conditions vary. The most useful model is likely to be one where m is assumed to be a material characteristic and α and β are determined by the conditions. Such a model is analysed in the next section.

ILLUSTRATION

Nine crack tests on specimens of steel for pressure vessels are analysed. The two steels are A533B and A503B (12). The exponent m of the Paris-Law is assumed to be a material property and α and β are assumed to depend on the conditions. The likelihood to be maximised is then a function

$$L = L (m, \underline{\alpha}, \underline{\beta}; \text{samples})$$

where $\underline{a}$ and $\underline{\beta}$ are vectors of parameters, α_1 corresponding to test 1 and so on.

The maximum likelihood estimators are found easily by setting

$$\hat{\alpha}_i = \frac{1}{(m-1)(t_{in_i} - t_{io})} \sum_{j=1}^{n_i} (x_{ij-1}^{1-m} - x_{ij}^{1-m})$$

$$\hat{\beta}_i^2 = \frac{1}{n_i(m-1)^2} \sum_{j=1}^{n_i} \frac{\left\{x_{ij-1}^{1-m} - x_{ij}^{1-m} - \hat{\alpha}_i(m-1)(t_{ij} - t_{ij-1})\right\}^2}{(t_{ij} - t_{ij-1})}$$

whereupon the log-likelihood for the i-th group becomes

$$L_i(m, \hat{\alpha}_i, \hat{\beta}_i) \quad = \quad - n_i \sum_{j=1}^{n_i} \ell n \, x_{ij} - n \, \ell n \, \hat{\beta}_i - \frac{1}{2} \sum \ell n (t_{ij} - t_{ij-1}) - \frac{1}{2} n_i .$$

Thus it is easy to obtain $\hat{m}$, the estimator of m, by a search method to determine the maximum of the log-likelihood

$$L = \sum_i L_i(m, \hat{\alpha}_i, \hat{\beta}_i) .$$

The estimates for the different groupings are given below.

Steel A508 $\hat{m} = 2.99$

$\hat{\alpha}_i$	$\hat{\beta}_i$	conditions
0.5547×10^{-9}	0.3039×10^{-7}	water @ 20°C; 0.1Hz; 55kN - 5kN
0.5709×10^{-9}	0.4852×10^{-7}	water @ 20°C; 0.1Hz; 50kN - 5kN
0.6996×10^{-9}	0.4498×10^{-7}	air at room temp; 0.5Hz; 55kN - 5kN

Steel A533B $\hat{m} = 3.64$ all in air and at 10Hz

$\hat{\alpha}_i$	$\hat{\beta}_i$	conditions
0.4652×10^{-7}	0.5029×10^{-7}	35kN - 3.5kN
0.1198×10^{-8}	0.9646×10^{-7}	9kN - 0.9kN
0.5362×10^{-8}	0.3992×10^{-7}	16kN - 1.6kN

Steel A533B $\hat{m} = 2.29$

$\hat{\alpha}_i$	$\hat{\beta}_i$	conditions
0.9058×10^{-8}	0.6442×10^{-6}	air; 0.5Hz; 55kN - 5.5kN
0.7514×10^{-8}	0.6814×10^{-6}	water; 0.1Hz; 50kN - 5kN
0.7741×10^{-8}	0.3934×10^{-6}	water; 0.1Hz; 55kN - 5kN

When the two groups of steel A533B are combined we find, ordering the data sets as above,

Steel A533B $\hat{m} = 2.69$

$\hat{\alpha}_i$	$\hat{\beta}_i$
0.9599×10^{-6}	0.1560×10^{-5}
0.2003×10^{-7}	0.1304×10^{-5}
0.1015×10^{-6}	0.1251×10^{-5}
0.2089×10^{-8}	0.1342×10^{-6}
0.1731×10^{-8}	0.1562×10^{-6}
0.1824×10^{-8}	0.8563×10^{-7}

To estimate the length of a crack we use the fact that for a given t the expected value of x^{1-m} is

$$\hat{x}^{1-m} = x_0^{1-m} - \hat{\alpha}_i(m - 1)(t - t_0) .$$

The results are shown in Figures 1 to 4 where it can be seen that the model works reasonably well.

74

Figure 1.

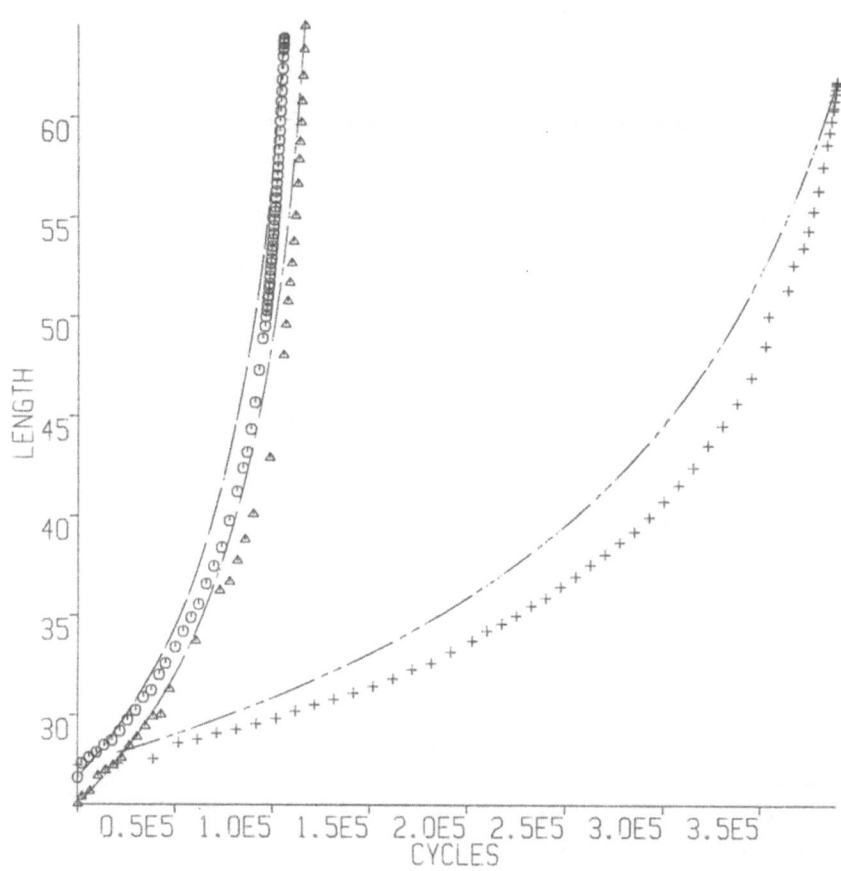

PWR1:   STEEL A508 CL.2 FCG DATA
PWR2:   STEEL A508 C1.2 FCG DATA
PWR3:   STEEL A508 CL.2 FCG DATA

Figure 2.

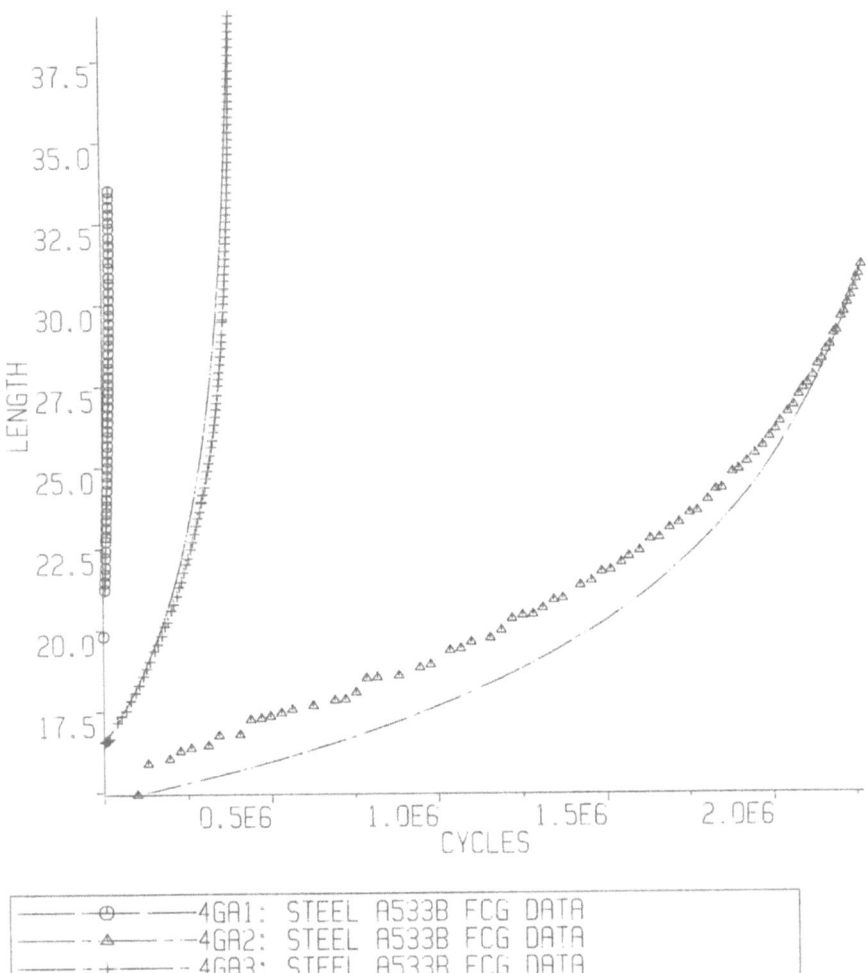

Figure 3.

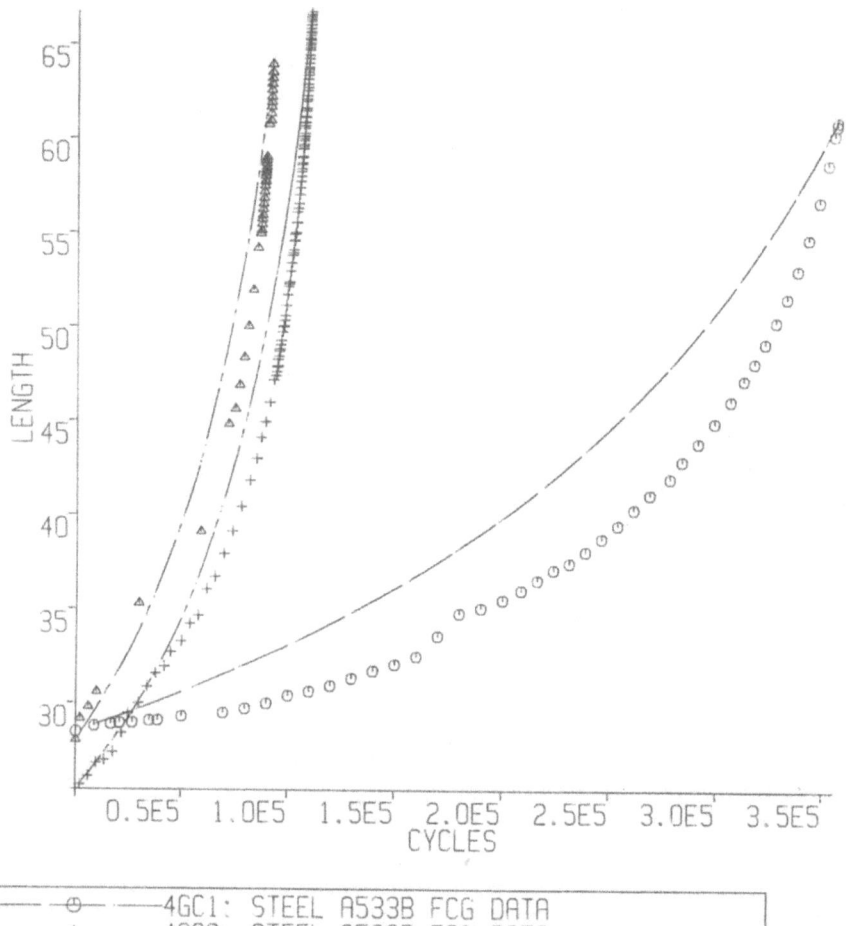

Figure 4.

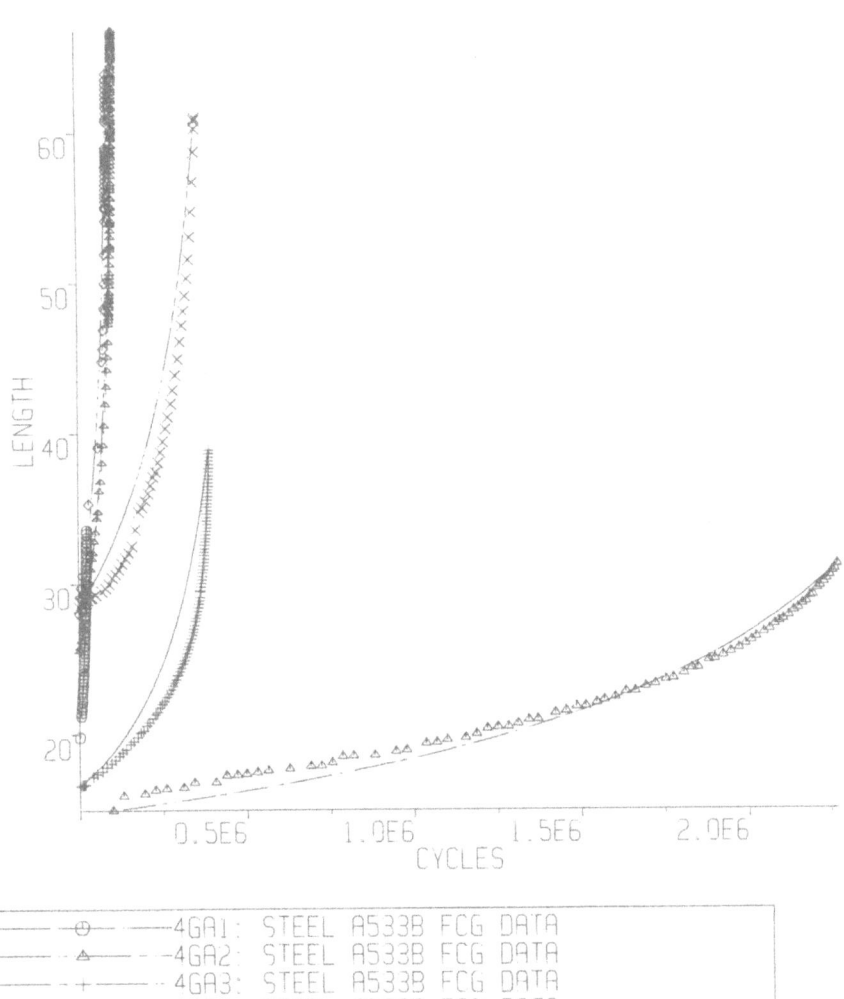

LENGTH

CYCLES

——⊕——4GA1:	STEEL A533B FCG DATA
——△——4GA2:	STEEL A533B FCG DATA
——+——4GA3:	STEEL A533B FCG DATA
——×——4GC1:	STEEL A533B FCG DATA
——⊘——4GC2:	STEEL A533B FCG DATA
——⊕——4GC3:	STEEL A533B FCG DATA

CONCLUSION

The models have been developed and some results obtained. From the results it can be seen that the model is effective and supports the argument that the exponent in the Paris-Erdogan law is indeed a material property whereas the parameters α and β are influenced by conditions. Moreover the models have allowed the pooling of data without the need for identical experimental conditions. The model also has the ability to estimate the variance of the estimated lengths through the parameter β, information which cannot be obtained from a least squares approach to estimation.

Acknowledgment

My thanks are due to Per Becker and colleagues in the Systems Analysis Group at Risø National Laboratory Denmark for introducing me to the problem and supplying data.

REFERENCES

1. Bogdanoff, J.L., A new cumulative damage model - Part 1. J. Appl. Mech., 1978, 45. 246-250.

2. Bogdanoff, J.L. and Krieger, W., A new cumulative damage model - Part 2. J. Appl. Mech., 1978, 45. 251-258.

3. Bogdanoff, J.L., A new cumulative damage model - Part 3. J. Appl. Mech., 1978, 45. 733-740.

4. Bogdanoff, J.L. and Koyin, F., A new cumulative damage model - Part 4. J. Appl. Mech., 1980, 47. 40-44.

5. Kozin, F. and Bogdanoff, J.L., A critical analysis of some probabilistic models of fatigue crack growth. Engng. Fracture Mechanics, 1981, 14. 59-89.

6. Cox, D.R. and Miller, H.D., The Theory of Stochastic Processes, Chapman and Hall, London, 1977.

7. Krausz, A.S., The random walk theory of crack propagation. Engng. Fracture Mechanics, 1985, 21. 1151-1168.

8. Newby, M.J., Markov models for fatigue crack growth. Engng. Fracture Mechanics, 1987, 27. 477-482.

9. Sobczyk, K., Modelling of random fatigue crack growth. Engng. Fracture Mechanics, 1986, 14. 609-623.

10. Wong, E., Stochastic Processes in Information and Dynamical Systems, McGraw-Hill, New York, 1971.

PREDICTING PIPELINE RELIABILITY
USING DISCRIMINANT ANALYSIS

Y. L. Bakouros
University of Bradford Management Centre, England

ABSTRACT

This paper describes an ongoing research programme for modelling
offshore pipeline failures using a technique known as 'Discriminant
Analysis'. Using data from the North Sea, this paper shows how this
technique can be employed to forecast the likelihood of a pipeline failing
and the probability of that occurrence. It is thought that this
methodology has potential for more cost effective design and operation of
pipelines taken singly and in networks throughout their life cycles.

INTRODUCTION

During the past two decades a comprehensive distribution network
comprising approximately 5800 kilometres of pipelines has been developed in
the North Sea. These pipelines have been designed to stringent engineering
standards to withstand the most severe stresses which can be imposed by the
harshness of this environment. High standards of reliability have been
needed to justify the very high capital investment in these pipelines and
to prevent extremely difficult and costly repairs and consequential revenue
losses, in the event of their failure. Despite the search and design for
high reliability however, North Sea pipeline failures have occurred with
consequences varying from relatively inexpensive repairs involving no
production loss or hazard, to those involving extremely expensive repairs
and substantial pollution.

Several authors and organizations have studied these failures in detail,
from a Reliability Engineering viewpoint, with the aim of improving our
knowledge of their inherent failure rates, to enable cost effective
reliability and safety improvements to be made. Uppermost in this respect has
been the work of Cannon (1985), Battelle (1983) and Det norske Veritas (1980).
Unfortunately all of these authors have been forced to classify pipeline
failures according to their 'average failure rate' in terms of 'failures
per 1000 kilometre-years'; terminology which has become imbedded in the
industry. These failure rates have to be used with caution and selectivity
otherwise they can lead to erroneous decision making. Furthermore they

lead to a very restrictive model, for estimating the probability of a pipe-line failing within a prescribed period of time, which is known to be wrong.

Because of these limitations, which are described in detail later, the objective of this ongoing research programme is to develop a methodology for overcoming these problems with a view to facilitating accurate pipeline reliability prediction.

The work reported here is concerned with North Sea pipelines only. A similar study of pipelines in the Gulf of Mexico is also ongoing.

WORK CARRIED OUT

In common with previous studies, this research was concerned initially with the collection and collation of data involving pipeline construction, commissioning and operating failures. This has resulted in one of the most extensive, comprehensive and accurate data systems of its kind albeit that it has certain deficiencies which will be discussed later. A comparison with other data bases has revealed gross deficiencies in their accuracy and coverage.

Two data banks have been developed, one describing the inventory of pipelines and the other 'failure events'. Typical examples of these data are presented in Tables 1 and 2 where the information has been partly censored in order to retain its confidentiality. The Pipeline Inventory data bank also includes other general information, which has no direct bearing on this study, such as the contractor's name, the lay method, the hydrostatic test pressure, the pipeline's maximum capacity and its frequency of inspection. A review of these examples shows that the two different kinds of data are linked by the 'pipeline number' which features in both data sets. These data can be analysed by a variety of means using our interactive suite of computer programs.

For reasons which will become obvious later, it was decided to investigate initially all those 'failure events' which resulted in a change in a pipeline's state (from its intended state) and thereby incurred unnecessary costs. Resulting from this definition of a 'failure' the scope of the work reported here is broader than most previous studies, which have confined their attention to post-commissioning failures.

So far this study has confined its attention to data available from 1968 to 1984. During that period 163 pipelines approximating 5871 kilometres in length were laid in the North Sea and they incurred 61 failures as shown by Table 3. This table illustrates how the frequency of failures has changed with time; a factor which we shall return to later.

Table 4 gives the locations of these failures and it highlights their high occurrence in the 1 kilometre 'platform zone' area. Table 5 gives the status of the pipelines at the moments of their failures and it demonstrates their vulnerability during the installation phase. It should be noted that neither of these last two tables contain 61 failures due to certain critical information being unavailable at the moment.

A large variety of analyses of these data has been done with a view to obtaining a crude but robust understanding of their implications for pipeline reliability, using the failure rate criterion 'failures per 1000 kilometre-years'. Only a few of the most relevant of these analyses are presented here.

Table 6 shows the overall failure rate and frequency as a function of pipeline length and it highlights a declining average failure rate with increasing length; a phenomenon which is well known and understood.

Table 7 shows the overall failure rate and frequency of pipeline failures as a function of their diameter and it illustrates the well known phenomenon of a declining failure rate with increasing diameter; at least for those pipelines with diameters exceeding 12 inches.

Table 8 suggests there is a substantive difference between the average failure rates for 'oil' and 'gas' pipelines in the North Sea. Previous research by Det norske Veritas (1980) showed, but did not prove conclusively, that this resulted mainly from a diameter dependent effect. Table 9 shows how the failure rates of pipelines with different commissioning dates have declined with their 'life-times'. This also is an extremely important matter which we shall examine later.

In all of these crude analyses, the average failure rate has been calculated according to the following method:

$$\text{Average failure rate} = \frac{\text{Number of failures in class x 1000}}{\text{Total experience of pipelines in class}}$$

Where the 'experience' of a single pipeline is its length, multiplied by the difference between calendar date and its date of commissioning.

In all of these calculations, common cause failures arising from the multiple damage of an individual or set of pipelines have been censored to ensure that only one failure is used in the calculations.

DISCUSSION OF INITIAL RESULTS

Table 6 shows how the average failure rate declines with the increasing length of the pipeline, a result which is completely at variance with intuitive reasoning, since one would expect that a longer pipeline would have a greater chance of being damaged. To some extent this anomaly can be explained by the fact that very short pipelines (mostly inter-platform connections) incur extremely high failure rates which are approximately one order of magnitude greater than for long distance pipelines. In effect therefore this phenomenon is partly explained by the very high failure rates in the 'platform zone' areas being diluted with increasing pipeline length. Another but connected reason is that the frequency of failures in the North Sea has changed throughout this investigation period and has tended to decline as the pipelines' experience has increased, thereby effecting both 'numerator' and 'denominator' in our calculations.

Table 7 also produces a result which is somewhat odd because one might reasonably expect a larger diameter pipeline to be exposed to a greater chance of damage by a variety of means. It could be argued that this result is directly related to the increasing thickness of larger diameter pipelines rather than to their diameters. Unfortunately previous studies have been frustrated by a dearth of information regarding each pipeline's thickness so the complex relationship which exists between length, diameter and thickness has tended to be overlooked. Outside of the special conditions which prevail in the 'platform' and 'surf' zones, it might be reasonable to hypothesise that the probability of a pipeline failing is related to its dimensionless group:

$$\frac{length \times diameter}{\left(thickness\right)^2}$$

which we are investigating in detail.

Table 1 shows that the number of failure events changed in a complex manner but that the rate of failures seemed to diminish, at least during this short interval of time, as lay-vessel activity declined. To some extent this phenomenon is re-inforced by the results of Table 9. In effect therefore it would appear that pipeline failure rates are not time invariant as the average failure statistic naturally assumes. Three different categories of factors affect pipeline failures:

1. Factors which reduce the probability of failure

   such as a reduction in lay-vessel activity in close proximity to existing platforms and more responsible fishing and anchoring procedures, as knowledge of each pipeline's location becomes better known.

2. Factors whose influence remains constant

   these involve random 'chance' occurrences such as earthquakes, storms, mudslides and sinking vessels.

3. Factors which increase the probability of failure

   such as the ageing effects of corrosion, erosion and fatigue.

Despite human intervention, it would seem reasonable to expect the relative strengths of these three different categories of factors to change with time thereby giving rise to pipeline instantaneous failure rates which are time dependent.

In conclusion therefore the rationale for using an average failure rate statistic based on the 'number of failures per 1000 kilometre-years' is suspect for the following reasons:

1. it assumes that the failure rate is time invariant, which is contrary to the empirical evidence and the theoretical arguments given here and elsewhere.

2. it cannot distinguish between the high incidence of failures in the platform zone and the lower incidence elsewhere in a pipeline's system.

3. it is unable to distinguish properly between the different effects of length, diameter and thickness.

In effect therefore the Exponential Distribution which is currently used to estimate the probability $F(t)$ of a pipeline failing by a prescribed age 't' based on an average failure rate (r) and the formula:

$$F(t) = 1 - \exp(-rt)$$

is also suspect. The objective of this ongoing research programme therefore is to formulate a means for overcoming these shortcomings using the methodology which is described below.

DISCRIMINANT ANALYSIS

Discriminant Analysis is a statistical technique for separating individual samples into different sets according to some measurement or combination of measurements of that sample's characteristics. For example, given a person's height we might like to decide whether they are male or female. In itself 'height' would be an inadequate way to discriminate between the sexes but some combination of 'height' and 'waist' measurement might suffice.

A discriminant variable (Z) is one which facilitates this separation process. Usually the discriminant variable is a linear function of the weighted characteristics of the sample but other forms exist which include, inter alia, quadratic, exponential and logistic functions. So far in this research we have dealt exclusively with the linear variety and discriminant functions of the following form:

$$Z = X_i \cdot W_i + X_2 \cdot W_2 + \ldots X_n \cdot W_n + C$$

where X_i = the measurement of the sample's ith characteristic (length, diameter, thickness, etc),

W_i = the best coefficient or 'weight' to attach to that measurement to get the best discrimination,

C = a constant,

$i = (1, n)$ where each individual sample can have 1, 2, 3 -----n different characteristics.

The objective of Discriminant Analysis is to discover those variables 'X_i' and their coefficients (W_i) which enable the separation of samples into distinctly different sets to be done as efficiently as possible. This requires the following procedure:

1. obtaining representative samples from the distinct populations,

2. obtaining their characteristic measurements $(X_1, X_2 \text{ --- } X_n)$

3. selecting those characteristics which are thought to give the best discrimination

4. calculating the corresponding Z and W_i values which provide the best separation based on the selection in (3)

5.   repeating steps (3) and (4) using other combinations of
     characteristics until an optimum separation of the samples into
     their distinct sets is achieved.

Because each individual sample (a pipeline) will have different
measurements (X_i) for the same characteristic (i) from any other sample it
follows that it will also have a different Z-score to other samples, except
by chance. As a result the Z-scores for different samples will be
distributed in some functional form which is related to the ways in which
their measurements are distributed. It is commonplace for these measurements to be distributed according to Normal (bell-shaped) Distributions,
similar to those shown in Figure 1, in which case the Z-scores will be
similarly distributed.

The result of such an analysis might be represented graphically by
Figure 1, which provides the Z-scores and their frequencies for two
separated populations. Quite obviously the purpose of the optimization
procedure is to separate the bell-shaped distributions as much as possible.
Perfect discrimination would be achieved in the event of these
distributions being completely separated without any overlap whatsoever.
In the event of an overlap, as shown in Figure 1, this means that the
discrimination is imperfect because those samples with Z-scores in the
overlapping area could be classified incorrectly. In effect discrimination
in these circumstances would involve some 'risk' of being wrong.

Having established the distributions of Z-scores for the separate sets
it follows that the Z-score of any further representative sample can be
calculated using the same coefficients and characteristics and it can then be
classified as belonging to one or other of the distinct sets without
knowing in advance its proper classification.

A further benefit of this method is that it is possible to calculate
the probability of an individual sample belonging to one or other of the
distinct sets.

DISCRIMINANT ANALYSES APPLIED TO PIPELINE FAILURES

This technique has been used to discriminate between those pipelines
which have 'failed' and those which have 'survived' (not failed) in the
North Sea. Table 10 shows that, up through 1984, a total of 61 out of 163
of these pipelines had failed by some means or other. All of the characteristics listed in Table 2 were used in the analysis and it is found that the
seven variables listed in Table 11 give the best discrimination to date.
The unstandardized coefficients of column A are those used to calculate the
Z-scores for the two sets of pipelines shown in Figure 1.

Because these unstandardized coefficients belong to variables with
different dimensions they cannot be compared until they have been
standardized: hence the reason for column B in Table 11. These
coefficients have been arranged in rank order according to their declining
absolute values; whether positive or negative. A positive coefficient
means that any increase in the dimension of the pipeline's corresponding
variable (length, weight cover, thickness, metal grade, diameter or
operating pressure) increases its Z-score and the chance of its belonging
to the 'failed' category. The reverse is true for variables with negative
coefficients (metal thickness and life-time).

It is instructive to note that the three variables length, diameter and thickness affect the probability of a pipeline failing in exactly the way predicted using intuitive reasoning. This conclusion is important because it lends credibility to the results and the method.

It is important to note that the 'life-time' has a negative coefficient. In this work the life-time is the time since the pipeline was commissioned up through 1984. This means that the longer a pipeline has worked the lower is its Z-score and the higher its chances of belonging to the 'survived' set. In some respects this result is intuitively reasonable but a deeper understanding belies its negative coefficient which represents the declining lay-vessel activity (which we would like to model directly) with calendar time.

We have certain reservations concerning the fact that the coefficients for weight cover thickness and metal grade have positive coefficients and we intend to research these matters in more detail.

To date we have not been able to discriminate between 'failed' and 'survived' pipelines as efficiently as would have liked. The summary of results given in Table 10 shows that our overall discrimination accuracy is 75% whereas we would have wished for 90-95% if this were possible. It is thought that the main reasons for this deficiency are:

1. limitations in the quantity of failure data pertaining to the North Sea,

2. missing data which is absent from our data sets,

3. limitations in the coverage of our data to include other factors which effect pipeline failures such as shipping activity, meteorological data and the effectiveness of pipeline surveillance and inspection routines.

Various actions are underway to reduce these limitations.

Effectively Discriminant Analysis attempts to maximise the distance between the Z-score centroids (peaks) of the distributions shown in Figure 1, and it discriminates between new samples using a criterion - 'the critical Z-score' which is given by:

$$Z \text{ (critical)} = \frac{\overline{Z}_F + \overline{Z}_S}{2}$$

where $\overline{Z}_F$ = the average Z-score for the pipeline 'failed' set,

$\overline{Z}_S$ = the average Z-score for the pipeline 'survival' set.

For North Sea pipelines the values are +0.91 and -0.55 respectively, in which case the value of Z(critical) is +0.18.

Providing subsequent samples of pipelines are representative of these North Sea pipelines, it follows that they can be classified into 'failed' and 'survived' sets in advance, without prior knowledge of the occurrence of their failing or surviving, using their Z-scores and the critical value 0.18. Pipelines with Z-scores greater than 0.18 would be predicted as failing within the prescribed time period whereas those with Z-scores less than 0.18 would be forecast to survive.

Using the statistical relationship known as 'Bayes Theorem" it is possible to calculate the probability of any pipeline with a known Z-score belonging to either set. In other words we can predict the probability of a pipeline 'failing' and 'surviving' using this technique. Two actual examples of this technique are presented in Table 12 which shows that not only would these predictions have been correct but their probability of being correct would also have been high.

CONCLUSIONS

Although there is much work needed to refine this technique, to improve its discrimination efficiency, nevertheless it overcomes all those deficiencies which exist in the use of any average failure rate criterion. The main reason for this is that it can accommodate the complex relationships (correlations) which exist between all those characteristics of a pipeline which influence its chances of failure - something which the criterion 'failures per 1000 kilometre-years' cannot do.

Of particular interest is the prospect of using this method to predict the probability of a pipeline failing within a prescribed period of time based on its Z-score. This matter is being investigated urgently because of its potential for more cost effective design of individual pipelines, networks of pipelines and means for their surveillance, inspection and repair.

RECOMMENDATIONS

It is recommended that oil operators, regulatory bodies, insurance companies and all those who have a vested interest in the life cycle implications of pipelines should consider using and developing this technique. Furthermore, they should consider seriously the release of better data on pipeline failures and pipeline characteristics which would permit this technique to be fully developed into an efficient tool.

REFERENCES

1. Cannon, A.J. and Lewis, R.C. (1985), 'The Reliability of Pipe Systems Operating in the British Sector of the North Sea". Paper 4A/R, Reliability Conference Proceedings N.E.C. July 11th.

2. Battelle (1983). Multiclient project 'Economic Implications of Pipeline Reliability'.

3. Det norske Veritas (1980). Report No. 80-0572. 'Pipeline Reliability'. de la Mare, R.F. and Anderson, O.

TABLE 1

An Example from the Pipeline Inventory Data Bank

Pipeline Number   =   30

Country	=	G.Britain
Greater Section	=	Frigg Area
Field	=	Frigg
Route	=	Frigg TCP2 - Inter.Platf.MCP01
Owner	=	Elf Aquitaine Norge
Product	=	Gas
Commissioning Date	=	1978
Length of Pipeline	=	188.5 KM
Diameter of Pipeline ..	=	32. IN
Thickness	=	.75 IN
Pipeline Grade	=	X65.
Corrosion Protection ..	=	Mastic
Cathodic Protection	=	Zinc Anodes
Weight Cover	=	Concrete
Cover Thickness	=	TR100
Operational Pressure ..	=	148.Bar

THE PIPELINE HAS NEVER FAILED

TABLE 2

An Example from the Pipeline Failure Event Data Bank

Failure Number   ..   =   12

Failure Code	=	36.
Pipeline Number	=	19.
Route of Pipeline	=	Ekofisk P - Inter.platf. 37/4A
Commissioning Date	=	1975
Date of Failure	=	1975
Length of Pipeline	=	123. KM
Diameter of Pipeline ..	=	34. IN
Thickness of Pipeline ..	=	.719 IN
Weight Cover	=	Concrete
Thickness of Weight Cover	=	TR100
Operational Pressure ..	=	114. Bar
Product	=	Oil
Pipeline Status	=	Hydrotest
Location of failure	=	Open Sea
Depth at the Location of Failure	=	70. M
Event of Failure	=	Ruptured Pipe
Cause of Failure	=	Construction
Repair Method	=	Replaced with New Section
First Source	=	Mats

TABLE 3

Frequency of Pipeline Failures in the North Sea 1968-1984

Year	Number of Failures
1968	1
1974	3
1975	8
1976	14
1977	12
1978	9
1979	3
1980	4
1981	4
1982	1
1984	2
TOTAL	61

TABLE 4

A Location Classification of Pipeline Failures

	Location		Number of Failures
1.	Near platform	..	24
2.	Open Sea ..	..	31
3.	Surf zone ..	..	2

TABLE 5

The Status of a Pipeline at Failure

	Status		Number of Failures
1.	Installation	..	25
2.	Hydrotest ..	..	10
3.	Operation ..	..	25

TABLE 6

Pipeline Failure Rate and Frequency Varying with Length

Pipeline Length (kms)	Combined Length (kms)	Operational Experience (km-years)	Number of Failures	Failure Rate
0.0 - 2.0	11.3	89.3	1	11.192
2.1 - 10.0	138.4	983.5	8	8.134
10.1 - 20.0	385.5	2194.6	13	5.924
20.1 - 50.0	636.5	4211.8	3	.712
50.1 - 100.0	937.2	7764.2	4	.515
100.0 -	3762.5	28431.8	24	.844
Total	5871.4	43675.2	53	1.214

TABLE 7

Failure Rate and Frequency Varying with Diameter

Diameter (in)	Combined Length (kms)	Operational Experience (km-years)	Number of Failures	Failure Rate
0.0 - 6.0	61.5	395.3	0	0.000
6.1 - 12.0	252.2	1166.7	2	1.714
12.1 - 18.0	553.3	4457.8	21	4.711
18.1 - 24.0	1026.0	4453.4	11	2.471
24.1 - 30.0	1103.4	10273.9	7	.681
30.1 -	2861.3	22905.2	16	.699
Total	5857.3	43652.2	57	1.306

TABLE 8

Pipeline Failure and Frequency Varying with Product

Product	Length (kms)	Operational Experience (kms-years)	Number of Failures	Failure Rate
Oil	1940.6	13109.5	30	2.288
Gas	3314.2	27837.2	26	.934
Other	616.5	2728.5	3	1.099
Total	5871.3	43675.2	59	1.351

TABLE 9

Pipeline Failure Rates Varying with 'Life-Time' for
Different Commissioning Dates

Pipeline Commissioning Date

Years of Operation	1974–1975		1976–1977		1978–1979		1980–1981	
	A*	B*	A*	B*	A*	B*	A*	B*
1	2	1.8	17	7.2	14	8.2	3	7.5
2	5	2.3	21	4.5	16	4.7	3	3.7
3	8	2.2	21	3.0	17	3.3	3	2.5
4	10	2.0	22	2.4	18	2.7	3	1.9
5	10	1.6	23	2.0	18	2.6		
6	10	1.4	23	1.6	18	1.8		
7	10	1.2	23	1.4				
8	10	1.0	23	1.2				
9	10	0.9						
10	10	0.8						

*NOTE: A = Number of failures
         B = Failure rates

91

TABLE 10

A Summary of the Discrimination Efficiency of
this Method to Date

Actual Pipeline Set	Number of Pipelines in Set	Z-score classification Failures	Survivors
Failed	61	48 (79%)	13
Survived	102	27	75 (74%)

Overall % of correct classification = 75%

TABLE 11

The Characteristics and their Coefficients which Discriminate
Best Between 'Failed' and 'Survived' North Sea Pipelines

Variable (Characteristic)	Unstandardised Coefficients (A)	Standardised Coefficients (B)
1. Length	0.13	+0.84
2. Thickness	-4.70	-0.73
3. Weigh cover thickness	0.39	+0.60
4. Life-time	-0.11	-0.37
5. Metal grade	0.26	+0.29
6. Diameter	0.02	+0.18
7. Operating pressure	0.003	+0.08
8. Constant	3.95	

TABLE 12

Two Examples of Discriminant Analysis as a Prediction Technique

Data Bank Pipeline Number	Actual Set	Discriminant Scores	Set Prediction Based on Z-Score	Probability of Belonging to Set
30	Survived	- 2.0194	Survived	96%
141	Failed	+ 1.5231	Failed	88%

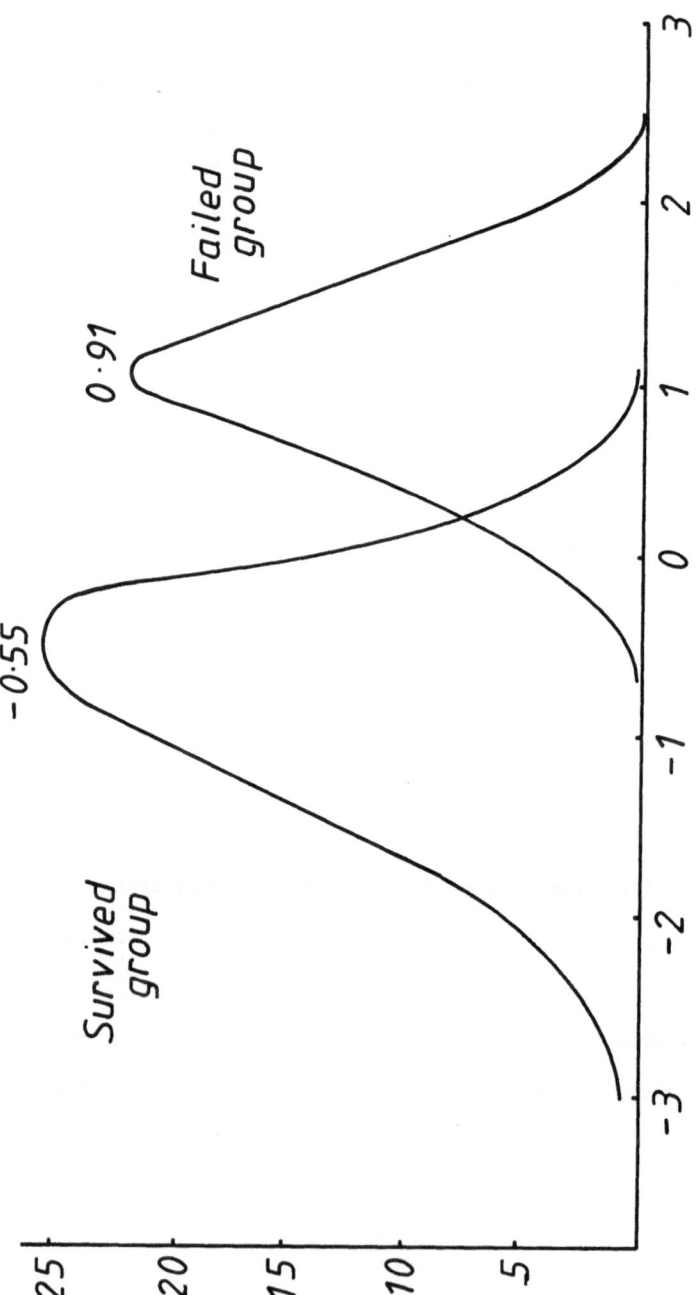

Fig 1 Probability density distribution of Z scores for North Sea Pipelines

ASPECTS OF MECHANICAL SYSTEM RELIABILITY

P. MARTIN

DEPARTMENT OF MECHANICAL ENGINEERING
THE UNIVERSITY OF LIVERPOOL

ABSTRACT

It is now well known that difficulties exist in the prediction of mechanical system reliability by theoretical analysis. The problems associated with different system levels are examined by considering a number of theoretical models which can provide guidance during the design and operation of mechanical systems.

A number of different system models are considered and their limitations are identified in relation to a number of practical cases.

It is suggested that the continued examination of such models and their application can provide guidance in understanding the factors to be taken into account during design and improved ability in the provision of appropriate monitoring methods during system operation. A number of these factors are identified.

INTRODUCTION

The modern technological systems which are now emerging as a result of the rapid and visible growth of the communication and information industries have emphasised the importance of the reliability of electronic and computer based systems. Rapid strides have been made in the understanding of the factors which govern the reliability of such systems and this coupled with improvements in technology has resulted in electronic and computer systems which demonstrate very high reliabilities in service operation. Such systems are now being integrated with mechanical systems to form highly complex technological systems. Major industries such as transportation, power generation, food production, chemical processing and advanced manufacturing involve examples of such highly integrated systems.

In many cases, the design constraints on such systems would involve the ability to satisfy three basic requirements:

- OPERATION AT ACCEPTABLE SAFETY LEVELS.

- REQUIRED PERFORMANCE AT MINIMUM CAPITAL COSTS.

- MINIMISED OPERATING AND MAINTENANCE COSTS.

The ability to satisfy these potentially conflicting requirements demands a deep understanding of the factors which lead to the failure of the system elements and the relationship between such failures and system reliability. Much progress has been made and a wide spectrum of engineering effort in reliability is directed towards this understanding and it has reached very advanced levels in dealing with systems where safety is of paramount importance in for example the aerospace and nuclear industries. Well established methods such as Failure Modes and Effects Analysis (FMEA) as described by Collacott (1) and Fault Tree Analysis as described by Fussell (2) are representative of methods used for ensuring safety.

MECHANICAL RELIABILITY CONSIDERATIONS

Early success based on probabilistic models held out the hope that it would eventually be possible to estimate system reliability during design but in the case of mechanically based systems, it is now apparent as discussed by Carter (3), that much further understanding is required before this becomes feasible to any significant extent for complex mechanical systems. The failure mechanisms are complex, often not statistically independent and in most cases are time dependent. In addition considerable difficulty can be experienced in dealing with generic component failure data, since failure mechanisms can be influenced by small differences in operating conditions. These points have been considered by Carter (3) who has demonstrated theoretically, based on the load-strength interference model, that the majority of mechanical components are intrinsically reliable and then become unreliable as strength reduction due to degradation with time in service takes place.

If these general propositions are accepted, then it is apparent that further understanding in the long term should be based on the behaviour of mechanical systems and their elements in service operation since degradation is going to be influenced by the service operating conditions of particular items of equipment and their constituent components.

To achieve this objective, it is suggested that an alternative approach to complement the general methods used to obtain generic failure data is required if the necessary depth of understanding is to be gained in order to satisfy the objectives of required performance at minimum capital costs and minimised operating and maintenance costs for complex mechanical systems.

Thus a radical departure from existing methods is required if progress is to be made in the long term.

In conducting projects aimed at reducing maintenance costs in a number of different regions of the process industries it has become possible to make a number of observations:

(1) There is often little if any direct communication between the design authority for the plant and the regions of the plant where mechanical reliability problems exist and hence where maintenance costs are occurring unless these are very high.

(2) Due to the requirement to keep the plant operating, maintenance is carried out extremely quickly in order to bring the plant back on line and it is often difficult to trace failures and running times for industrial components and items of mechanical equipment.

The vital evidence for improved understanding of service behaviour is thus lost. Clearly there are also many exceptions to these conditions.

Progress would be possible if the vital evidence relating to failure in operation could be fed back directly to the design authority for interpretation. This is clearly not easy with paper based communication systems but the advent of low cost computer systems can provide a way forward. It is suggested that such systems should not simply substitute for paper based systems but should aim to exploit the capability of advanced computer graphics systems using system models appropriate to the level of the system being considered.

COMPUTER MODEL CONCEPT

The increasing use of computer aided design systems for the design of highly integrated systems now opens up the possibility of linking the management of failures in systems directly back to the elements of the system in the design data base in order that the operating conditions can be examined relative to the design assumptions. The direct aim would be the reduction of failures and hence maintenance costs on the particular system and the indirect aim would be improved understanding of operating conditions and their relationship to failure with the objective of improved design procedures in the future.

It is already feasible to establish graphically based models of complex systems at the design stage and then link these models to the plant. Such models are standard for plant control purposes in some areas of industry. The approach can be found for example in chemical processing plant where measured values are represented on a computer model of the system diagram as in Figure 1.

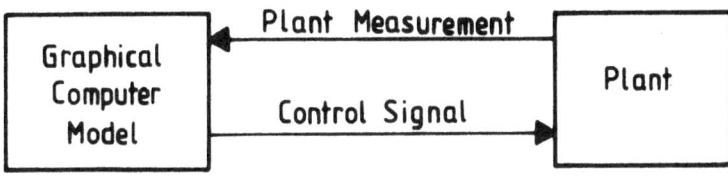

Figure 1   Computer model of plant

The measurement of failure conditions and the occurrence of failures in systems as opposed to system control is a much more formidable undertaking and in its initial stages it is clearly an advantage to relate any approach to existing models for systems to allow existing experience to be taken into account.

The feasibility of this approach is now considered by examining a number of existing theoretical models relevant to mechanical equipment.

MECHANICAL SYSTEM RELIABILITY MODELS

A wide range of theoretical models have evolved to deal with specific aspects of reliability but for the purposes of this examination just three models are considered:

1   THE COMPLEX EQUIPMENT MODEL

2   THE LOAD CHAIN MODEL

3   THE LOAD-STRENGTH INTERFERENCE MODEL.

These three models have been selected as they represent models derived to provide guidance at representative system levels in mechanical systems.

THE COMPLEX EQUIPMENT MODEL

This model is usually based on the proposition that it is possible to identify an item of equipment of sufficient complexity that its behaviour can be examined in isolation. Examples could include: an aircraft engine; a motor vehicle; a materials handling vehicle; a functional region of a more complex plant such as a boiler and its associated fittings in a power station.

An item of complex equipment would normally be part of a much larger system and would in itself contain other subsystems. It represents an identifiable level in the system around which precise boundaries can be defined.

Such a complex item of equipment could be regarded as an entity which has a number of INPUTS and a number of OUTPUTS as shown in Figure 2(a). In the case of an aircraft engine the INPUTS would include: air at velocity v_1 (flight speed); fuel at flow rate $\dot{m}$; control signals. The OUTPUTS could include: air and products of combustion at velocity v_2; compressed air for cabin pressurisation; high pressure hydraulic oil; electricity. Clearly the inputs and outputs would have to be established in practice for the particular engine type in its specific installation.

Similar sets of inputs and outputs could be established for the other representative complex equipment examples. As Green and Bourne (4) have explained any inputs and outputs to a system will have distributed values and there will be acceptance limits on both.

A **failure** in such a complex item of equipment could therefore be defined as the occurrence of any condition in the equipment which results in an output value which is **out of acceptable limits** when all input values are within limits.

This definition avoids the difficulties involved in considering whether particular failures result from particular material failure mode. The failure definition must, however, rely on the ability to measure or monitor the input and output values.

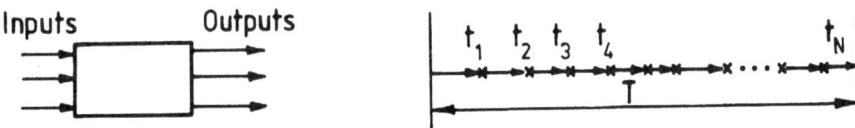

Figure 2(a) Complex equipment; (b) Failures in time period T

If the **operating** times between such **failure** occurrences which result in equipment shutdown are monitored during a time period, T, and recorded then if these times; t_1, t_2, t_3 ...., t_N are regarded as distributed randomly, then it follows that The Mean Time Between Failures (MTBF) is given by:

$$m = MTBF = \frac{t_1 + t_2 + t_3 + ... t_N}{T}$$

and the failure rate λ for the monitored time period, T, is given by:

$$\lambda = N/T = 1/m = FAILURE\ RATE$$

It follows that for these conditions, the Poisson process is applicable and the probability of the occurrence of x failures in some small time period, t, where t << T is given by:

$$P(x,t) = \frac{e^{-\lambda t} (\lambda t)^x}{x!}$$

If x = 0, then the reliability for time t is given by

$$P(0,t) = R(t) = e^{-\lambda t}$$

which is of course the well known EXPONENTIAL RELIABILITY LAW for complex equipment.

The assumption of random occurrence of failure events is usually based on arguments such as:-

(1) Successive failure and repair gradually results in a randomised distribution of operating times between failures.

(2) The equipment gradually stabilises to a reduced constant failure rate
     after falling due to: improvements in the equipment; the effects of
     learning; the operating conditions or a combination of all three,
     which is normally regarded as RELIABILITY GROWTH.

(3) The occurrence of failure incidents within the monitoring period T is
     so rare that no other assumption is feasible.

     If a single item of equipment is monitored, any other finding could
be interpreted as meaning that basic design defects exist which should be
immediately rectified. This assumes that the monitoring period T is
appropriately chosen. One year or 1000 hours would be typical periods.

     If multiple items of equipment were to be monitored for successive
periods T, then, three possible characteristics would result as shown in
Figure 3.

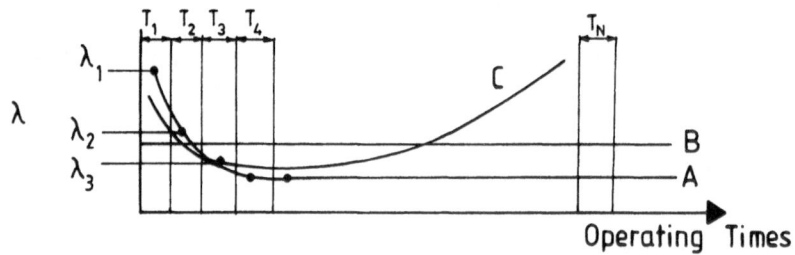

Figure 3 Failure rates for monitored operating time periods T

     Characteristic A represents the type of pattern to be expected from
either assumption (1) or (2) above or a combination of both.

     Characteristic B represents the type of pattern to be expected from
assumptions (1) or (3). For (3) to be valid the value of λ would be a
very low value. A high value of λ would suggest that assumption (1) would
be valid.

     Characteristic C would suggest that the equipment being monitored
could not be regarded as a complex item of equipment with constant failure
rate as the rising value of λ suggests that any degradation processes are
not randomised and wearout processes are beginning to predominate.

     The above illustration indicates that this method of monitoring can
result in some meaningful engineering guidance on complex systems.
However it will be appreciated that the illustration is essentially
dealing with particular examples of the non-homogeneous Poisson process.
A generalised theory of this process has been presented by Keller (5) who

has demonstrated theoretically and by simulation that the general theory for this process encompasses the well known model for reliability growth due to Duane (6).

Guidance on the behaviour of a carefully defined item of complex equipment with boundaries as depicted in Figure 2(a) can thus be obtained by monitoring operating conditions at the boundaries and running times between failures.

Clearly this model could be examined further to include the effects of maintenance time and thus lead to availability estimates. However it is not necessary to develop this argument further as information relating to this approach is well documented. However the model in itself provides only indirect guidance for design purposes. Such guidance can only be obtained by means of a careful examination within the boundaries of the equipment if the model reveals that the failure rate values are unacceptable.

Such a complex item of equipment in itself may be part of a much larger system which could be represented, for example, in the form of a reliability block diagram of classical form. However, this does not usually allow for the multiple input and output situation already defined in Fig. 2(a) and it may be necessary to establish a complete system model which deals with each input and output across carefully defined boundaries. This argument is not pursued further at this stage.

Any item of complex equipment by definition will also contain other systems. If these are mechanical systems careful treatment is necessary for the reasons advanced in (7 - 8). The arguments presented here are essentially concerned with mechanical systems within the complex equipment boundary with the load-chain model representing the next level down in a complex system.

THE LOAD CHAIN MODEL

The essential argument advanced in identifying this model is that an item of mechanical equipment should not be regarded as comprising components in series even though most items of mechanical equipment do not contain parallel paths or redundancy. Mechanical equipment consists typically of four basic system types: Thermodynamic, Fluid, Kinematic and Structural systems which operate within carefully defined environmental boundaries both internal and external. The first three of these systems can fail and these failures would result in failure at the complex item of equipment boundary. Each of these systems would feed loads into various points in the structural system elements as well as into any bearings and gears in the kinematic system.

The load-chain model then consists essentially of the structural elements and the associated bearings, gears or drive belts. Failure of the system would then occur at inherent flaws in the load chain. Examples would include stress concentrations, the surfaces of gear teeth, bearing tracks, etc.

Such analysis would enable the points of potential or incipient (8) failure in the equipment to be identified in the design data base. Such information in principle could be defined at the design stage or through failure modes and effects analysis. It might, however, only emerge from examination of failures in service. In this case the model would need updating or the design would need to be changed. A high incidence of failure in a particular region of the chain could indicate the requirement for careful condition monitoring in the region. For example, it may be necessary to record operating loads over an extended period.

It is apparent that the model can be applied to the analysis of engines, gearboxes, etc. A representative load-chain for a single reduction gearbox is shown in Figure 4.

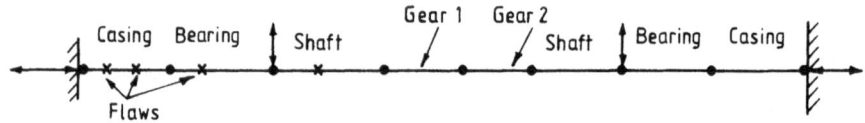

Figure 4   Load-chain for single reduction gearbox

Basic monitoring of such a model would involve a record of the running times to failure at each of the inherent flaws in the chain.

The various failure mechanisms relating to such load chains have been considered in (8) but these depend essentially on whether the "potential" or "incipient" failures are detected before they become "actual" failures. It has been shown in (8) that the basic reliability of such a chain depends essentially on the effectiveness of detection of "incipient" failures before they become "actual" failures. An "actual failure" occurs when the load exceeds the strength at any flaw in the chain.

The arguments leading to this model were based essentially on the premise that although mechanical equipment elements operate essentially in series, the survival events are not independent but usually involve conditioned or dependent failures, (8).

The mechanisms involved in failure at any particular flaw in a load chain are essentially, the province of the load-strength interference model.

LOAD-STRENGTH INTERFERENCE MODEL

This model is based on the realisation that the loads experienced by mechanical components are distributed due to variations in operating loads and the strengths of component materials are distributed about mean values each of which may change with time in service operation. Failure then occurs when a high load encounters a low strength material at a flaw in the load chain.

Based on the load-strength interference model, Carter (3) has explored comprehensively failure mechanisms which are inherent in mechanical elements. He has shown theoretically that mechanical components are intrinsically reliable and that it is essentially degradation of material in service operation which leads to reduced strength and eventually to failure. The measurement of times to failure in service at flaws in the load chain may provide useful guidance on these mechanisms in the long term.

The problem with prediction using this model is that recording values for the load and strength distributions is not easy. Loading distributions must be measured during service operation and such information is not easily obtained.

It is however vital to ensure that critical components are indeed intrinsically reliable.

An example from the motor industry demonstrates the possibilities. Based on a method suitable for use with a microcomputer, Everitt and Hill (9) have described a method for the long term collection and analysis of service loads for transmission components. The method demonstrates that it is feasible to measure actual service operating loads over a long period in situations where this is vital.

Values for the actual times to failure obtained from service operation of the load chain will begin to identify the material degradation conditions in critical regions. But it is not apparent that a generalised approach is possible. This clearly requires much careful examination. The availability of times to failure for the different flaw types at the point of design can aid this problem.

CONCLUSIONS

The main argument advanced is that the design constraints on advanced technological systems depend on a deep understanding and quantification of the failure behaviour of such systems in service operation.

It is suggested that the advent of modern computer based design systems provides an opportunity for an advance to be made in handling this information in a form suitable for feedback to design. Such a framework would also allow optimum service monitoring to be achieved. A framework for the proposed approach has been identified. It is based on three models relevant to different levels in complex mechanical systems. The continued examination of such models and the identification of departures from the assumptions of the models can provide a basis for guidance during the design and operation of future systems.

It also suggests that methods or arrangements for recording times to failure should be incorporated in the product or system definition at the design stage. Such arrangements are not necessarily complicated. For example, the incorporation of a distance recording device in a materials handling vehicle, immediately introduced some meaning into the maintenance records.

Clearly this is an initial outline of an approach which requires much further work before its general advantages could be confirmed.

REFERENCES

1. Collacot, R.A., Mechanical Fault Diagnosis, Chapman and Hall, Ltd., London, 1977, pp. 404-405.

2. Fussell, J.B., Fault Tree Analysis : Concepts and Techniques, Generic Techniques in Systems Reliability Assessment, NATO Advanced Study Institute Series, Noordhoff-Leyden, 1976, pp. 133-162.

3. Carter, A.D.S., Mechanical Reliability, Macmillan, 1986.

4. Green, A.E. and Bourne, A.J., Reliability Technology, John Wiley and Sons, 1972.

5. Keller, A.Z., A Generalised Theory of the Non-Homogeneous Poisson Process. Proc. 8th Advances in Reliability Technology Symposium, National Centre of Systems Reliability, UKAEA, Warrington, 1984, pp. B4/2/1 - B4/2/7.

6. Duane, J.T., Lerning Curve Approach to Reliability Monitoring, IEEE Transactions on Reliability, Vol. 2, No. 2, pp. 563-566.

7. Martin, P., A Systematic Approach to Interactive Failures, Mechanical Reliability (1980), IPC Science and Technology Press, London, 1980, pp. 36-54.

8. Martin, P., Consequential Failures in Mechanical Systems, Reliability Engineering, 3, (1982), pp. 23-45.

9. Everitt, D.R. and Hill, S.J., Development of an Onboard Data Processor for Transmission Components, Proc. International Symposium on Automotive Technology and Automation, Graz, Austria, 1985, pp. 95-108.

HUMAN FACTORS ANALYSIS
OF AUTOMATION REQUIREMENTS

- A METHODOLOGY FOR ALLOCATING FUNCTIONS

J. C. Williams
Sizewell 'B' Project Management Team,
Central Electricity Generating Board,
Knutsford, Cheshire, WA16 8QG
UK

ABSTRACT

This paper describes the development of a practical methodology for allocating automatic and manual functions at the design stage, whilst encouraging the development of a heightened awareness of man/machine system design implications in the areas of training and procedures development. The methodology employed was a combination of human factors analytical techniques, adapted and refined, for direct application to a specific set of automation decisions. It provided early insight into potential problems associated with human reliability and the impact of automation and quickly highlighted the portions of a manoeuvre in which automation should be considered. The methodology also identifies what it is about tasks that creates the need for automation, and facilitated the striking of an appropriate balance between operator action and automated operation.

INTRODUCTION

In contemporary complex man-machine system design and development the manner in which functions are allocated to automated systems or man is now subject to increasing levels of scrutiny.

Whilst some of the functional allocation can be made without major investigation, simply on the basis of feasibility or the application of checklists, the striking of an appropriate balance to maximise overall man machine system reliability has to date remained something of a "black-art".

This paper describes in detail the development of a practical methodology which has not only succeeded in allocating function in a highly face-valid fashion but has encouraged the development of awareness of man/machine system design implications in the areas of training and procedures development.

The methodology employed was the Function Analysis System Technique (FAST) which had been borrowed from the Value Analysis/Engineering community. This technique permits the generation of a hierarachical form of task analysis, if applied from the human perspective with respect to system goals.

The second method was a time-line analysis to depict the sequencing and duration of each genuine operator task in relation to each other and their relationship to the elapsed time of an important plant operating manoeuvre.

The third method employed was aggregate task difficulty and criticality rating by a multi-discipline group of experts using an abbreviated task analysis worksheet.

In order to obtain an assessment of potential workload a time-averaged task rating was calculated and calibrated by reference to a known task for which fixed staffing, and instrumentation requirements had been established.

The fourth method involved examination of the source of task loading with respect to the plant manoeuvre trajectory and an evaluation of counter-measures such as task re-ordering, additional staffing and automation by reiteration of the analysis in the affected areas.

TASK ANALYSIS

Because detailed operational procedures did not exist for the plant at the design stage it was necessary to ascertain which tasks would have to be performed and in what order. This was accomplished by conducting a form a hierarchical task analysis in order to establish the extent and nature of the tasks so that an assessment of mission complexity and candidacy for automation could be made.

The technique used for the task analysis was the Function Analysis System Technique (FAST). This method of describing where, why and how system functions relate to each other was developed by Charles Bytheway (1) in the USA in the late '60s/early '70s. It was originally developed as a tool to assist the value engineer in understanding (a) how systems really work and (b) where it might be possible to modify their properties in order to achieve cost savings by omitting functions or combining them in previously unconsidered ways.

The method works by requiring the analyst to ask two basic questions at every stage of his analysis. These questions are "HOW ?" does a sub-ordinate function achieve a super-ordinate function, and "WHY ?" is a super-ordinate function accomplished with respect to its sub-ordinate functions.

When a function is properly analysed its super-ordinate and sub-ordinate functions can only be described by logic which will satisfy both directions of questioning. It is normal to use an action verb and a noun for each functional description. When individual functions are described in these terms by iteration it becomes clear where particular functions fit in relation to others, ie in situations in which there is known to be a lot of activity, the goals and their reasons for existence will quickly make themselves known in ways that were not obvious at the outset.

A modification to the basic method devised by Bytheway is to use conventional "AND" and "OR" gates to connect the super-ordinate and sub-ordinate functions. When all these features are incorporated into the analysis the function analysis system technique facilitates the creation of a hierarchical task analysis which is very robust.

FAST is especially valuable when the situation to be depicted has no procedures but nevertheless would benefit greatly from a hierarchical type of analysis. It is also a good method to use when the tasks are very complex and, in the first instance, not easily described.

The major advantage of FAST is that it is totally logical, and if used illogically this will show up when the basic HOW ? and WHY ? questions are asked of each function description.

The bottom level of tasks finally depicted consists of genuine tasks and shows very strongly the extent of human interaction.

The principles of the method are a little difficult to learn at the outset, and this may deter the potential user. In addition the technique is very resource-intensive in the initial stages.

For complex technical systems analysis from a human perspective FAST works best with a multi-disciplinary team, and so such a team was assembled. This team comprised one operator, one human factors engineer, one procedures engineer, one training engineer, two commissioning engineers, three systems engineers and additional experts called upon to assist the analysis as needed.

An important plant manoeuvre was selected for analysis. This manoeuvre was known to involve significant operator actions, had high safety significance and was required as a natural consequence of a variety of fault conditions.

The analysis was designed and intended to result in a high likelihood that all genuine operator tasks would be identified. It was also intended that the method would be usable by different analysts in a consistent manner in isolation, facilitate effective review and minimise reliance on the availability of operating procedures.

In order to ensure that all analysts had a common technical knowledge-base a Functions-Based hierarchical task analysis was performed, in parallel, but slightly in advance of Operations-Based analysis. The Functions-Based analysis was based on the plant design and represented the interconnections between the technical functions of the plant to give a graphical representation of the plant design basis for the manoeuvre.

The Operations-Based hierarchical task analysis then followed in order that the tasks which collectively constituted the total manoeuvre could be depicted. The goal of this manoeuvre was to place the plant in a condition which would minimise the potential hazard to personnel gaining access to a controlled area for extended periods of time. The overall goal was broken into two basic phases, one in which the heat removal function was served by latent heat removal and the second in which heat removal was via conductive heat exchange. These two phases were subsequently amalgamated to depict the total manoeuvre.

After each analytical interaction the resulting task hierarchical tree was reviewed by the working group for veracity, completeness and logical consistency. A grid identifier system was also developed to facilitate the rapid location of tasks within the tree.

The Operations-Based analysis was developed to an advanced state and after many critical reviews it became clear that continued interaction was unlikely to result in significant structural or detailed change.

In order to establish the bottom-level "genuine" tasks a set of stopping rules was employed to determine whether a task had been analysed to the extent necessary to facilitate an accurate work-load assessment. The stopping rules were:-

o   The set of bottom-level tasks had to be complete (ie any route top-down had to terminate at a bottom-level task).

o   Bottom-level tasks had to be discrete (ie no bottom-level task was allowed to be the objective of any other bottom-level task)

o   No box could describe a bottom-level task if one of the tasks supporting it had other objectives as well.

o   Given the above 3, bottom-level tasks where possible were to be defined at functional control level such that the task uniquely defined:

The function being controlled.

The control feature being used.

The mode of operation of the control feature.

TIME-LINE ANALYSIS

Time-Line Analysis (TLA) facilitates the examination of time-related features within major missions. It is usual to set out a time axis and depict the identified tasks on this axis as analogues of their temporal positions with respect to one another and their duration. The time estimates are indicated by bar lines and give a good overall impression of task intensity. Time-Line diagrams also give good indications of task/time incompatibilities and the likely problems associated with staffing and task sharing.

The "bottom-level" tasks from the Operations-Based Hierarchical Task Analysis (HTA) defined the set of tasks for incorporation onto the TLA. However, not all bottom-level tasks were included on the TLA because they were either:

o   Duplicates of other bottom-level tasks identified elsewhere in the HTA.

o   Non-normal tasks which had been included in the HTA for completeness.

o   Monitoring tasks which were considered to be encompassed by an associated control action task.

o   Tasks which were no longer applicable to the manoeuvre due to further information concerning the task becoming available during the course of the analysis.

o Tasks which were identified as being frequent background monitoring tasks not related to a specific operator action(s). These tasks were removed from the TLA and replaced by a single continuous task to represent all background monitoring requirements specific to the manoeuvre being analysed.

Additionally, background control-room operational tasks ie those tasks that are always required to be performed and hence which are not specific to the manoeuvre, eg communications duties etc, were represented on the TLA as a single continuous task.

Having defined the set of tasks for examination via the TLA, these tasks were then located on the time-line to specify both the timing of the task and the time taken to perform TLA by using:

o The operational experience of the working group of the sequence and timing of tasks.

o The functional relationship of tasks on the HTA (ie knowing why a task is being performed) to enable the timing of the task to be determined.

o The detailed task information including the reasons for performing a task and the relationship between tasks identified on the worksheets used for identifying priority tasks.

o Design information where applicable.

The TLA was developed through an iterative review process with progress being discussed at regular working group meetings throughout the analysis.

TASK DIFFICULTY
AND CRITICALITY RATING

When assessment of task difficulty and criticality are to be made on a total of about 100 tasks the two problems of resources and accuracy become somewhat acute.

Assessment of task difficulty would normally involve judgements against at least 20 performance-related attributes, and if many analysts were to be involved, such assessments might be expected to produce considerable inter-analyst variability, as well as thousands of individual assessments, which would have to reconsidered and aggregated in some meaningful way.

It quickly became apparent that such an information explosion could be precipitated without much effort, and so a concerted effort was made to contain the potential problem whilst still achieving the principal allocation of function insights desired of the study. To facilitate the process a simplified task assessment worksheet was devised which contained 8 principal items requiring only key task-relevant details to be recorded and 4 assessment categories which required a 3 level judgement against each category.

A scoring scheme was adopted to serve as an index, and marks were awarded to each of the assessment categories to reflect the high, medium and low judgements as 3, 2 or 1 mark(s) respectively.

Task assessment worksheets were structured as follows:-

Number Grid   Task Description

Ref   Performance Factor

1   BACKGROUND INFORMATION
1.1   Reason for carrying out task
1.2   Method of control (C & I)
1.3   Relationship to other tasks
     (in terms of plant behaviour)
1.4   Personnel required to perform task
2   TASK DIFFICULTY
2.1   Task critical cues
2.2   Time constraints
2.3   Required accuracy
2.4   Task complexity
2.5   Skills and knowledge required
2.6   Difficulty rating (/3)
3   TASK CRITICALITY
3.1   Likely errors
3.2   Potential conflicts
3.3   Task failure consequences
3.4   Criticality rating (Safety) (/3)
3.5   Criticality rating (Availability) (/3)
     Overall rating (/6)
     [=2.6 + greater of 3.4 or 3.5]

The scoring was kept simple because the group did not consider that anything other than approximate judgements could be made.

Criticality was assessed on two bases, safety and availability. Safety referred to both personnel safety and plant safety while availability referred to plant availability. Tasks were assessed on the effects an error in carrying them out might have on these aspects of plant operation.

Difficulty was assessed using the factors of complexity, accuracy, special skills and knowledge required, time constraints and the critical cues required to commence or terminate the task.

The overall rating of the task was assessed by summating the scores for difficulty and for criticality. Since availability criticality and safety criticality are two aspects of essentially the same factor, only the greater of the two criticality scores was used.

Clearly the minimum scoring for any task would be 2 and the maximum 6. If in assessing the difficulty or criticality of a task an assessor considered that there were no features in the task raising an individual score above 1, then he was not required to complete any of the descriptive entries in that part of the worksheet. Conversely, if he assessed that there was a feature which raised the scoring to 2 or 3 in any category, he had to include the reason why he felt such scoring would be appropriate. Application of this rule in completing the worksheet resulted in economy of effort for both assessor and reviewer.

Any task awarded an overall marking of 4 or more was identified as a 'p' task, ie a task which by reason of difficulty or criticality or both might impose in itself a significant workload on the operator. Where two or more of these tasks occurred simultaneously in the TLA, closer analysis of the operator workload was made in order to see if it could be reduced. Also, any task achieving a rating of 5 or 6 was examined, irrespective of whether or not it occurred in combination with others, to see if means were available to reduce the workload.

Additional performance factors not immediately relevant to the workload assessment of the task were included in section 1 of the worksheet. This section, which always had to be completed in full, demonstrated when completed that the assessor had analysed the task in sufficient detail. In addition, comparison between worksheet entry 1.2 Method of Control (C & I) and entry 2.1 Task Critical Cues made it possible to identify problems with pre-existing C & I provisions with respect to a particular task and enabled recommendations for additional or modified C & I to be made. Entry 1.3, 'Relationship to Other Tasks' (in terms of plant behaviour), assisted in making the criticality assessment since it provided an indication as to the behaviour of the parameters which the particular task affects and if applicable, how they might behave if the task were not carried out correctly.

A systematic process of marking and review was adopted to optimise the reliability of the task assessment phase whilst making most efficient use of experts' time. This process consisted of:

o   Allocating the scoring of the worksheets amongst the members of the group and subjecting these scores to review by different group members.

o   Choosing a sample of five tasks for scoring by every member of the group to assure all-round consistency.

With minor differences, generally resolved in brief discussion, the scores were agreed by initial assessors and reviewers. The scores awarded for the five common tasks were found to be generally consistent, although the reasons given for the assessments tended to differ occasionally. Thus a fair degree of inter-analyst judgement reliability was assured and it would appear some reliance could be placed on the judgement of a small number of expert assessors.

TASK LOADING ASSESSMENT

In order to obtain an assessment of the potential workload pattern for the manoeuvre the time sequence was split into 12 minute periods and a time-averaged task rating calculated for each period. This overall rating was calculated on the basis of the proportion of each time sector in which the task was performed multiplied by the original task rating. These ratings were then summated over each affected 12 minute period and the resulting information used as a workload indicator with respect to time.

The plant manoeuvre trajectory was then plotted on the same axis as the workload indicator and it was possible to observe the manner in which workload could be expected to be affected by key features of the manoeuvre. This indicator was calibrated by reference to a known task for which fixed staffing and instrumentation requirements had been established.

In detail the method was as follows:

Summated task ratings were calculated for:

o   The overall task rating (ie the sum of the difficulty task rating and the higher of the two ratings for criticality with respect to safety or availability.

o   The difficulty task rating only (this was used to check that trends in the workload indicator were not being spuriously influenced by changes in task criticality). The above two summated task ratings were plotted in the form of histograms on a timescale comparable with both the TLA and the plant manoeuvre trajectory to enable ready comparison between them. The workload indicator chosen for use was that calculated from the overall task rating.

The workload indicator was used to determine the sensitivity of the overall workload to various assumptions about the degree of automation and the staffing levels. This enabled automation requirements for various staffing levels to be determined.

In order to obtain an absolute indication of workload, the workload indicator was calibrated against an operational condition where the workload was known. This was taken to be a steady shutdown state which did not involve any plant transients and which was judged to be within the capability of one operator with some spare capacity, for example, to respond to alarms.

Following workload indicator calibration, it was possible to determine from the histogram, phases during the manoeuvre where one operator would be insufficient and subsequently investigate the various options available to reduce the workload.

Before automation was considered to reduce operator workload other possibilities were explored. These included:

o   Where local peaks in workload or a large number of simple but coincident tasks occurred, the workload was, where possible, evened out by re-locating one or more tasks with respect to time.

o   Where the high difficulty rating of a given task contributed significantly to the workload, an assessment was made to identify possible MMI or system design changes which could reduce the workload. Where two tasks were considered incompatible with performance by a single operator MMI changes were investigated.

111

o   An assessment of control location was made at various points throughout the manoeuvre to identify coincident tasks requiring control from different sections of the Control Suite.

Having attempted one or more of the above if the workload was still found to be unacceptably high, automation was investigated in relation to a variety of possible staffing levels. Automation was also investigated when it was identified that manually performing the task was an inappropriate requirement on the operator (eg due to the accuracy required or the length of time spent performing the task).

The workload was seen to increase significantly during the period from the initial steady state to the intermediate steady state. Consequently this period was assessed as perhaps requiring additional general assistance in addition to some automation provisions.

The period of highest workload appeared to occur during transition to intermediate state with apparently demanding tasks associated with latent heat removal to achieve the required trajectory.

Though a major aim of the study was to identify automation provisions considered necessary for satisfactory operation, the group consciously regarded automation as only one of several means of reducing workload and improving reliability. It therefore examined other options which might prove cheaper or simpler before turning to automation. While automation is a powerful method for reducing workload, and achieving adequate reliability, it is likely to prove expensive, and a case has to be made that in the particular area under examination the workload reduction is necessary and that automation is the only satisfactory method for reducing it.

During the period of highest workload, four of the tasks were found to dominate the workload, three of which had to be performed continuously. It was clear that workload reduction via automation was likely to offer maximum benefit in this area.

It was evident that there were tasks which require high accuracy and/or reliability of execution, and for which automation would be the most effective (both in terms of cost and safety) way of ensuring that exacting task requirements are achieved.

Particular attention must also be paid to the presentation of the instructions where parallel activities are specified. If such activities are performed by more than one operator the dependencies between the operations must be clearly identified and the distinction between the operators' roles must be identified. This analysis showed where such dependencies might arise.

Where the ratings on the task sheets indicated that a task was particularly critical either with respect to safety or availability this was used to identify the need for a 'caution' to be included in the instruction.

Similarly, where the comments in the task sheet identified that care was necessary with regard to the manner in which the task would be performed, the instruction could be worded to reflect this.

When procedures are developed from task analysis of the type described the development is unambiguously described and is therefore auditable.

The analysis and rating of the bottom level tasks also aids training by identifying the more significant tasks in terms of difficulty and criticality. Particular attention can then be paid to these tasks during training to ensure the student understands the tasks and can successfully perform them.

The time-line analysis illustrating both the straightforward task density and then modifying it to account for the difficulty and criticality of the individual tasks (i.e. the workload indicator), had both a direct and an indirect training benefit.

The direct benefit was that it provided an early identification of the sequencing, timing and intensity of tasks before operational experiences would be available. This information can be incorporated within the training programme to ensure the training is as realistic as possible.

The indirect benefit was probably the more significant of the two. The time-line analysis and workload indicator provided a means of identifying staff requirements and aids the writing of workable realistic procedures, both necessary for effective training to be promoted.

<u>SUMMARY</u>

The methodology described shows that a human factors analysis of automation requirements can assist with the allocation of functions.

Two applications of FAST were made, one to create a functions based approach in order to depict the systems goals, and the other an operations-based approach in order to show the human content. The functions-based approach quickly captured the requisite knowledge of system function so that different analysts could work from a common data source, and the operations-based approach created the framework from which genuine operator tasks could be identified and confirmed.

FAST permitted the creation of a hierarchical analysis at a time when it would be normal for most projects at a design stage not to have operational procedures from which to work. An additional benefit of the analysis is, of course, that the information gathered could then be used in the detailed development of procedures, and as mentioned previously, the development of training curricula.

When genuine operator tasks had been identified at the bottom of the task hierarchy they were stripped off and subjected to a time-line analysis. This technique required that the sequencing and duration of each genuine operator task be depicted in linear fashion with respect to the elapsed time of an important plant manoeuvre.

Each genuine operator task was then evaluated to assess reliability in the context of difficulty by a number of analysts using an abbreviated task

113

analysis worksheet. In order that consistency was assured the abbreviation only took place once the principal factors associated with task load had been identified and incorporated in the shortened worksheet.   Secondly checks were carried out on a sample basis to ensure reasonable inter-analyst reliability was achieved.

When these analyses were complete it was a requirement that two basic judgements regarding the rating of task difficulty and task criticality be made and an overall rating of each task was made by summating the difficulty and criticality ratings.   Task criticality was assessed on two bases, safety and availability, and task difficulty was assessed on the basis of complexity, accuracy, special skills and knowledge, time constraints and task-critical cues.   Again, checks were carried out to ensure a high degree of inter-analyst judgemental reliability.

Where points of high workload were indicated it was possible to examine the source of the loading and deploy effective counter measures such as task re-ordering, additional staffing and automation.   Where automation was applied it was then possible to assess the predicted impact by reiterating the analysis in the affected local areas.

CONCLUSION

This methodology provides early insight into potential problems associated with human reliability and the impact of automation.   It quickly highlights the portions of a manoeuvre in which automation should be considered and it identifies what it is about complex tasks that creates the need for automation and facilitates the striking of an appropriate balance between operator action and automated operation.

ACKNOWLEDGEMENTS

The following made significant contributions to the development, documentation and execution of this methodology:- G Pendlebury and W B Sargeant of the CEGB Sizewell 'B' Project Management Team, Dr D P D Whitworth, A Fewins, K Mitchell and J W Marriott of the National Nuclear Corporation, A Biggs and S Petts of the Operational Engineering Division (South) of the GEGB and J Bennett of the CEGB Nuclear Power Training Centre.

The author wishes to thank Miss J Askey and Mrs M Hitchen for typing this paper and Mr B V George, Project and Technical Director of the Sizewell 'B' Project Management Team for approving its publication.

REFERENCE

1. Churchill, R. (ed.) 1980)

SAVE Proceedings International Conference, (Society of American Value Engineers) 1980 Dallas, Texas, May 14-17, 1980, published by SAVE (Society of American Value Engineers), Irving, Texas, 1980, pp. 338.

MINIMISING THE CAUSE OF HUMAN ERROR

SP Whalley
Lihou Loss Prevention Services Ltd
Grays Court
1 Nursery Rd, Edgbaston,
Birmingham B15 3JX
UK

ABSTRACT

Accidents or incidents due to human error do not 'just happen' and it is rare for the cause to be as simple as human incompetence or negligence. This paper commences with a practical consideration of why we need to move away from this stereotyped answer and culminates with the reassurance that a comprehensive analysis is a useful and possible pursuit.

As a simple introduction to various concepts and terminology relating to Human Reliability, the terms Error Type and Error Cause are defined and illustrated through reference to equipment failure modes and failure causes. Extending this introduction to the source of human error, the underlying contributory factors "Performance Shaping Factors" are linked into the model. The nature of PSFs is explained with reference to their main classsification types and the extent of their influence is illustrated through the use of explicit examples taken from industrial incident reports.

One method a company may use to minimize human error is that of analysing past events, using this information as a starting point for considering potential problems in the same or a different area. Alternatively a PSF assessment booklet or ergonomic guide can be used as an audit tool for existing work places. Both methods are introduced.

Finally, a new predictive technique; PHECA, the Potential Human Error Cause Analysis, will be demonstrated as a method to assist the design engineer with minimising the potential causes of human error from the concept stage onwards.

INTRODUCTION

THE NATURE OF THE PROBLEM

On analysis it can be seen that all major incidents are due to a prior sequence of events yet a traditional accident investigation looks for the ultimate position of **blame**. This view point has led to the type of categorization scheme demonstrated within the IChemE Loss Prevention Bulletin (1), and reproduced in Table 1. Within this scheme the causes of accidents directly associated with the **'operator'** are;

Cause 5. Worker / operator works in an unsafe manner,

Cause 8. No use or improper use of work permits,

Cause 13. Physiological, psychological causes

and

Cause 14. Poor concentration, inattention, negligence

What is more, when assessing the distribution of causes, this report places almost 50% of the accidents examined as resulting from Cause 14 and approximately 15% due to Cause 5 with Cause 13 being assessed as the fourth highest implicated reason for an accident.   At best such a report is extremely frustrating and at worst a very negative inditement which could lead to ineffectual reprimands replacing the need to suggest improvements based on the root cause.

The analysis of information is primarily dependent upon the type of classification used; if a classification is coarse grained then the results reflect this limitation, if information has to be fitted to a preconceived set of categories then the results will reflect the biases of those who produced the scheme and those who attempt to use it.   Similarly when viewing the behaviour of others, what the casual observer may consider to be a slip stemming from negligence may be interpreted very differently by someone with an informed knowledge of human behaviour and error mechanisms.   Fortunately our understanding of human error has progressed beyond the point demonstrated within table 1.

Table 1.   Causes of Accidents
(as presented in the IChemE Loss Prevention Bulletin, no.072)

```
1.   Technical Deficiency
2.   Design Error
3.   Change which was not thoroughly examined
4.   Unforeseen, unknown
5.   Worker / Operator works in an unsafe manner
6.   Toleration by the supervisor
7.   Lack of supervision
8.   No use or improper use of work permits
9.   Poor training/instruction
10.  Poor communication / co-ordination
11.  Decision of superiors concerning technical matters
12.  Decision of superiors concerning organization
13.  Physiological, psychological causes
14.  Poor concentration, inattention, negligence
15.  Remaining risk
16.  Unknown or undetermined causes
```

USE OF THE EVENT SEQUENCE CHART

In order to achieve the most benefit from historical data, an incident needs to be examined in terms of the preceeding series of events. This can be achieved by producing an **event sequence chart** (2) for each incident.   The chart acts as an event summary and should tie together many aspects that in some way could have influenced the outcome. In addition to indicating the 'point of no return' it enables the assessor to consider the combination of factors and events surrounding the incident. Of specific importance is the inclusion of aspects termed 'causal factors' and 'systemic factors'.

An example of an event sequence chart is provided in Figure 1. The main flow diagram gives the immediately related sequence of events leading to the incident, the contributory boxes provide secondary information (Causal Factors) directly affecting the outcome of the incident and the bubbles indicate policy aspects (Systemic Factors) that are constantly present, there by providing the latent conditions that could turn a performance error into an incident. People make mistakes all the time it is **when** and **where** they occur that turns them from inoccuous, unnoticed and self corrected performance discrepencies into reportable events.

It is immediately obvious that such an approach to incident analysis leads to the identification of factors that would benefit from improvement. Systemic factors by their very nature could have an equally undesirable effect on another task or plant area if left uncorrected.

How would the example scenario have been classsified in terms of the error cause classification previously quoted ? Out of context the error could have been seen as yet another example of Cause 5 "worker / operator works in an unsafe manner" or perhaps Cause 14 "Poor concentration, inattention, negligence", depending upon the assessor. Once the background to the incident is established causes 2, 3, 6, 9, 10 and 13 can all be seen to play their part, yet this scheme insists upon identifying one main cause. It must be contended therefore that such a classification of incident causation is entirely inappropriate; not only is it ineffectual at determining what went wrong when an accident has occurred, it has no place within accident prevention.

This paper contends that a number of alternatives do in fact exist that assist with minimising the cause of human errors and hence accidents. These methods are placed in context by initially considering the theory of human reliability in relation to that of equipment reliability.

HUMAN RELIABILITY

When examining the terminology of the engineer, a fault ie Failure Mode is "what went wrong" with the system or piece of equipment - this equates with the **Type** of human error. Extending this analogy, the term Primary Failure is used to describe one of the 'ultimate reasons' for an equipment fault, this equates to the **Cause** of human error. In human factor terms, this would suggest that both a functional (external) assessment and a psychological (internal) assessment of human error would be required in order to provide a qualitative analysis of human reliability.

TYPES OF HUMAN ERROR

There has been a lot of confusion over the years with respect to error type and error cause, the functional deviation or 'what went wrong' (the valve was left open) has been insufficiently separated from the 'why' (the task was interrupted by a telephone call). A suitable error type classification must conform to the following requirements:

1. Include all unwanted deviations in performance output
2. Be simple to apply and determine
3. Suggest possible important failures to the user
4. Be generic in nature
5. Use simple language

117

Figure 1. Event Sequence Chart

H_2S CYLINDER INCIDENT

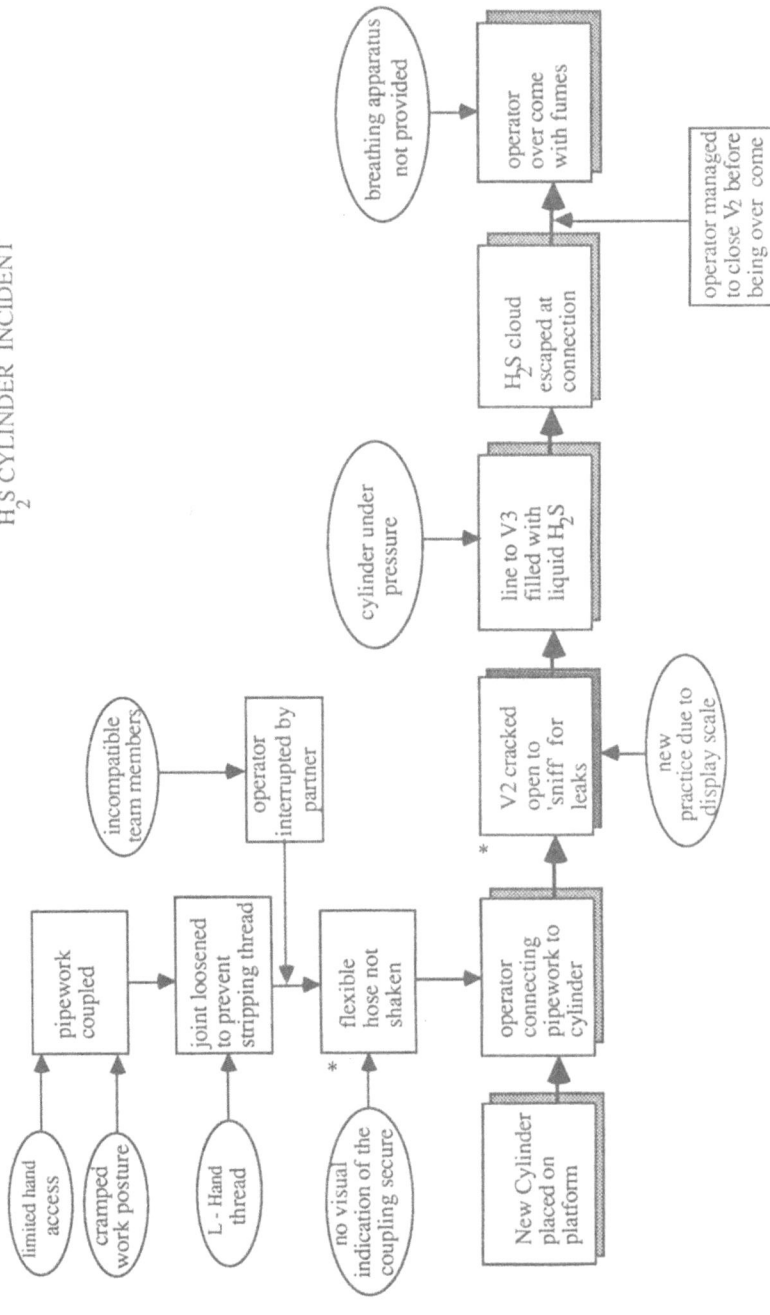

It has been suggested by Whalley (3) that Hazard and Operability study keywords form an ideal basis for error types, especially due to their familiarity to the engineer, these are presented in Table 2.

Table 2.   Error Types related to Hazop key words

ERROR TYPES				
HAZOP	ERROR	HAZOP		ERROR
1. No	1. Not Done	------		6. Repeated
2. Less	2. Less Than			7. Sooner Than
3. More	3. More Than	6. Reverse	<	8. Later Than
4. As Well As	4. As Well As			9. Mis-ordered
5. Other Than	5. Other Than	7. Part Of		10. Part Of

ERROR CAUSES

Hazops, hazard analysis, fault tree analysis, failure mode & effect analysis and cause consequence diagrams all consider the causes of an equipment / process failure or deviation to some degree.   Similarly with human reliability analysis the causes of human error must be considered as a route to reducing error likelihood.

Henley and Kumamoto (4), WASH 1400 (5) and others distinguish between primary failure, secondary failure and command faults as three groups of causes of component deviation.   Primary failure can be considered as due to **Internal** causes, command faults are associated with **External** causes, whilst secondary failures are due to a **Mismatch** between internal capabilities and external demands.   An obvious analogy exists with human error cause.

Distinguishing between internal, external and mismatches immediately suggests whether the error cause can be minimized (internal) or even removed by improving design (external and mismatches).   The major classification of error causes suggested by the author (6) are termed error mechanisms.   These in their turn cover psychological factors and physical factors, Figure 2, which if identified can suggest areas for improvement.

Henley & Kumamoto (4) also demonstrate the electronic engineers interest in identifying the potential causes of component failure in relation to potential component deviations.   It is interesting to see that **plant personnel** are cited twice as a secondary cause of component failures.   To pursue this model; component failure mode can be linked to the primary failure cause and the component performance shaping factor - that is human error and then extended to the cause of human error and ultimately the personnel performance shaping factors, figure 3.

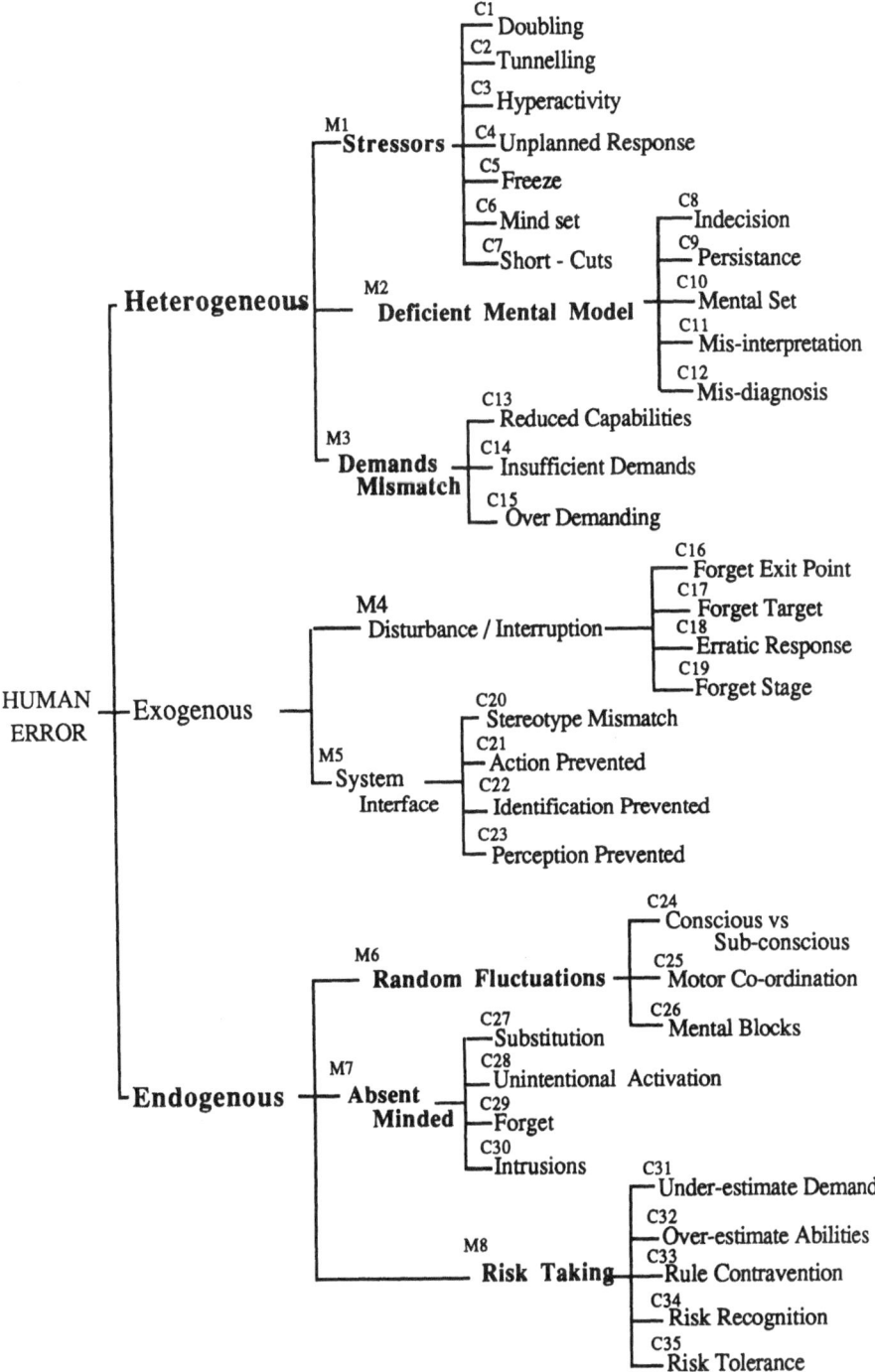

Figure 2. Error Causes Grouped by Error Mechanisms

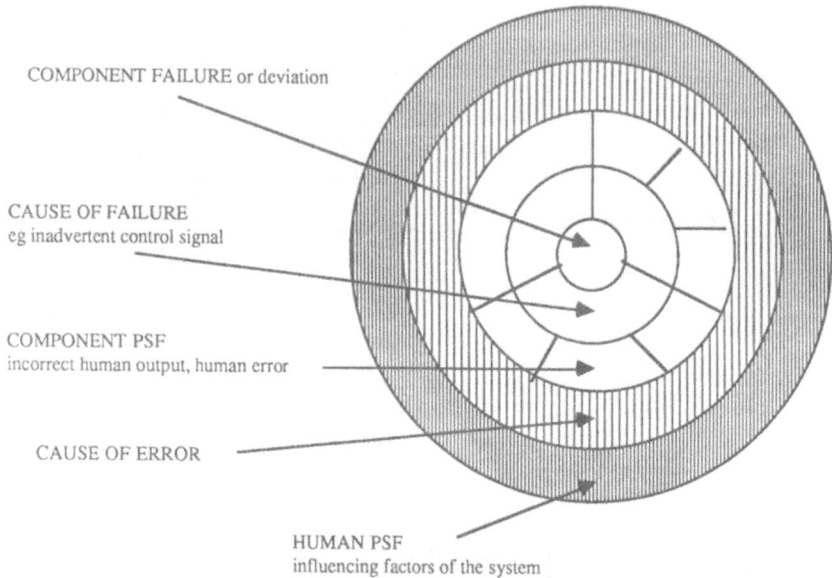

COMPONENT FAILURE or deviation

CAUSE OF FAILURE
eg inadvertent control signal

COMPONENT PSF
incorrect human output, human error

CAUSE OF ERROR

HUMAN PSF
influencing factors of the system

Figure 3. The relationship between human failure and component failure

PERFORMANCE SHAPING FACTORS

Performance shaping factors (PSFs) as their name suggests are individual aspects (factors) within the system that can affect (shape) output (performance). PSFs link directly to error causes; a specific PSF may be associated with many causes and each cause will be influenced by several PSFs, a many to many mapping. If these PSFs can be identified for a specific situation and optimised, ensuring that they no longer have a negative influence on performance, then the potential cause of human error can be reduced.

Personnel performance can be viewed as a PSF associated with secondary failure and command fault causes of system **component** failure. It is equally important to identify what factors may constitute personnel PSFs. Human PSFs have been hierarchically classified (7) with an increasing level of detail, such detail should assist improvements to design when justified. This type of classification has also been considered helpful for auditing existing plant. The structure of the PSF classification scheme is contained within Figure 4.

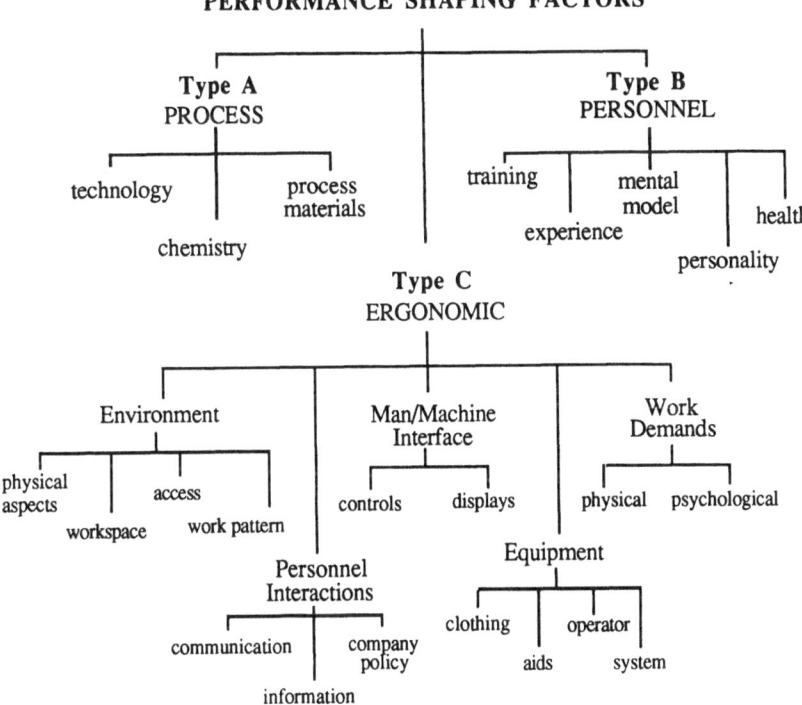

**Figure 4. Major sections of the performance shaping factors
classification structure**

CASE EXAMPLES

The value of these three main concepts; error types, error causes and performance
shaping factors, can be demonstrated by considering how they relate to specific incidents.

INCOMPLETED BATCH

What Happened ?
The supervisor informed the shift team that an order had to be completed by the end of
the shift. This meant filling the normal amount of drums at the normal rate. However an
hour before the end of the shift the leading hand reported that another man was needed to
ensure that the order was completed. The supervisor believed that they were 'trying it on'
since the required work rate was the same as usual and told the leading hand that there was
not a man spare. At the end of the shift the order was incomplete.
Error Type = PART OF

Why ?
What the supervisor had not taken into account was the **change in work environment**. He had been unaware that this could affect performance.

Normally the filling operation takes place in an enclosed building but due to plant modifications the two end walls had been removed leaving a covered area in the shape of a tunnel. This particular day was exceptionally **cold** with the first snow flurries of the year. The operators were wearing their **jackets** and **windcheaters** (due to the tunnel increasing the effect of the wind). They were wearing their usual protective gloves but their hands were **numb** and the smaller jobs were difficult (labelling, fastening and unfastening the bungs). The combinatorial effect of the cold windy environment and the extra protective clothing meant that **movement was more difficult** resulting in higher energy expenditure to achieve the same work and hence a lowering of the work rate. They found that even working through their normal rest breaks did not improve the situation. Although this would increase the time available for completing the job it would remove any possible chance of **recuperation**. The work rate would continue to drop rather than recommencing at a higher level following a **rest break**.

```
ERROR CAUSES - Mental Model (mis-diagnosis) - supervisor
               Demands Mismatch (reduced capabilities) - operators

INFLUENCING FACTORS -
   (PSFs)              physical environment, location, workspace,
                       physical demands (power, dexterity), clothing
                       work pattern, team work , communication,
                       management attitude
```

Lessons: 1. Work rate and energy expenditure are effected by environment as well as by the specific work demands (nb heat as well as cold, humidity and atmosphere).

2. 'Protective' clothing is vital in terms of peoples health and safety but will reduce their capability, therefore make allowances - normal jobs may take longer.

3. When making changes to plant, consider the environmental impact on those people working in the area, not just on the equipment and process.

AN EXPLOSION RESULTING FROM A PRESSURE RISE

What Happened ?
The operator of a computer controlled plant was **unexpectedly** having problems with its control, parameters were moving into **alarm**. In an attempt to fulfil the supervisor's request to keep the plant running as long as possible, he placed the plant into manual control. He then found that he was unable to re-stabilise the plant and had also averted an automatic shutdown. This resulted in a rapid pressure build up leading to an explosion.

Error Type = Other Than

Why ?

Normally the operator had **little positive interaction** with the plant, his job was to monitor the process parameters via two VDU screens and to make manual control modifications if required. In fact this **plant ran very smoothly** therefore the operator rarely needed to intervene. On this particular occasion his **arousal was low**, nothing unusual was expected, it was **well into the shift** with **no visitors** and no maintenance crews to monitor.

Whilst checking one VDU page with its set of parameters another group was altering (though not sufficiently to cause an alarm) yet remaining in a similar ratio (**pattern**). When this page was subsequently examined the operator noticed nothing different. A while later one of these parameters went into alarm closely **followed by several others**. This was completely **unexpected** and the operator's mind **went blank**. Due to the operators limited involvement with the plant his **understanding of the process was hazey**, a lot of what he had been taught had been **forgotten**. The operator contacted his supervisor who asked him to **keep the plant running** as long as possible. The operator started to try and correct the situation placing **control into manual**. His **attention was fixed** upon the original problem. In his attempts to correct this situation other parameters started to alter and the plant became unstable suffering a number of pressure surges. The operator was **unaware** that the overall pressure in the system was rising rapidly and that due to the manual setting an automatic shut down would not occur.

```
ERROR CAUSES -  Stressors (tunnelling, freeze)
                Mental Model (indecision, persistance,
                                     misinterpretation)
                Demands Mis-match (under demanding,
                                     followed by over demanding)
                System Interface (perception retarded)
                Risk Taking (risk recognition)


INFLUENCING FACTORS -
                      personnel involvement (occasional),
        PSFs          technology, chemistry, controls, displays (location,
                      response time,alarms), communication, supervision,
                      fail safe, training, experience,knowledge,personality
```

Lessons:

1. If people have very little to occupy them within their jobs they become "under aroused" ie their level of concentration drops, their ability to notice small changes diminishes and their capacity to react suitably and promptly to difficulties reduces.

2. If skills and knowledge are irregularly used or tested they can become reduced and forgotten. Operators in charge of plants needing little interaction should have simulation or self test facilities plus regular assessment and retraining.

3. Sudden emergencies, which are very rare, cause an increase in arousal and if the individual is not trained to cope with the situation this can result in high stress and a rapid drop in functioning "over arousal". Note that in this example stressors led to the following:

THE USE OF AUDIT TOOLS

1967 saw the first ergonomics check list for plant designers, this was developed by the International Ergonomics Society and two years later a more specialised design guide was made available by Meister & Sullivan (10) for visual displays. Despite this early progress a specific guide to reducing human error in process operation (11) was still considered necessary in 1985 in order to assess existing plant design. Process plant and associated systems were still being designed with insufficient consideration of the people expected to run them.

The Human Factors in Reliability Group short guide to reducing human error was produced to cover the spectrum of Ergonomics (interface design, procedures, work place, environment, training, task design and job organisation) and is presented as a check list of questions, each to be answered Yes or No. If the system is well designed all the answers should be Yes. Alternatively a PSF check list can be used as a prompt to examine factors that could have an affect on those working within the system, however in this case if a PSF is relevent, recording a tick, then the designer must check that it is designed to suitable standards. In both cases there is a problem associated with deciding whether a particular aspect is adequately designed or not and if not, what to do about it. The handbook of perception and human performance (12) may provide the answer but to date it is still an ongoing venture.

Note that although auditing is typically associated with standing plant, it is also feasible to consider its use as an assessment of drawing board designs.

One final problem associated with auditing techniques is that there is no method of ensuring that a limited budget is most effectively spent. It is necessary to know which aspects are the most **important** in terms of their influence on human error.

POTENTIAL HUMAN ERROR CAUSE ANALYSIS
- PHECA

The most positive means of minimising the cause of human error is by identifying the most dominant factors that could influence performance, prior to a system's design. The basic philosophy of the PHECA model (13) is that the cause of error is dependent upon three variables; the type of task being performed (Task Type), the expected form of response (Response Type) and any deviations in performance (Error Types) considered to be important given the specific circumstances. Each of these variables has an associated set of error causes. Only if an error cause is identified by all three routes can it be registered as a **potential error cause** and receive a likelihood weighting for inclusion in the error cause chart.

At the end of an analysis the error causes cascade forward via established links to the potential performance shaping factors. These receive weightings in their turn based upon their own importance and that of the associated error causes plus the frequency with which each was implicated. By this mechanism it is possible for the final output to present a prioritised list of individual PSFs associated with that particular set of circumstances. Hence system designers and managers can receive consistent and relevant guidance for achieving the most cost effective reduction of the causes of human error without solely depending upon implicit expert judgement.

3.1 "Freeze" -   initially the operator was unable to make any response, he couldn't think what to do.

3.2 "Tunnelling" -   the operator fixed all his attention on the original problem area, ignoring other aspects of the plant (this is a normal way of trying to cope with too much information, we concentrate on one thing at a time)

3.3 Mis-interpretation -   the operator had not thought through the possible effects of his actions and the current problem on other parts of the plant.

3.4 "Risk Recognition" -   the operator failed to realize that by keeping the plant on manual control no automatic shut down would occur

4. Be careful if making a request to an operator to keep the plant running, this can alter his perception of the situation and may lead to unnecessary risks, particularly if the operator likes to oblige.

ERROR REDUCTION METHODS

HISTORICAL DATA

The case studies presented in the previous section indicate how it is possible to use the qualitative human reliability principles in order to benefit from an analysis of past events. An additional way in which historical data can be used as a guide to reducing the causes of error is by examining the dynamics or the 'error profile'. The three classical patterns of errors are: Systematic errors, Random errors and Spurious errors (8).

If all the error records show a similar performance discrepancy, ie the same error is repeated, then the error profile is said to be **systematic**. This means that the cause of the errors is a permanent feature of the system, whether internal or external to the task performer. If the same error consistently occurs independently of who performs the task, then the designer or manager can be confident that the fault is due to external factors and can examine such PSFs as the environment, the interface, the job demands etc. Conversely if the errors are dependent upon the performer, then internal aspects should be considered, for example; training, the individual's understanding (mental model), experience, personality, ability, health.

Alternatively if the errors are frequent and of no consistent type this suggests a truely **random** pattern of errors. This is characteristic of the novice and is due to a lack of experience or training, or if obviously not the case it is probable that the task itself is too demanding for consistent performance.

Finally the more refined group of random errors are those that occur intermittently and for no apparent reason - the **spurious** error. This is the form most common to the expert and is the most difficult to prevent. The designer needs to consider what may trigger these errors, for example; a disturbance, sudden noise, schema cross-overs (the task being attempted contains features similar to a more frequently performed task which takes over - 9), and stereotype contravention. Only then can the system design be changed to keep their likelihood to a minimum or to counter-act their threat.

USER INPUTS

PHECA relies on three user inputs as initiators of an error cause analysis. These inputs are the Generic Task Type, the expected Response Type and a set of important Error Types.

Generic Task Type - whether a task analysis, hazop or fault tree analysis registered the need to consider causes of human error, there must be some form of performance requirement that can be considered as a single task or a set of associated tasks. Once these have been identified their specific characteristics require reformulating in terms of generic principles. The seven generic task types, table 3, account for changes in the type and number of psychological processes that have to be performed during a task, these have been seen to directly influence the possible causes of error.

Table 3. Generic Task Types

1. Stimulus / Response
   A direct instinctive response to a well known situation with no requirement made of the higher level mental capacities - no conscious decision making.
   For example: Stopping at a Red Light

2. Integration / Response
   A number of inputs are attended to and integrated, they are all comprehensible producing a recognisable pattern that relates directly to an internal model. A known response to a known group of variables.
   For example: Following diversion signs

3. Interpretation / Response
   A picture of the current situation is developed and the individual must relate this to an internal model and previous experience, decision making.
   For example: Route Planning

4. Requirement / Response
   A predetermined activity that must be completed at some point in time. Often time dependent, it is internally triggered rather than relying on external stimulus.
   For example: Stopping for petrol (prior to the warning light coming on!)

5. Self Generation / Response
   A self determined activity sometimes requiring planning and decision making to provide a method for obtaining a goal.
   For example: Putting the car radio on

6. Choice / Response
   The individual must select a particular plan of action or goal when more than one could be correct or there are several alternatives.
   For example: Choosing when and where to stop for lunch

7. Correction Required / Response
   It is realised that something other than the planned response is being carried out or that the original aim was inappropriate for the situation, this is established by self monitoring and system feedback. It is during this type of situation that compounded errors may emerge.
   For example: Driving the wrong way down a one-way street and reversing back up.

Response Type   - in the same way that task types require classification, response types also need some degree of distinction. The most obvious preliminary distinction is between discrete responses and those that form part of a sequence or can be sub-classified into a sequence. From working on chemical plants and studying incident reports it also appeared important to separate communication activities from those that actively changed the system in some way and indeed to identify situations in which the only correct response would be to wait and do nothing. This resulted in a set of seven possible responses;

> Discrete Tasks: Get Information, Give Information, Action and No Action,
> Sequence Tasks: Get Information, Give Information and Action

Error Types - this set of ten has already been discribed in relation to the Hazop keyword system, table 2. This is the only variable input where the user can select more than one response. PHECA addresses the 'what if' scenario therefore if the user is attempting to assess the potential situation, rather than using the technique to guide an investigation of an actual incident, it is quite probable that more than one type of error could potentially interfere with the success of the activity under review, therefore all those error types considered to be important must be covered.

THE ANALYSIS

Once PHECA has received all the relevant inputs for a sub-task the program's linkages are traversed to error causes. The mathematics involved is simple since it is associated with set theory. Based on a universal set of error causes each task type, response type and error type has an associated sub-set, these are overlapping but no two alternatives link to exactly the same set of causes. Initially the program makes a union of the error causes associated with the identified error types, variable three. Secondly it establishes the intersection of the error cause sets for each of the three variables. Only error causes within the intersection are registered for a likelihood weighting. Once weightings have been established these are normalised, prior to their recording in tabular and chart form. The error causes provide an overview of understanding which highlights both the unlikely as well as likely reasons for errors.

Error causes remain one step away from assisting the designer, what is required is detailed information relating to features of the system. This is provided by the second step within the analysis. The weighted error causes link forward to individual PSFs which have their own in-built weightings expressing their general level of influence. More than one error cause can link to a PSF therefore the final PSF importance weighting reflects; the PSF general weighting, the importance of the error cause implicating the PSF and the frequency with which the PSF has been registered. Once again after establishing the weightings for all the PSFs these are normalised in order to assist comparisons between designs. The PSF output is in the form of a prioritised list and the user may request to view the top 50%, 25%, or 10% of PSFs with weightings above the mean.

The final stage of the technique will be the elimination of irrelevant (remember that this is a generic technique) and well designed features from the PSF list. By back tracking through the links this deletion will be used to assess the extent of error cause reduction achieved through good design. This facility will enable the system assessor to check the extent of improvement possible due to a different design of the system.

PHECA is still being validated as a technique but is already demonstrating its potential for indicating likely problem areas within a system.

TO CONCLUDE

Ergonomics and the human reliability specialists have moved a long way forward towards the understanding of the causes of human error and away from the type of classification demonstrated within the introduction to this paper. It is only through increased understanding that real improvements can be made. Note that in 1975 the independent reports following Three Mile Island (14) recommended the continued use of probabilistic risk assessments due to their ability to increase awareness and understanding of potential problems. Similarly a technique that can provide understanding of potential human problems and their causes should be considered vital.

REFERENCES

1. Causes of accidents (the results of a study). Loss Prevention Bulletin, 1986, 072.

2. Johnson, W.G., MORT Safety Assurance Systems, Marcel Dekker, 1980.

3. Whalley, S.P., Types and causes of human error. Risk Assessment of Marine Hazards course manual, Lihou Loss Prevention Services Ltd, Birmingham, 1987.

4. Henley, E.J. and Kumamoto, H., Reliability Engineering and Risk Assessment, Prentice Hall, New Jersey, 1981.

5. US Nuclear Regulatory Commission, WASH 1400 Reactor safety study: an assessment of accident risk in US commercial nuclear power plants, US atomic energy commission, Washington, 1974

6. Whalley, S.P. and Maund, J.K., Improving human reliability by design. Hazards in the Process Industries: Hazards IX, IChemE symposium series no. 97, IChemE,Rugby, 1986, pp 235-248.

7. Whalley, S.P. and Maund, J.K., The influence of performance shaping factors on operator interactions with process control. Human Decision Making and Manual Control, ed H-P. Willumeit, Elsevier Science, North Holland, 1986, pp 195-201.

8. Edwards, E., Human Error. VNV Dutch airline pilots association symposium, Safety and Efficiency: the next 50 years, The Hague, September 1979.

9. Senders, J.W., On the nature and source of human error. Second Symposium on Aviation Psychology, Colombus, Ohio, April 1983, pp 421-426.

10. Meister, D. and Sullivan, D.J., Guide to Human engineering design for Visual Displays, Contract no. N00 014-68-C-1278, Bunker-Ramo corp, California, 1969.

11. Human Factors in Reliability Group members, Guide to Reducing human error in Process Operations - Short Guide, SRD UKAEA, Warrington England, 1985.

12. Boff, K.R.and Lincoln, J., Handbook of Perception and Human Performance, Wiley and sons, New York, 1986 onwards.

13. Whalley, S.P., Factors affecting Human Reliability in the Chemical Process Industry, PhD thesis, Aston University, 1987.

14. Rogovin, M. and Frampton, G.E., Three Mile Island a report to the Commissioner and to the Public, Vol 1, US NuReg Commission, Washington, 1976.

ACHIEVING CONSISTENCY IN THE USE OF HUMAN FACTORS ANALYTIC METHODS FOR RELIABILITY ASSESSMENT

J.A. Astley
Applied Psychology Division
Aston University
Birmingham B4 7ET
GB

ABSTRACT

In highly complex systems, the human has a role to play in optimising system availability, maintenance and safety. However within modern technological systems it is also recognised that human error can be a major contributor to system failures. There are a range of techniques available to quantify and predict human error. Many of these methods are used in parallel with human factors task analysis to identify the performance goals and objectives of the human. A problem exists in the use of these methods. Often there are inconsistencies in the analysis due to variations in interpretation by analysts. So analyses can result in different, but equally valid, inputs into the human factors design of the system. This paper aims to explore the reasons why variability exists and to look at potential solutions to overcome discrepancies between analysts. Case studies from the high risk industries and defence industries are used to illustrate. Analysts with varying ranges of experience are examined. The difficulties encountered are assessed and the approaches used to resolve them outlined.

INTRODUCTION

Human error is widely recognised to be a contributory factor in many industrial system failures. This has led to increased emphasis being placed on the analysis of the role of humans in a system, both for identifying the human's capacity for error and assessing reliability. There are a range of methods available to quantify and predict human error. Most of these techniques employ a task analysis either directly or indirectly. Task analytic techniques provide information on the different elements of a task in the system and can allow the potential for error to be assessed quantitatively or qualitatively.

It is important that the analysis carried out is effective, both in its thoroughness in analysing the task and its potential for error and in its application to design. To achieve this the technique used should have a sound theoretical basis and produce consistent results in practical application. Due to the resources needed to carry out analytic activities, the opportunity to examine analyses for consistency as a measure of their validity rarely occurs. This paper outlines case studies where consistency has been examined and explores reasons why inconsistencies occur, suggesting steps that can be taken to overcome them.

THE IMPORTANCE OF CONSISTENCY

Human factors experts have traditionally carried out task analysis themselves. Now the trend is for engineers and other personnel to become involved in the process. Due to this a new range of problems arising from the application of techniques is becoming evident. Analysis is a time and person intensive activity and so must be as efficient as possible. Consistency is important to avoid large discrepancies in the quality and type of analysis produced by the analysts and to avoid the need for repeated reiteration.

The opportunity to assess the consistency of the results of a technique rarely occurs, even in large scale systems where there are several analysts working. The time and effort expended in carrying out an analysis means that analyses are rarely allowed to overlap significantly, so there is little opportunity to compare the results of several analyses on one system or part of a a system. Consequently the validity of techniques has tended to be shown by the results of their use and inputs into the design process.

Ensuring consistency helps to ensure that a method is valid. Validity is a broad concept, but has 2 main aspects. Theoretical validity is based on the underlying psychological theory of tasks and of the causes of human error. Empirical validity relates to the use of the technique in practice, for example whether its ouput gives a justifiable contribution to design and whether the results of the analysis are repeatable.
One means of assessing the practical validity of a method is by 'inter-rater reliability'. This is the degree of agreement between analysts independently anaysing the same task. It does not quantify consistency but helps to give an identifiable measure of the method's repeatability. Aiming at an analytical technique that is perfectly reproducable would need a high degree of training and a method with a complex and constraining rule structure. Fine (1974) took this approach by training in the use of language for analysis as well as in the method itself. At present formal training is not common for analysts and analysis tends to be carried out be human factors experts.

Inconsistencies in the way analysis is carried out at this stage could lead to errors in design and often these will be difficult to trace. The quality of information input into design from the analysis reflects the qulaity of the information input into the analysis itself. To achieve reliability in analysis the information collection stage is as important as the analysis itself. There are many sources of information available existing systems, subject matter experts, documentation and simulation.

In a new system analysing the humans role and the possibilities for error can be particularly problematic. The information concerning the human can be fragmented . In this case it is particularly important to identify who the stakeholders in the new system will be and to use their expert knowledge as a source of analytical information.

Previous successes in practical applications and the extent to which a method has been tested in the field are good guides to the effectiveness of a method in practice. If the method produces consistency across a wide variety of applications then it can be considered fairly robust.

CASE STUDIES

Two case studies have been carried out to highlight some of the problems encountered when trying to ensure consistency in analysis. Some solutions and their usefulness are discussed.

The two case studies looked at the problems in use of an analytical method, and also at the consistency of end results of an analysis of the same complex control task by several analysts. The first study looked at the application of FAST (Functional analysis systems technique Creasy 1980) by a group of engineers to the analysis of part of a control room operators task in a high risk industry. The analysts were given no formal training in the method and the consistency of 5 analysts each looking at the same 5 tasks was considered.

The second case study looked at Hierarchical task analysis (HTA) which is used as a basis for the human reliability assessment technique SHERPA (Embrey 1986). Both software and hardware engineers were trained in the use of HTA. The consistency of analysis was measured on a complex sonar operator's task set in the context of the defence industries. Consistency in applying HTA was also assessed using a group of trainee ergonomists who had had some training in the use of task analysis. In this case the task analysed was a complex monitoring and procedural control task.

It is not appropriate to discuss the results of the studies in detail here, but some of problems encountered relating to consistency are outlined and solutions suggested.

PROBLEMS OF ACHIEVING CONSISTENCY

Categories of problem were encountered that accounted for variability amongst analysts, inluding problems resulting from misconceptions about applying the technique, how to represent the analytical information and problems of conflict with existing knowledge.

Many of the inconsistencies evident in the FAST analysis could have been overcome by some training in the application of the technique, and the explanation of some of the concepts. In some places confusion arose, as the method left ambiguities in the required content of the analysis at a more detailed level. Just where to start and finish the analysis is a common problem, including how detailed to make an analysis and how to balance detail between the levels of the hierarchy.

The final result of an analysis is usually represented in a diagrammatic, graphical or tabular format, the main content of which is text rather than symbols or numerals. By its very nature this is less exact than than a mathematical formula or algorithm, and so the subjective content may seem to some extent to limit validity and repeatability. Some expert judgement in an analysis is inevitable and consistency can be maintained provided the meaning of the analyses are the equivalent of one another when judging consistency.

The hierarchical structure of both techniques (Figures 1 illustrates the structure of HTA) was reported to help in structuring the analysis and ensuring it was both systematic and thorough. Problems were encountered with the description of the conditions and specificaton of ordering that governed the task elements (Plans in HTA and logic symbols in FAST).

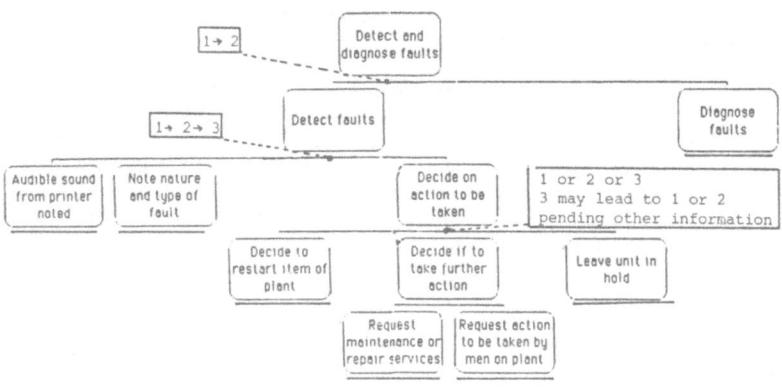

Figure 1.
The structure of a Hierarchical Task Analysis

Some difficulty was encountered with the content of the text using HTA as it was orientated towards human design (For example training) rather than system design. However this was overcome by practice and the use of examples.

Within human tasks there can be many equally valid routes to achieving the same functional goal, as within a system there can be many possible algorithms that achieve the same result. In analysis all the sensible routes can be reported (But here the analysis must be exhaustive) or a route can be selected which is optimal for that particular system function. Some of the routes identified may be more prone to human errors than others and identification of all the possible routes and opportunities for human error can help in the choice of procedure for training etc. A common related problem is that inconsistencies amongst analysts resulted from incomplete analyses failing to show all the possibilities that were appropriate. This led to error in analysis which could have ramifications for errors in design. Often this can be rectified by iteration in the analytic process and gaining comprehensive task information prior to the analysis.

In the first case study, where the analysts largely depended on themselves as the source of expert information, more gaps in analysis resulted. This is largely because experts have much knowledge that has become part of their bahaviour pattern and so they find it difficult to verbalise how they perform a task without outside questioning. Other methods of information collection provide a more acceptable alternative. Further difficulties arose with the analysis and documentation of 'cognitive' task components such as problem solving and decisison making for engineers who were unfamiliar with psychological concepts and analysis of humans.

CONCLUSIONS

Several vulnerable areas within analysis were highlighted by the case studies. The need for collection of comprehensive task information before embarking on the analysis should not be underestimated. The analysts should avoid trying to use themselves as a subject matter expert, it is far more effective to elicit knowledge from elsewhere.

Assuming that a method is valid, consistency relates strongly to both the documentation surrounding the method and how complex it is to apply. If a method forces an analyst to be systematic and thorough then the analyses tend to be more consistent in their outcome. One of the first steps in ensuring consistency is to select a method that is systematic and has a method clearly outlined step by step, which does not allow ambiguity. Both the structure of the representation and also taxonomies can help give structure and form, provided of course that the boundaries between categories are clear.

Stopping rules tend to very context specific , however Shepherd (1982) developed a set of stopping rules for HTA that can be generally applied: a) firstly an analysis is sufficiently detailed if performance of the same task on the current system is satisfactory; b) if there is some means to ensure the performance can be carried out with acceptable reliability, such as by the use of job aids or additional training; c) if a particular task element has been analysed elsewhere; d) finally the analysis can be terminated if there is no further information available on that particular task.

As there is a likelihood of personnel other than human factors experts carrying out task analyses, the method should be clear and aviod the need for knowledge of complex psychological thoeries or user models.

Regarding this, the greatest difficulties highlighted by the case studies were conceptual problems with the analysis. Problems with both the terms used and of how the structure of the analysis could be used to represent the task were common. Especially where the analyst had to deal with cognitive elements of the humans task and not just the overt physical aspects If an analysis method is chosen that uses concepts that are compatible with engineering terms and do not conflict with those used by the person doing the analysis (For example the concept of a system can differ in its human factors and engineering meanings.), then training time is reduced and the analyses are found to be more consistent.

Subjective variations within the language used for the content of the analysis do not affect consistency if the overall meaning of the analysis remains constant. Most analysts experience some difference in the organisation of the content of the analysis but training and guidelines can help to overcome this.

If an analysis is complex, consistency can be a problem not just between, but within individual analyses. This could be overcome by the use of computer aids either to computerise some of the analysis or as a tool for checking consistency in the structure and relationships between task elements in the hierarchy. For example the inputs and outputs of the task elements could be checked automatically.

If a method is reliable it will generate similar information for human error probabilities, or for operating procedures from any application of the

134

analysis to the same system. Much hinges on selection of the methodology, how well documented it is, its proven validity and reliability and its usability as a technique .

Ulimately the method should provide an auditable account of the humans role in the system highlighting possibilities for error ,feeding into the design process and providing a justification for design.

REFERENCES

1. Astley, J.A.and Stammers, R.B., Adapting Hierarchical Task Analysis for user-system interface design. In New Methods in Applied Ergonomics, eds. J.R. Wilson, E.N. Corlett, I. Manenica, Taylor and Francis, London, 1987, pp. 175-184.

2. Creasy, R, Problem solving the fast way. Proceedings of the SAVE conference, 1980, pp 173-175.

3. Embrey, D.E., SHERPA: A Systematic Human Error Reduction and Prediction Approach. In Proceedings of the International topical meeting on advance in human factors in nuclear power systems, European Nuclear Society, Knoxville, Tenessee, 1986.

4. Fine, S.A., Holt, M., Hutchinson, M.F.,Functional Job Analysis : How to standardise task statements. Kalamazoo M.I., 1974.

5. Shepherd, A., Carrying out Hierarchical Task Analysis- a course manual. Chemical and Allied products ITB, 1982.

INTERVAL AND BAYESIAN AVAILABILITY MODELLING

A.Z. Keller
N. Nemat-bakhsh

University of Bradford

ABSTRACT

Two models which enable availability to be calculated using observed data
are given. The first is classical in nature and is appropriate for the
calculation of interval availability. The second model is Bayesian and
has particular relevance in assessing uncertainties involved with the
prediction of instantaneous and asymptotic availability. Monte Carlo
techniques are used to validate both models and to establish their
relative accuracies.

INTRODUCTION

For the purpose of the present paper the following definitions are used
[2, 1985]:

1. Instantaneous availability (or pointwise availability) is defined as
   the probability that a system is operational at any random time t
   under stated conditions.

2. Asymptotic (steady state) or limit interval availability is the above
   availability when the time interval considered becomes indefinitely
   large.

3. Interval availability (average uptime) is the ratio of uptime to total
   time corresponding to N cycles of operation and repair. The definition
   is seen to depend upon N.

Hosford, J.E. [11, 1959] defines interval availability as a measure of
system dependability and gives formulae for the case where both failure
and repair are exponentially distributed.

Kabak, I.W. [12, 1959] gives formulae for the mean availability and
variance for a given number of cycles of operation and repair in a given
time interval assuming a steady state availability and shows how the
variance of the availability can be used to support an economic evaluation
of start-up operations. Kabak assumes an exponential failure time and a
deterministic repair time.

Martz, Jr. H.F. [13, 1971] investigated single-cycle availability and
showed that the median cycle availability is equal to long run or
asymptotic availability provided that the failure and repair distributions
are related by the following expression:

$$F_1\left(\frac{E_1}{E_2}\,t\right) = F_2(t) \tag{1}$$

where $E_i = \int_0^\infty f_i(t)t\, dt$

$i = 1$, 2 and where $f_1(t_1)$ and $f_2(t_2)$ are failure and repair time distributions and $F_1(t)$ and $F_2(t)$ are the cumulative distributions respectively. (1) is true when both failure and repair times belong to the same parent distribution having a common shape parameter but different scale parameters. Providing the previous conditions hold (1) can be shown to still apply for N cycles. Nakagawa and Goel [14, 1973] gave a definition to cover a finite interval of time. They also present asymptotic solutions for interval availability. Mine and Nakagawa [15, 1977] discuss interval availability further and apply the concept to the formulation of optimum preventive policies. They also assume that failure and repair times are exponentially distributed.

Whilst the Bayesian approach has been applied to a wide range of problems, it has principally been directed at questions of statistical inference [5, 1971]. Methods for obtaining measures of system performance such as availability and reliability have been largely developed using non-Bayesian (classical) approaches [6, 1975]. The parameters of distributions describing the underlying processes are generally obtained by using classical estimating procedures such as Maximum Likelihood [6, 1975]. In these classical approaches the parameters, whilst unknown, are assumed to be constants. Bayesian methodology departs from the classical by assuming that the uncertainty in the true values of the distribution parameters can be described by probability statements [6, 1975]. This concept is particularly valuable in obtaining measures of uncertainty in predicted values based on observed data. Furthermore, Bayesian methods allow prior knowledge or even subjective judgement to be incorporated and hence such methods have the ability to handle small data sets as is often the case in practical applications.

Brender [1, 1968] was the first to predict and measure system availability using a Bayesian method. Again Gaver and Mazumdar [7, 1969] consider long run availability in a Bayesian sense whilst Thompson and Springer [8, 1972] analysed system availability on the basis of snapshot data obtained for each of N component subsystems. Using the Mellin integral transform the above authors derived an overall system p.d.f. from the p.d.f.'s of the series components comprising the system. In considering the same problem, Thompson and Policio [9, 1975] present a numerical procedure for computing Bayes intervals for the availability of a series of parallel systems. Kuo [3, 1980] formulated a Bayesian availability function for a system whose on-time and cycle time are described by gamma distributions. Kuo has more recently [4, 1986] applied this Bayesian method to the assessment of a digital radio transmission system.

Model - 1

It is assumed that the system studied alternates between two states, up and down, according to unknown identically distributed random processes; the form of which is known but whose governing parameters are unknown. Let $f(t)$ and $g(t')$ be the p.d.f.'s for failure and repair respectively. Define $u = t/t'$ then the probability of $t \geqslant ut'$ is:

$$R_1(ut') = \int_{ut'}^{\infty} f(t)dt \tag{2}$$

As all times of failure T' are possible it is necessary to weight $R_1(ut')$ with the failure p.d.f. and integrate over the time domain. Thus:

$$R(u) = \int_0^{\infty} R_1(ut')g(t')dt' \tag{3}$$

The p.d.f. q(u) governing u can now be obtained using:

$$q(u) = -dR/du$$

This gives:

$$q(u) = \int_0^{\infty} t'f(ut')g(t')dt' \tag{4}$$

Availability Modelling

Assume a system with N cycles of breakdown and repair; let t_i and t_i' be uptime and downtime for the i'th cycle.

It is then convenient to define

$$u = \frac{T_1}{T_2} \quad \text{where } T_1 = \sum_{i=1}^{N} t_i \quad T_2 = \sum_{i=1}^{N} t_i' \tag{5}$$

T_1 and T_2 are total uptime and downtime respectively.

The interval availability is now given as:

$$A = u/(1+u) = T_1/(T_1 + T_2) \tag{6}$$

If uptime and downtimes are exponentially distributed with failure parameter λ and repair parameter μ, i.e.

$$f(t) = \lambda \exp(-\lambda t) \quad g(t') = \mu \exp(-\mu t') \tag{7}$$

The p.d.f.'s for T_1 and T_2 are given by:

$$f^*_1(T_1) = \frac{\lambda^N T_1^{(N-1)}}{(N-1)!} \exp(-\lambda T_1) \tag{8}$$

$$f^*_2(T_2) = \frac{\mu^N T_2^{(N-1)}}{(N-1)!} \exp(-\mu T_2) \tag{9}$$

Using equation (4), one obtains:

$$q_N(u/N) = \frac{\theta^N u^{(N-1)} (2N-1)!}{(u+\theta)^{2N} [(N-1)!]^2} \qquad \text{where } \theta = \mu/\lambda \qquad (10)$$

Recalling equation (6), which gives:

$$du/dA = \frac{1}{(1-A)^2} \qquad \text{and letting} \qquad (11)$$

$$q^*_N(A)dA = q(u/N)du$$

$$q^*_N(A) = q_N[A/(1-A)]du/dA$$

It follows that the p.d.f. for A is given by:

$$q^*_N(A) = q_N\left(\frac{A}{1-A}\right) \frac{1}{(1-A)^2} \qquad (12)$$

(12) expresses the p.d.f. for A in analytical form defined as the ratio of uptime to total time for N cycles.

If the time interval T is sufficiently large so that N >> 1 (number of cycles much greater than one) then:

$$N = [T\lambda\mu/(\lambda+\mu)]$$

If N is small then it is necessary to compute the individual probabilities of cycles occurring in time T for all N. A composite interval availability can then be obtained as follows:

$$q(A) = \sum_{N=1}^{\infty} q_N^*(A) P_N(T)$$

The calculation of the probabilities $P_N(t)$ are governed by a non-homogeneous Poisson process [10, 1984], these calculations will be presented in a later paper. However, approximate values can be obtained by assuming a Poisson distribution with a mean $\mu = T\lambda\mu/(\lambda+\mu)$.

Model - 2

Bayesian Availability

It is assumed as before that the failure and repair times t_i and t_i', for the i-th cycle can be described by independent and identically distributed variates. For simplicity and purposes of illustration these distributions are taken as exponential and as given by equations (7).

The likelihood function describing the above failure observations can be written as:

$$L = \lambda^N \exp\left(-\lambda \sum_{i=1}^{N} t_i\right) \tag{13}$$

Multiplying by a diffuse prior $1/\lambda$, then after normalisation the Bayesian p.d.f. for λ is given by:

$$f_1(\lambda) = \frac{1}{(N-1)!} \lambda^{N-1} \left(\sum_{i=1}^{N} t_i\right)^N \exp\left(-\lambda \sum_{i=1}^{N} t_i\right) \tag{14}$$

Similarly a Bayesian p.d.f. for the repair rate can be derived as follows:

$$f_2(\mu) = \frac{1}{(N-1)!} \mu^{N-1} \left(\sum_{i=1}^{N} t_i'\right) \exp\left(-\mu \sum_{i=1}^{N} t_i'\right) \tag{15}$$

Asymptotic Availability

This as given by [2, 1965] can be written as:

$$A_\infty = \frac{\mu}{\lambda+\mu} = \frac{1}{1+\mu/\lambda} \tag{16}$$

If the Bayesian p.d.f. for $v = \mu/\lambda$ can be calculated from equations (14) and (15) then the p.d.f. for A can readily be computed using (16). Following the same procedure as in the derivation of equation (10) the p.d.f. for v, f(v), is given by:

$$f(v) = \frac{v^{N-1} (2N-1)!}{(v+\theta)^{2N} [(N-1)!]^2} \theta^N \tag{17}$$

where again

$$\theta = \frac{T_2}{T_1} = \sum_{i=1}^{N} t_i \Big/ \sum_{i=1}^{N} t_i'$$

T_2 and T_1 are the observed total up and downtimes respectively. Recalling equations (5) and (12) the p.d.f. for A_∞ is now given by:

$$f^*(A_\infty) = \frac{v^{(N-1)} (2N-1)!}{(v+\theta)^{2N} [(N-1)!]^2} \frac{1}{(1-A_\infty)^2} \theta^N \tag{18}$$

since

$$f^*(A_\infty) = f\left(\frac{A_\infty}{1-A_\infty}\right) \frac{1}{(1-A_\infty)^2} \tag{19}$$

Equation (19) expresses in Bayesian form a p.d.f. for the asymptotic availability depending only upon observed data. Although (19) is of the same mathematical form as (12), with $\theta = T_2/T_1$ replacing $\theta = \lambda/\mu$, their interpretation and application are completely different, insomuch that (12) expresses the p.d.f. for the interval availability as defined by equation (6) over an interval comprising N cycles. On the other hand (19) expresses the uncertainty of the asymptotic or steady state availability A_∞, and is calculated from observed data covering N cycles.

Instantaneous Availability

The analytical formulation for the instantaneous availability of a system with exponentially distributed failure and repair times following [2, 1965] can be written as:

$$A(\lambda,\mu|t) = \mu/(\lambda+\mu)\exp(-(\lambda+\mu)t)+\lambda/(\lambda+\mu) \tag{20}$$

where λ and μ are again the failure and repair rates.

Treating λ and μ as random variates governed by equations (14) and (15), sample values of λ and μ can be generated by Monte Carlo simulation methods. These can then be substituted into equation (20) to provide values of $A(\lambda,\mu|t)$. Repeating this procedure a sufficiently large number of times a p.d.f. for $A(\lambda,\mu|t)$ can be then obtained from the simulation.

Applications

Interval Availability Model

Taking values of λ and μ corresponding to exponential failure and repair rates, p.d.f.'s for varying number of cycles were calculated using formula (12). These results are given in Figure 1.

Bayesian Model

Asymptotic Availability

Initial values of $\lambda_0 = .3$ and $\mu_0 = .7$ for the failure and repair rates were again assumed. Corresponding to these values, sets of times t_i and t_i' were generated by Monte Carlo simulation using equation (7). P.d.f.'s for λ and μ were obtained together with a p.d.f. for A by using (14), (15) and (19). The results are given in Figures 2 and 3 respectively. It is seen that the modes for the p.d.f.'s in Figures 2 and 3 compare favourably with the values initially assumed for λ and μ. Similarly the mode for the p.d.f. in Figure 4 compares favourably with the value of the asymptotic availability calculated using equation (16) provided $N \geqslant 7$. The cumulative distribution

$$Q(A_\infty) = \int_{A_\infty}^{\infty} f^* (A_\infty) \, dA_\infty$$

was also evaluated and the results are given in Figure 5.

Instantaneous Availability

Because of the mathematical complexity involved in calculating a mean, analytically mean values of $A(\lambda,\mu|t)$ values were estimated, instead, by Monte Carlo methods using equations (14) and (15) as follows:

Random values of λ,μ corresponding to the distributions (14) and (15) were generated. The algorithms used were

$$\lambda = -\frac{1}{T_1} \sum_{i=1}^{N} \ln Xi \qquad \mu = -\frac{1}{T_2} \sum_{i=1}^{N} \ln Xi$$

where the Xi are random numbers.

Using (20) and the values of λ,μ as generated above, mean values of $A(\lambda,\mu|t)$ were calculated. A comparison of values of $E(A(\lambda,\mu,t))$ generated by simulation with those given by equation (11) is given in Figure 6. Agreement again is seen to be good.

142

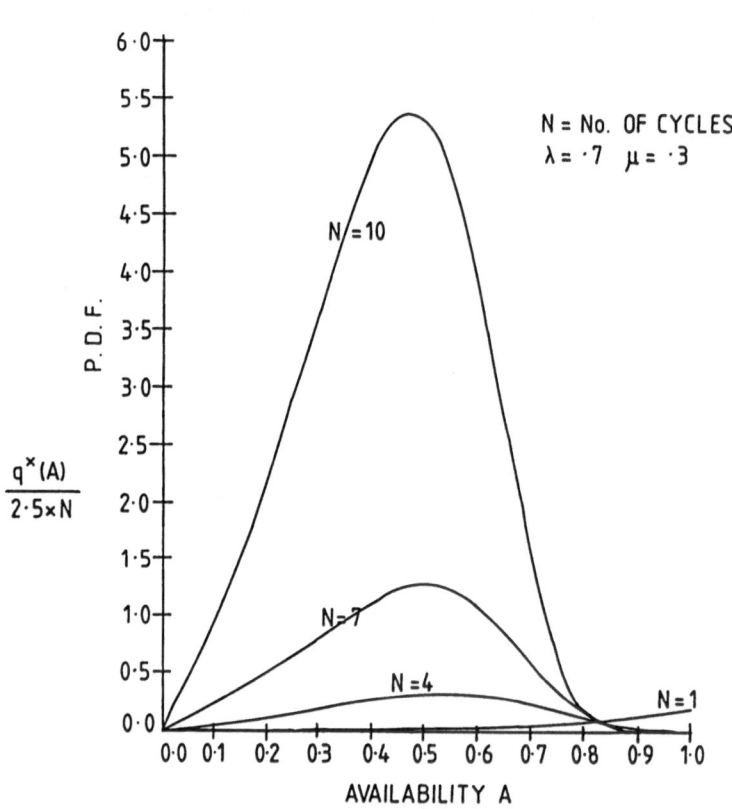

FIG. 1

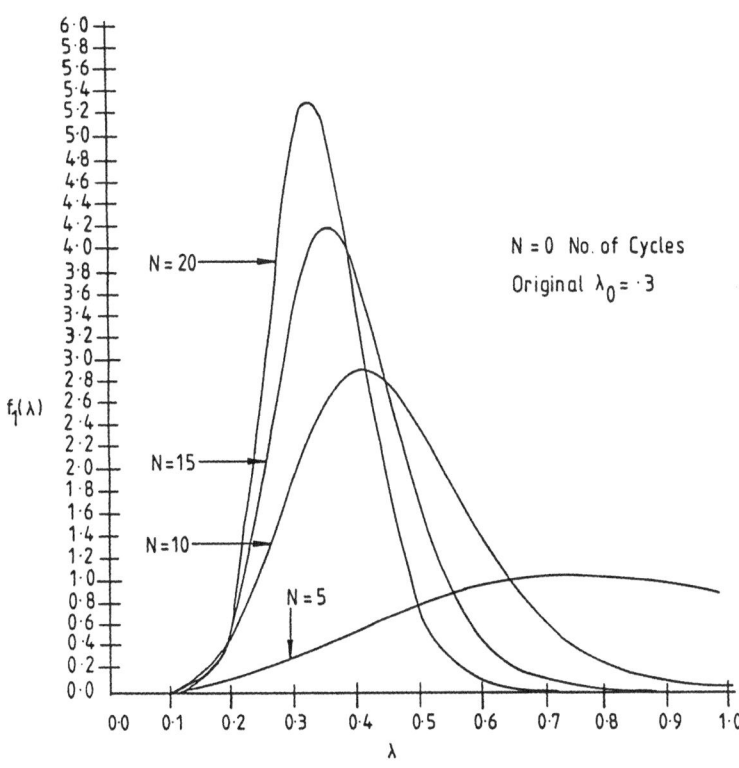

BAYESIAN Pdf $f_1(\lambda)$

$f_1(\lambda)$

N = 20

N = 15

N = 10

N = 5

N = 0 No. of Cycles

Original $\lambda_0 = \cdot3$

λ

FIG. 2

144

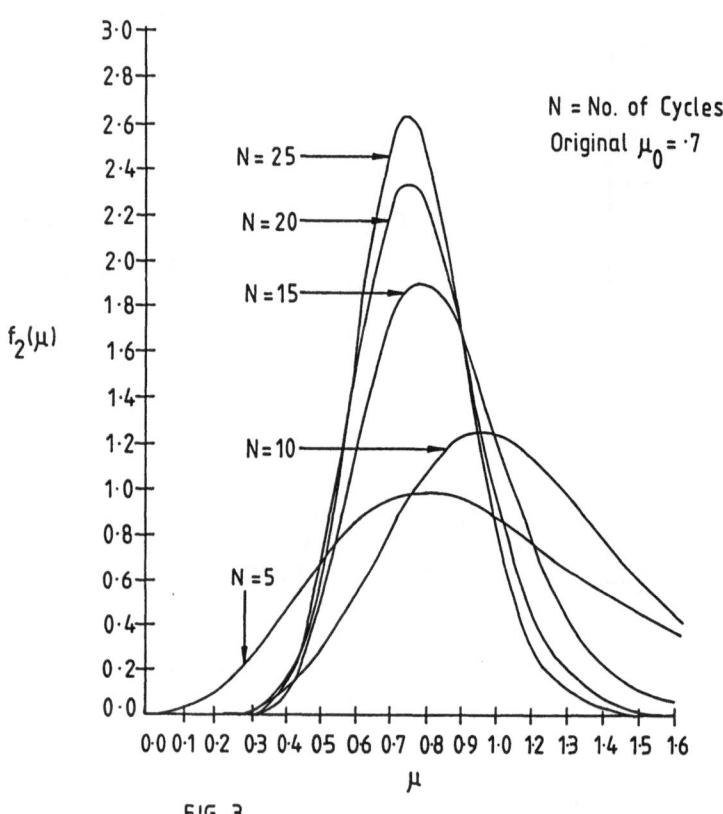

FIG. 3

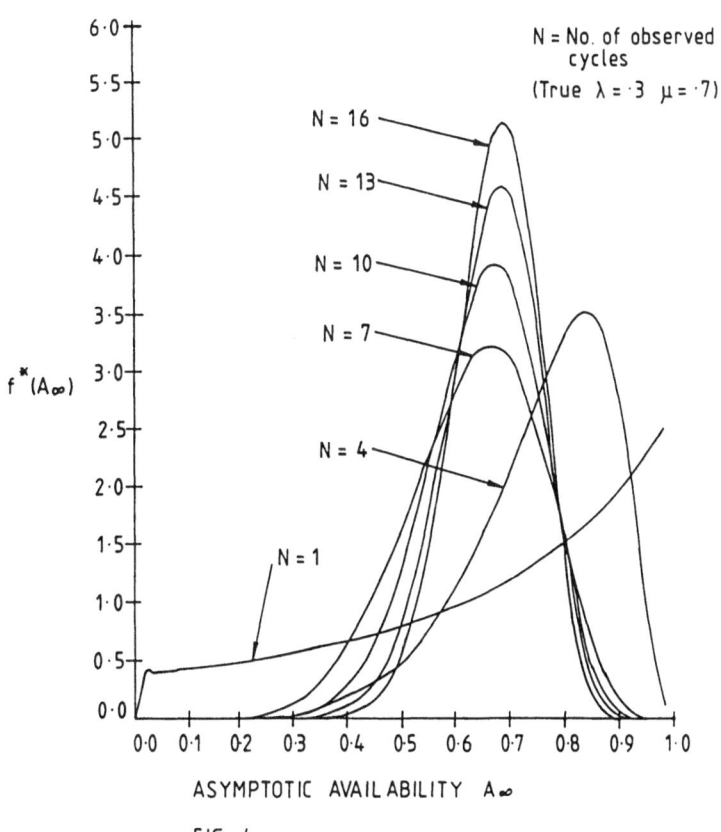

BAYESIAN PDF ASYMPTOTIC AVAILABILITY
VARIATION WITH NUMBER OF OBSERVED CYCLES

N = No. of observed
cycles

(True $\lambda = \cdot 3$ $\mu = \cdot 7$)

N = 16

N = 13

N = 10

N = 7

N = 4

N = 1

$f^*(A_\infty)$

ASYMPTOTIC AVAILABILITY A_∞

FIG. 4

CUMULATIVE DISTRIBUTION FUNCTION Q(A∞)

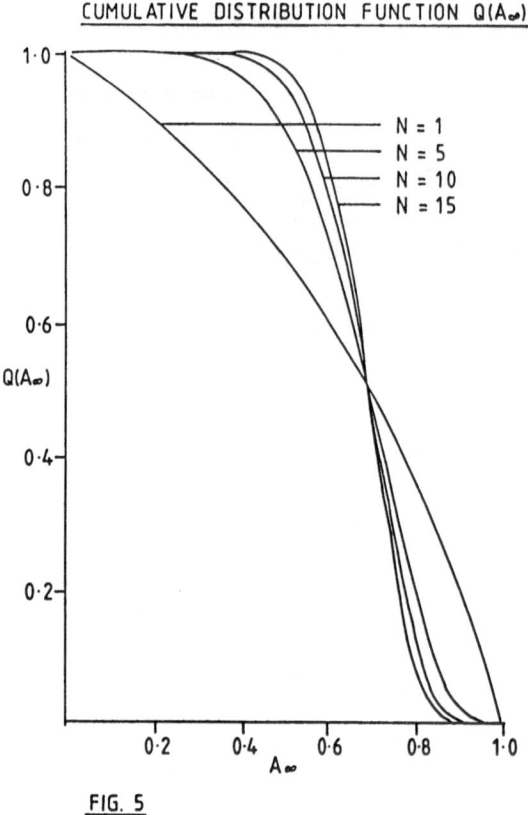

FIG. 5

FIG. 6   AVAILABILITY COMPARISON

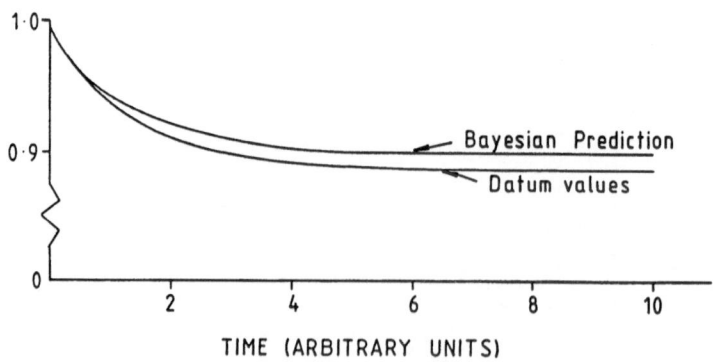

CONCLUSIONS

1. In cases where utilisation over a fixed time period is a parameter of major interest, this parameter should be described by a probability distribution rather than a constant value.

2. Instantaneous and other availabilities can be obtained from small observation data sets using Bayesian methods.

3. For risk situations, the methods presented here allow uncertainties arising from availability prediction to be taken into consideration.

4. The methods presented should be extended to deal with failure and repair rates which are other than constant.

REFERENCES

1. Brender, D.M., "The prediction and Measurement of System Availability: a Bayesian Treatment", IEEE Trans. Reliability, Vol.R-17, No.3, pp 127-138 (Sept. 1968).

2. Barlow, R.E. and Proschan, F., (1965), "Mathematical Theory of Reliability", Wiley and Sons Inc., New York.

3. Kuo, W., (1980), "System Effectiveness Models Via Renewal Theory and Bayesian Inference", PhD Dissertation, Kansas State Univ., Manhattan, Kansas, USA.

4. Kuo, W., (1986), "Bayes Weighted Availability for a Digital Radio Transmission System", IEEE Trans. Rel. Vol.R-35, No.2.

5. Zellner, A., (1971), "An Introduction to Bayesian Inference in Economics", New York, Wiley.

6. Barlow, R.E. and Proschan, F., (1975), "Statistical Theory of Reliability Models", Holt, Rinehart and Winston Inc., New York.

7. Gaver Jr., D.P. and Mazumar, M., "Some Bayes Estimates of Long Run Availability in a Two State System", IEEE Trans. Rel., Vol.R-18, Nov. 1965, pp 184-189.

8. Thompson, W.E. amd Springer, M.D., (1972), "A Bayes Analysis of Availability for a System Consisting of Several Independent Subsystems", IEEE Trans. Rel., Vol.R-21, pp 212-214.

9. Thompson, W.E. and Palicio, P.S., (1975), "Bayes Confidence Limits for the Availability of Systems", IEEE Trans. Rel., Vol.R-24, pp 118-120.

10. Keller, A.Z., (1984), "A Generalised Non-Homogeneous Poisson Process", 8th Advances in Reliability Symposium, Bradford, UK.

11. Hosford, J.E., (1960), "Measures of Dependability", Operational Research, Vol.8, No.1.

148

12. Kabak, I.W., (1969), "System Availability and Some Design Implications", Operational Research, Vol.17, No.5.

13. Martz Jr., H.F., (1971), "On single Cycle Availability", IEEE Trans. on Rel., Vol.R-20, No.1.

14. Nakagawa, T. and Goel, A., (1978), "A Note on Availability for Finite Interval", IEEE Trans. on Rel., December.

15. Mine, H. and Nakagawa, T., (1977), "Interval Reliability and Optimum Preventive Maintenance Policy", IEEE Trans. on Rel., Vol.R-26, No.2, June.

MONTE CARLO SIMULATION APPLIED TO
POWER SYSTEM RELIABILITY EVALUATION

R.N.Allan   Y.A.Jebril   A.Saboury   J.Roman

Department of Electrical Engineering and Electronics
UMIST
Manchester M60 1QD
UK

ABSTRACT

This paper considers the application of Monte Carlo simulation techniques to the reliability evaluation of electrical generating systems. It describes the basic concepts and evaluation techniques used. The range of indices, including expected values, standard deviations and frequency distributions, that are evaluated are discussed. The techniques are applied to the IEEE Reliability Test System. Some of the simulation results are compared with those evaluated using analytical methods. Although discussed in terms of power systems, the general concepts and results are equally applicable to other engineering systems.

INTRODUCTION

An electrical power system serves one function only and that is to supply electrical energy to its customers as economically and as reliably as possible. These two constraints, economics and reliability can conflict and hence lead to difficult managerial decisions in both the planning and operating phases. These problems have always been recognised and various design criteria have been specified which attempts to satisfy both the economic and reliability constraints. The criteria initially developed however were all deterministically based.

The need for probabilistic evaluation of power system reliability was recognised and initially developed in the 1930's. However, it is only in more recent times that significant developments and applications have been made [1-4]. Generally, the vast majority of the evaluation techniques [5-6] have been analytically based and simulation techniques have taken a minor role. The main reason for this is because Monte Carlo simulation generally requires large amounts of computing time and analytical models and techniques have been sufficient to provide system planners and designers with the results needed to make objective decisions. This is now changing and increasing interest is being shown in modelling the system behaviour more comprehensively and in evaluating a more informative set of system reliability indices. This implies the need to consider Monte Carlo simulation.

This paper considers the application of Monte carlo simulation in the reliability evaluation of one area of power systems, namely that associated with the generation of electrical energy and its ability to satisfy the system demand adequately. This area of power system reliability is known [7] as hierarchical level I (HLI). The only concern in HLI studies is to estimate the necessary generating capacity to satisfy the system demand and to have sufficient reseve capacity to perform corrective and preventative maintenance and to cope with unexpected load increases. The ability to move the generated energy to the consumers is therefore ignored in these studies.

The Monte Carlo simulation techniques described in this paper are applied to the IEEE Reliability Test System (RTS) [8]. It illustrates the range of results that can be evaluated and compares some of the results with those obtained by direct analytical techniques.

Although discussed in terms of power systems, the general concepts and results are equally applicable to other engineering systems.

RELIABILITY EVALUATION TECHNIQUES

Reliability evaluation techniques make use of probabilistic methods which are therefore able to take into account the stochastic nature of power system behaviour, of consumer demand and of component failure. The majority of these methods are direct analytical techniques which generate system or subsystem risk indices [1-4].

Analytical techniques [5,6] represent the system by a mathematical model and evaluate the reliability indices from this model using direct numerical solutions. They generally provide expectation indices in a relatively short computing time. Unfortunately, assumptions are frequently required in order to simplify the problem and produce an analytical model of the system. This is particularly the case when complex systems and complex operating procedures have to be modelled. The resulting analysis can therefore lose some or much of its significance. The use of simulation techniques are very important in the reliability evaluation of such situations.

Simulation methods estimate the reliability indices by simulating the actual process and random behaviour of the system. The method therefore treats the problem as a series of real experiments. The techniques can theoretically take into account virtually all aspects and contingencies inherent in the planning, design and operation of a power system. These include random events such as outages and repairs of elements represented by general probability distributions, dependent events and component behaviour, queuing of failed components, load variations, variation of energy input such as that occuring in hydro-generation, as well as all different types of operating policies.

If the operating life of the system is simulated over a long period of time, it is possible to study the behaviour of the system and obtain a clear picture of the type of deficiencies that the system may suffer. This recorded information permits the expected values of reliability indices together with their frequency distributions to be evaluated. This comprehensive information gives a very detailed description, and hence understanding, of the reliability of the system.

STOCHASTIC SIMULATION

The first step in any simulation process is to define the system to be studied. The next step is to derive a model that represents the behaviour of this system. Simulation is then the process of experimenting with the model. These three steps are described below.

A system can be defined [9] as a set of related entities called components or elements. These elements have certain characteristics, or attributes, that have logical or numerical values. A number of activities (relations) exist among the elements, and consequently the elements interact. These activities cause changes in the system. External and internal relationships are also considered. The internal relationships connect the elements within the system, whilst the external relationships connect the elements with the environment.

After defining the system in terms of its elements and the relationships between these elements, it is necessary to derive an appropriate scientific model. This can be defined [9] as an abstraction of some real system, an abstraction that can be used for prediction and control. The purpose of a scientific model is to determine how one or more changes in various aspects of the modelled system may affect other aspects of the system as a whole. For the purpose of most studies, it is not necessary to consider all the details of a system. Therefore a model is not only a substitute for a system, it is also a simplification of the system [10]. For this reason care must be taken while building a model to ensure that it remains a valid representation of the problem [9].

Once the system and its model have been defined, the process of simulation can commence. This can be defined [9] as a numerical technique for conducting experiments on a digital computer which involves the mathematical and logical models that describe the behaviour of a system over extended periods of real time. Simulation is, in its wide sense, a technique for performing sampling experiments on the model of the system [9]. Stochastic simulation is experimenting with the model in time and includes sampling stochastic variates from probability distributions.

Sampling from statistical distributions implies the use of random numbers. It is for this reason that sometimes stochastic simulation is called Monte Carlo simulation. Since Monte Carlo simulation is widely used as a synonym of stochastic simulation, the name Monte Carlo will be used for the procedural stochastic simulation discussed and described in this paper.

The process of Monte Carlo simulation is fully described in several texts [9-12]. Its use is considered when one or more of the following conditions apply [12].

a)   A complete mathematical formulation of the problem does not exist or analytical methods for solving the mathematical model have not yet been developed

b)   Analytical methods are available but the mathematical procedures are so complex that simulation provides a simpler method of the solution

c)   Where analytical solutions necessitate gross assumptions and simplifications, simulation methods may be used as a means of verification or the basis for comparison

d)   It is desired to observe a simulated history of the process over a period of time in addition to estimating certain parameters

e)   Simulation may be the only possibility because of the difficulty in conducting experiments and observing phenomena in their actual environment

f)   Time compression may be required for systems or processes with long time frames. Simulation offers complete control over time.

RANDOM NUMBERS AND VARIABLES

The input data in Monte Carlo simulation is sampled from the appropriate probability distribution and the behaviour of the system represented by its mathematical model is observed. An adequate number of samplings then gives an estimate of the system indices with a desired level of certainty.

Since Monte Carlo simulations require a sequence of random numbers which must be drawn from distributions that in general are not uniform, a procedure for generating random numbers is an integral part of any Monte Carlo simulation study. Computers are deterministic machines and hence can not generate random numbers. Therefore the so-called 'pseudo-random numbers' are used in simulation instead of genuine ones.

A pseudo-random number generator is an arithmetical algorithm for producing pseudo-random number sequences. These generators are normally designed to generate numbers that are uniformly distributed between zero and one inclusive. This means that every real number within this range is equally likely to be the result of the algorithm. The desirable properties of a sequence of random numbers can be classified as follows [9,13,14] :

a)   The numbers must be uniformly distributed, that is all numbers are equally likely to occur

b)   The numbers must be statistically independent. This means that previous numbers must have no or as little effect as possible on the new number

c)   The sequence of numbers must be reproducible. This allows programs using random numbers to be tested. It allows a program to be run several times with the same sequence of numbers but with different input data and the results to be compared

d)   The sequence of numbers must have a sufficiently long period . The sequence will eventually repeat itself because pseudo-random numbers are generated by mathematical operations. The length of this cycle is called a 'period' and for any application it must be sufficiently long to prevent any repetition.

There are several different algorithms for producing sequences of pseudo-random numbers. The one used in these studies is the multiplicative congruent method. Although other and more complicated methods do exist, there seems to be little to be gained under most circumstances from using anything other than one of the multiplicative congruent algorithms [12].

A brief summary of this method is described below. The method is based on the equation:

$N_{i+1} = MOD (a N_i , m)$

where N_i and N_{i+1} are the i-th and (i+1)-th random numbers of the generated sequence

a is the positive constant integer multiplier

m is the positive constant integer modulus

MOD represents the mathematical operation of taking the remainder of dividing $(a N_i)$ by m.

The choice of the multiplier a and the modulus m depends on the word size of the computer being used. For a particular binary computer with a word size of B digits, the modulus m is set equal to 2^B. The multiplier a and the starting value N_0 (known as the seed) must be chosen so that the maximum period is obtained and the correlation between the numbers is minimised. These conditions are met for a binary system if:

$a = 8r \pm 3$

and a is close to $2^{B/2}$

where r is any positive integer

and N_0 is any positive odd integer less than m.

GENERATION OF RANDOM VARIATES

The general requirement in simulation studies is for a sequence of random numbers with a non-uniform distribution. The most common way of deriving such numbers is based on the principle of transforming a uniformly distributed sequence into the required sequence.

There are three basic methods for generating variates from probability distributions. These are:

a)   inverse transform method

b)   composition method

c)   acceptance-rejection method

Let x be a random variate with a probability density function (pdf) of f(x) and a cumulative probability distribution function (cdf) of F(x), that is:

$$F(x) = \sum_{t \leq x} f(t) \qquad \text{(discrete)}$$

$$F(x) = \int_{-\infty}^{x} f(t)\ dt \qquad \text{(continuous)}$$

The inverse transform method is a straightforward method and can be used if F(x) has a known inverse function $F^{-1}(x)$. Distributions in this category are exponential, uniform, Weibull, logistic and Cauchy. If f(x) is bounded and x has a finite range, the acceptance-rejection method may be used. The composition method can be used for generating variates from some important distributions. Also, by using this method, random variates with complex distributions can be generated from simpler distributions that are themselves generated by inverse transform or acceptance-rejection techniques.

A full description of the methods can be found in References 9, 11 and 15.

SIMULATION PROCEDURE

The simulation process can be carried out after the elements of the system have been modelled with all the characteristics deemed important for the study. The simulation can follow one of two approaches:

a)   sequential - this examines each basic interval of time of the simulated period in chronological order

b)   random - this examines the basic intervals of time after choosing them in a random manner

The basic interval of time must be chosen according to the type of system under study, as well as the length of the period to be simulated in order to ensure a certain level of confidence in the estimated indices.

The choice of simulation approach depends on whether the history of the system plays a role in its behaviour. The random approach would be used if the history had no effect but the sequential approach would be used if the past history affected present conditions. This is the case of a power system particularly those containing hydro-plant since the past use of energy resources, e.g. water, affects the ability to generate energy in subsequent time intervals. This was the method used therefore in the present studies.

If the model is simulated over a sufficiently long period of the system lifetime, the results will provide records of the events and deficiencies which the system can encounter. These records contain information for calculating expected values, moments about the mean and complete frequency distributions.

APPLICATION TO POWER SYSTEMS

A computer model has been developed to evaluate the reliability of thermal and hydro-thermal generating systems using Monte Carlo simulation. This is based on the techniques described in previous sections of this paper. The deficiencies encountered during the simulation process are recorded and analysed. The deficiencies associated with a power system are the required curtailment of load and energy due to the demand exceeding the available generating capacity in the interval of time being simulated. The deficiences are recorded in terms of their magnitude, frequency of occurrence and duration. All the commonly used reliability indices [6] can be evaluated from this recorded information together with the frequency distributions. The following indices are evaluated in the present model:

- loss of load expectation (LOLE) in hr/yr

- loss of energy expectation (LOEE) in MWh/yr

- energy not supplied per interruption (ENSI) in MWh/int

- load curtailed per interruption (LCI) in MW/int

- load curtailed per year (LCY) in MW/yr

- frequency of interruption (FOI) in int/yr

- duration of interruption (DOI) in hr/int

- energy index of reliability (EIR)

The expected value, standard deviation and frequency distribution is evaluated for most of these indices.

APPLICATION TO THE IEEE-RTS

The techniques have been applied to the IEEE Reliability Test System (RTS). This was established in 1979 as a reference system for the purpose of testing and comparing alternative techniques. The complete data is given in Reference 8. Briefly it consists of 32 generating units having a total installed capacity of 3405 MW which supplies a load having an annual peak of 2850 MW.

In these studies, each simulated year of the IEEE-RTS consisted of 8736 hours and each simulated hour was considered in chronological order, i.e. the sequential method was used. The number of simulated years needed depends on the level of accuracy required and the characteristics of the system. Some preliminary studies were performed in order to assess the required number of years for the IEEE-RTS. Typical plots of risk as a function of number of simulated years are shown in Figures 1 and 2. It was decided from these results to use 400 years of simulation.

Reliability indices of the IEEE-RTS have been calculated [16] by analytical methods without any approximations. Results obtained from the simulation technique are compared with these exact results [16] in Table 1. The similarity between these two sets of results justify the simulation model and its ability to assess the reliability of generating systems at

156

the HLI level.

Table 1.   Comparison of results with exact indices

index	simulation result	exact result
LOEE(GWh/yr)	1.182	1.176
LOLE(hr/yr)	9.212	9.394
EIR	0.999923	0.999923

The simulation method can evaluate a much greater number of indices than those shown in Table 1. The expected values and standard deviations of those indices defined earlier are shown in Table 2. Each of these indices represents a different facet of the system and the problems encountered with it. None of them individually can be considered to be the all-important one; they all have a role and some significance in the decision making process, which is usually to decide how much additional generation is required. The ability therefore to calculate a wide range of indices is very important.

Table 2.   Results for IEEE-RTS

index	expected result	standard deviation
LOEE (MWh/yr)	1182.9	3177.8
LOLE (hr/yr)	9.21	16.06
ENSI (MWh/int)	646.4	1103.1
LCI (MW/int)	82.7	78.3
LCY (MW/yr)	151.3	289.8
FOI (int/yr)	1.83	
DOI (hr/int)	5.03	3.94
EIR	0.999923	

One of the most significant features of the results shown in Table 2 is the wide dispersion evident in many of the indices, e.g. the expected value of LOEE is 1182.9 MWh/yr whilst its standard deviation is 3177.8 MWh/yr. These dispersions show the benefit of calculating, not only expected values, but a measure of this dispersion. They further demonstrate the benefit of deducing frequency distributions such as the histograms shown in Figures 3-8 since standard deviation alone does not indicate the shape of a distribution. The results shown in Figures 3-8 all exhibit considerable skewness which means that, although an expected value may be tolerable, the extreme upper values may not. An example which could indicate this is that, although the expected value of LOEE is 1182.9 MWh/yr, values in excess of 8000 MWh/yr are clearly evident from Figure 3. This understanding of range as well as expected values can be very important in making relevant decisions.

The simulation model also accounts for the restricted nature of water inflows to hydro-plant and the dependent nature between them. An example illustrating the change in indices due to limited water energy is shown by the results in Table 3. This example assumed the six hydro-units of the IEEE-RTS were energy limited due to restricted water input whereas the previous results assumed they were energy unlimited. A comparison between Tables 2 and 3 clearly show that the risk is significantly greater in the energy limited model, for instance, the value of LOEE is increased by a factor of 4.

Table 3. Effect of hydro-plant

index		expected result
LOEE	(MWh/yr)	4642.2
LOLE	(hr/yr)	26.2
ENSI	(MWh/int)	1054.6
LCI	(MW/int)	115.9
LCY	(MW/yr)	510.5
FOI	(int/yr)	4.40
DOI	(hr/int)	5.94
EIR		0.999697

CONCLUSIONS

Monte Carlo simulation is used primarily for reliability problems in which the system under study is either too complex or too large to be solved realistically in any other way. The studies described in this paper show that an electric generating system is one area of application in which Monte Carlo simulation can have some advantages. These systems are frequently complex, particularly hydro-systems, and analytical techniques are then limited. Also the ability to evaluate a wide range of indices including standard deviations and frequency distributions can play a significant role in the decision making process associated with expansion planning.

It can be concluded that Monte Carlo simulation offers the following advantages:

a) the feasibility of taking into account theoretically any random variable, any operation policy, any environmental effect, etc.

b) it offers planning engineers a synthesis of the final results and a detailed description of the events that cause the results. Therefore, it provides planning engineers with some "operating experience" of the system, as well as an insight into its structure and behaviour.

c) it can handle some reliability problems which cannot be solved by analytical methods.

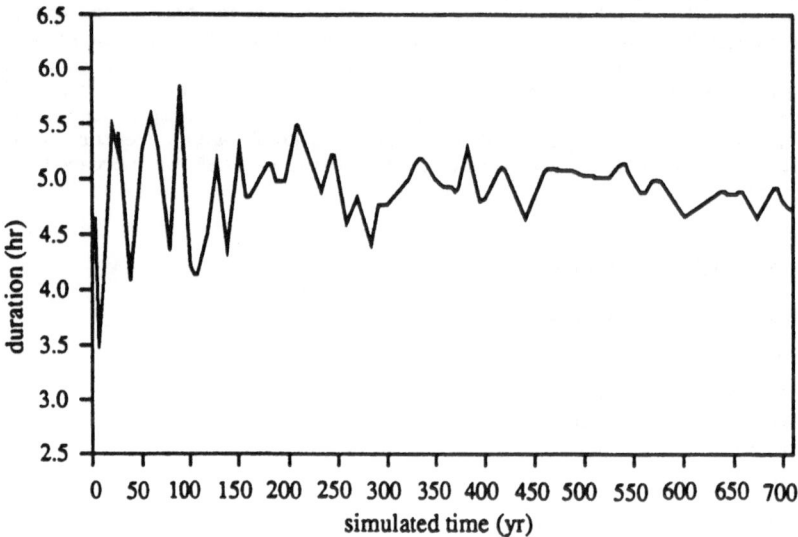

Figure 1. Capacity deficiency duration as function of simulated time

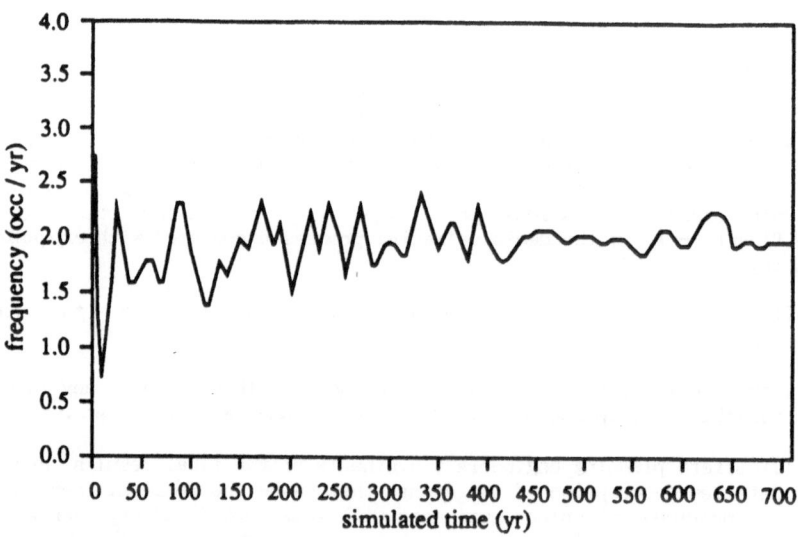

Figure 2. Capacity deficiency frequency as function of simulated time

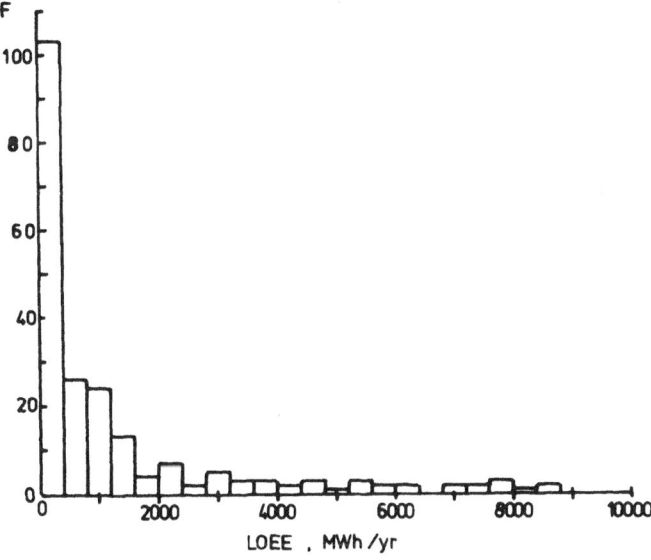

Figure 3. Frequency distribution of LOEE

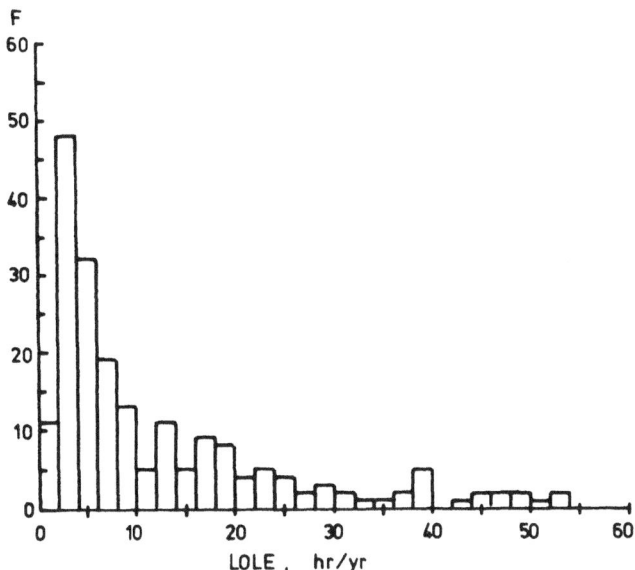

Figure 4. Frequency distribution of LOLE

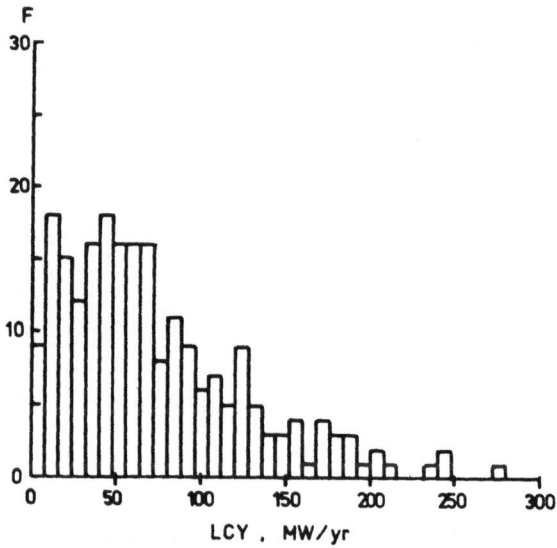

Figure 5. Frequency distribution of LCY

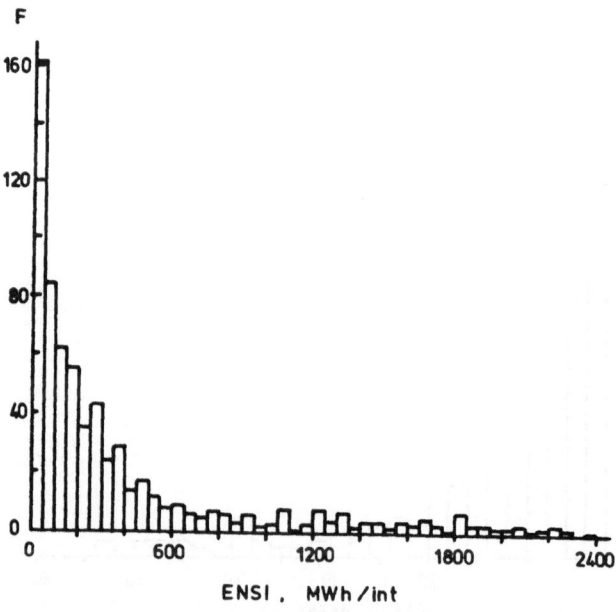

Figure 6. Frequency distribution of ENSI

161

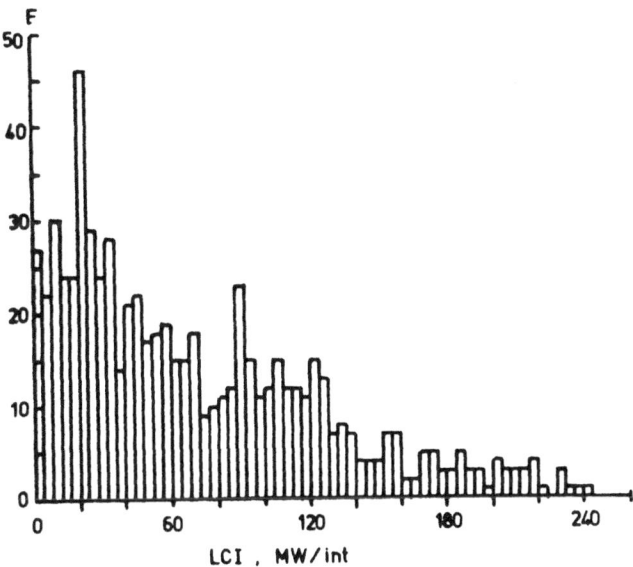

Figure 7. Frequency distribution of LCI

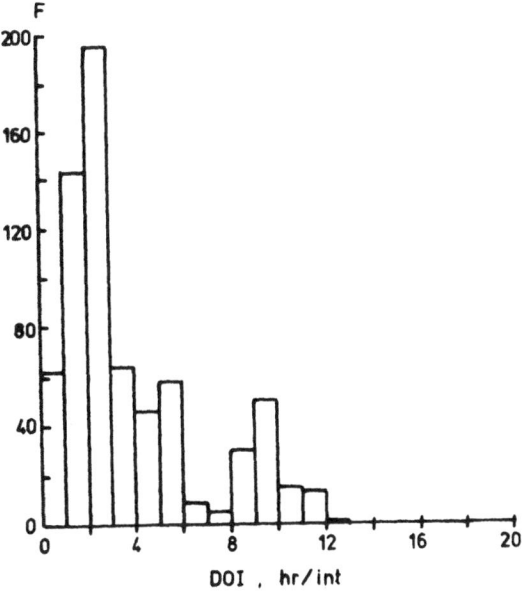

Figure 8. Frequency distribution of DOI

162

REFERENCES

1. Billinton, R., Bibliography on the application of probability methods in power system reliability evaluation, IEEE Trans, PAS-91, 1972, pp 649-660

2. IEEE Committee Report, Bibliography on the application of probability methods in power system reliability evaluation, 1971-1977, IEEE Trans, PAS-97, 1978, pp 2235-2242

3. Allan, R.N., Billinton, R. and Lee, S.H., Bibliography on the application of probability methods in power system reliability evaluation, 1977-1982, IEEE Trans, PAS-103, 1984, pp 275-282

4. Allan, R.N., Billinton, R. and Shahidehpour, S.M. and Singh, C., Bibliography on the application of probability methods in power system reliability evaluation, 1982-1987, IEEE Winter Power Meeting, New York, February, 1988

5. Billinton, R. and Allan, R.N., Reliability Evaluation of Engineering Systems: Concepts and Techniques, Longman, London/ Plenum, New York, 1983

6. Billinton, R. and Allan, R.N., Reliability Evaluation of Power Systems, Longman, London/ Plenum, New York, 1984

7. Billinton, R. and Allan, R.N., Power system reliability in perspective, IEE Journal on Electronics and Power, 30, 1984, pp 231-236

8. IEEE Committee Report, IEEE Reliability Test System, IEEE Trans, PAS-98, 1979, pp 2047-2054

9. Rubinstein, R.Y., Simulation and the Monte Carlo Method, Wiley, New York, 1981

10. Gordon, G., System Simulation, Prentice-Hall, New Jersey, 1969

11. Hammersley, I.M. and Handscomb, D.C., Monte Carlo Methods, Wiley, New York, 1964

12. Shannon, R.E., System Simulation: The Art and Science, Prentice-Hall, New Jersey, 1975

13. Naylor, T.J., Balintfy, J.L., Burdick, D.S. and Chu, K., Computer Simulation Techniques, Wiley, New York, 1966

14. Bulgren, W.G., Discrete System Simulation, Prentice-Hall, New Jersey, 1982

15. Fishman, G.S., Principles of Discrete Event Simulation, Wiley, New York, 1978

16. Allan, R.N., Billinton, R. and Abdel-Gawad, N.M.K., The IEEE reliability test system - Extensions to and evaluation of the generating system, IEEE Trans, PWRS-1, No.4, 1986, pp 1-7

FAULT TREE UNCERTAINTY ANALYSIS USING A MONTE CARLO METHOD

A M Irving

Rolls-Royce and Associates Limited
P O BOX 31
DERBY
DE2 8BJ

Abstract

Uncertainties in the initiating events of a Fault Tree can be propagated
through the tree using Monte Carlo sampling. The sampled values of the
top event are fitted to one of the Johnson probability distributions.
The fitted Johnson distributions can then be used to estimate confidence
limits, or for further analyses eg. Accident Sequences.

Introduction

When the probabilities of the initiating events in a Fault Tree are
subject to uncertainty, the probabilities can be considered to be random
variables described by some probability distribution. The probability
of the top event in such a Fault Tree will then also be a random
variable, and the form of its probability distribution will depend on
the tree structure and the probability distributions of the initiating
events.

This paper will describe a Monte Carlo method for the propagation of
uncertainties on the probabilities in a Fault Tree. The Monte Carlo
method is used to obtain a probability distribution to fit the top
event probability in a Fault Tree using information on the tree structure
and the assumed probability distributions of the initiating events in
the Fault Tree. The probability distribution of the top event is chosen
from the Johnson family of probability distributions. For large Fault
Trees, the tree may be disaggregated into smaller trees each of which is
then simulated. The Monte Carlo method is then used again to simulate
the original Fault Tree from its disaggregated parts. The top event
probability distribution can then be combined using the Monte Carlo
method with other such top event distributions to obtain, for example,
the probability distribution of particular Accident Sequences. These
Accident Sequences may, for example, represent particular initiating
event/fission product release categories. From such distributions the
Monte Carlo method can be used to obtain probability distributions for
a Farmer curve. A major advantage of using a Monte Carlo method for
propagation of uncertainty through a Fault Tree is that the sensitivity
of the final event (eg. top event, accident sequences or Farmer curve)
to the assumptions on the underlying probability distributions for the
initiating events can be investigated quickly and easily.

Background

Traditionally, the method used for the propagation of these uncertainties
through a Fault Tree has usually taken one of three forms:

a) the initiating event probabilities are represented by their best
    estimate or median values, and the resultant top event probability
    derived from these values.

b) the initiating event probabilities are represented by ranges of
    probabilities, and the bounds of these ranges are propagated through
    the Fault Tree to obtain bounds on the top event probability.

c) the initiating event probabilities are represented by ranges
    obtained by taking a median value (M) and an error factor F, to
    obtain bounds M/F and M*F. These ranges are then assumed to be
    confidence intervals from lognormal probability distributions. The
    parameters of these probability distributions are obtained from the
    values of M, F and the assumed confidence level for the ranges (eg.
    90%). Monte Carlo sampling is then performed to obtain samples of
    the top event probability from the chosen samples of the initiating
    event probabilities propagated through the Fault Tree. Percentiles
    of the top event probabilities are then obtained by ranking the
    simulated top event probabilities.

The advantage of method a) is that it is quick and easy to do.
However, the top event probability will just be a point estimate
probability to which it is impossible to attach any other statistical
interpretation eg. median. Similarly, method b) is quick and easy
to do, but the inherent pessimism in the calculation is difficult
to quantify. For example, if the initiating event bounds correspond
to 95% confidence intervals, the resultant top event interval will
not correspond to a 95% confidence interval. It will have associated
with it a much higher confidence value. Method c) depends initially
on the assumption of lognormality of the initiating probabilities,
which in some situations may not be a reasonable assumption to
make. This method allows the top event probability interval to be
obtained more realistically. The top event probability may be
required to be used in further probabilistic analyses eg. in
estimating particular Accident Sequence probabilities. However,
just knowing the percentiles of the top event probability will make
it difficult to construct the combination probability or confidence
intervals in such further analyses.

The problems in interpretation of the results of the above three
methods are illustrated for the simple Fault Tree in Fig 1. The
probabilities of the 4 initiating events are all assumed to have
median values of .01 and ranges .001 to .1. For method c) these
ranges are assumed to correspond to a 90% confidence interval, and
the results are based on 10000 Monte Carlo simulations.

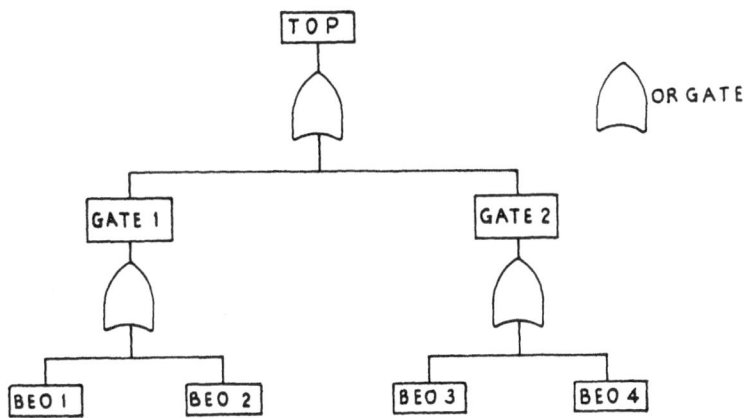

FIG 1. SIMPLE EXAMPLE.

The results are as follows:

method a)   probability   = .04

method b)   bounds   = .004 to .34

method c)   median   = .074

               90% confidence interval = .0186 to .294
               95% confidence interval = .014 to .385
               99% confidence interval = .0084 to .598

These results illustrate the optimism that is inherent in method b)
compared to the confidence intervals derived by method c), and the
large difference between the point estimate value of a) and the median
value of method c).

Method

The initial step in the proposed Monte Carlo method is to obtain
probability distributions to characterise the uncertainties on the
initiating event probabilities in the Fault Tree. Where field data
exists for the initiating events it may be possible to derive
probability distributions to describe the probabilities using the
data.   In many situations little or no data may exist, and the
probabilities may have to be expressed by some subjective estimates
eg. a median value and an error factor. The assumption then made
would be that these probabilities follow a lognormal distribution.
The parameters of such distributions can be estimated as in Ref 2.
The justification on the choices of probability distributions used
in any situation will not be covered further in this paper. However,
the advantage of using a Monte Carlo approach to the error propagation
is that the sensitivity of the top event probability to the
assumptions on the initiating event probabilities can be readily
investigated.

For large Fault Trees it may be more convenient to disaggregate the large Fault Tree into smaller Fault Trees (eg. using a method like Ref 3), performing the Monte Carlo method suggested in this paper, and reforming the larger Fault Tree from its component modules using further Monte Carlo operations.

Given a Fault Tree (whether the overall system or some module of it) the tree is represented by a Boolean equation for the top event in terms of the initiating event probabilities. For example, in the Fault Tree of Fig 1 if the initiating event probabilities are denoted by Xi (i=1, --,4) then the top event probability would be approximately $\sum_{i=1}^{4} X_i$. For any general Fault Tree the following steps for setting up the top event Boolean function work well in practice, and uses a 'bottom-up' approach:

a) for each initiating event information on the gate it feeds into and the type of gate it is, is used to set up the Boolean function for each of those gates.

b) given information on the gate each gate feeds into, and the type of gate each of the latter is, then the Boolean functions for these gates can be set up using the Boolean functions set up in stage a), if the stage a) gates have no other gates feeding into them. This process is repeated for all levels up the tree until all gate functions are completed (including any gate functions remaining from stage a)).

c) the final gate function will correspond to the top event.

The efficiency of the above algorithm depends on the care that is taken in setting up the gate information. Badly constructed information may necessitate several iterations of the algorithm, before the top event Boolean function is finally obtained. Splitting the Fault Tree into levels and working upwards from the initiating event gates generally works efficiently.

When the top event function is available the random choices from the initiating event probability distributions can be propagated through the function to obtain samples of the top event probability. After all the simulations have been done, a probability distribution can be sought which fits (under statistical criteria described later) these sampled values. The type of probability distribution chosen is one of the Johnson family of probability distributions. These probability distributions are transformations of the normal probability distribution, and are:

a) normal distribution.

b) lognormal distribution (Johnson S_L).

c) Johnson Su distribution.

d) Johnson S_B distribution.

The Johnson distributions are described by the parameters γ η λ ε and the transformations:

S_L : $z = \gamma + \eta \log_e (\lambda(x-\varepsilon))$ $\qquad\qquad$ $x > \varepsilon$

S_U : $z = \gamma + \eta \sinh^{-1} \left(\dfrac{x-\varepsilon}{\lambda}\right)$ $\qquad\qquad$ $x > \varepsilon$

S_B : $z = \gamma + \eta \log_e \left(\dfrac{x-\varepsilon}{\lambda + \varepsilon - x}\right)$ $\qquad\qquad$ $\varepsilon + \lambda > x > \varepsilon$

where z = standardised normal deviate

$\qquad$ x = top event variable

To determine which Johnson distribution is to be chosen for the particular top event sample, the following two statistics are calculated:

$\sqrt{b_1}$ = measure of skewness = $\displaystyle\sum_{i=1}^{N} \dfrac{(x_i - \bar{x})^3}{NS^3}$

b_2 = measure of kurtosis = $\displaystyle\sum_{i=1}^{N} \dfrac{(x_i - \bar{x})^4}{NS^4}$

where N = number of samples

$\qquad$ $\bar{x}$ = mean of the sample

$\qquad$ S = standard deviation of the sample

The particular Johnson distribution chosen for the sample is determined by referring to Fig 2, and the values of b_1 and b_2.

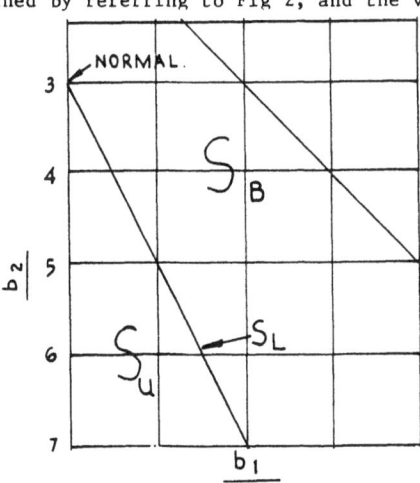

FIG 2.   JOHNSON DISTRIBUTION SELECTION.

Once the type of distribution has been chosen, the parameters
γ, η, λ, ε, have to be estimated. An initial estimate of the values
of γ, η, λ, ε, can be obtained by the method of moments, where the
functions of the parameters corresponding to $\sqrt{b_1}$, b_2, mean and
standard deviation specific to the chosen distribution are compared
to the sample values. Algorithms exist for solving these equations
for each type of Johnson distribution (see Ref 4). The adequacy
of the fitted distribution can be assessed using the Chi-Squared
Goodness of Fit Test using the histogram of the simulated values,
where the test statistic is

$$G = \sum_{i=1}^{m} \frac{(O_i - E_i)^2}{E_i}$$

where m = number of intervals the simulated values have been
split up into

O_i = observed number of values in interval i

E_i = expected number of values in interval i

G will be distributed as a Chi-Squared variable with m-5 degrees of
freedom. If the value of G is not statistically significant, then
one could conclude that the fitted distribution adequately represents
the simulated values. In many situations the method of moments does
not provide an adequate fit to the simulated values. Another method
that may be adopted is the method of maximum likelihood which will
use an iterative method to obtain the parameter estimates (eg. Ref 5).
This method may still not guarantee that the calculated value of G
using the optimised parameters will give a non-significant G value.
However, in general the maximum likelihood estimates will be an
improvement on the method of moments estimates. A refinement on
the maximum likelihood method is to impose a constraint on the
choice of parameters such that the calculated value of G is minimised.
This will be described as the minimisation of the Goodness of Fit
method.

Fig 3 illustrates the use of the Monte Carlo simulation method and
the Johnson distribution fitting using the minimisation of the
Goodness of Fit method. Each initiating event is assumed to be
lognormally distributed with the median values shown in Fig 3, and
each having an error factor of 5. The number of simulations
performed was 10000. The chosen Johnson distribution was Su with
parameters:

$$\gamma = -2.6746 \quad \eta = 1.5561 \quad \lambda = .00134 \quad \varepsilon = .00173$$

Using this probability distribution the upper .05 probability estimate
for the top event is .0124. The bounds on the top event probability
using the bounds on the initiating events propagated through the
Fault Tree are .00077 to .0193. The estimate of the median was
.0054 compared to a value of .0039 obtained by propagating the
median values through the Fault Tree. This example again illustrates

169

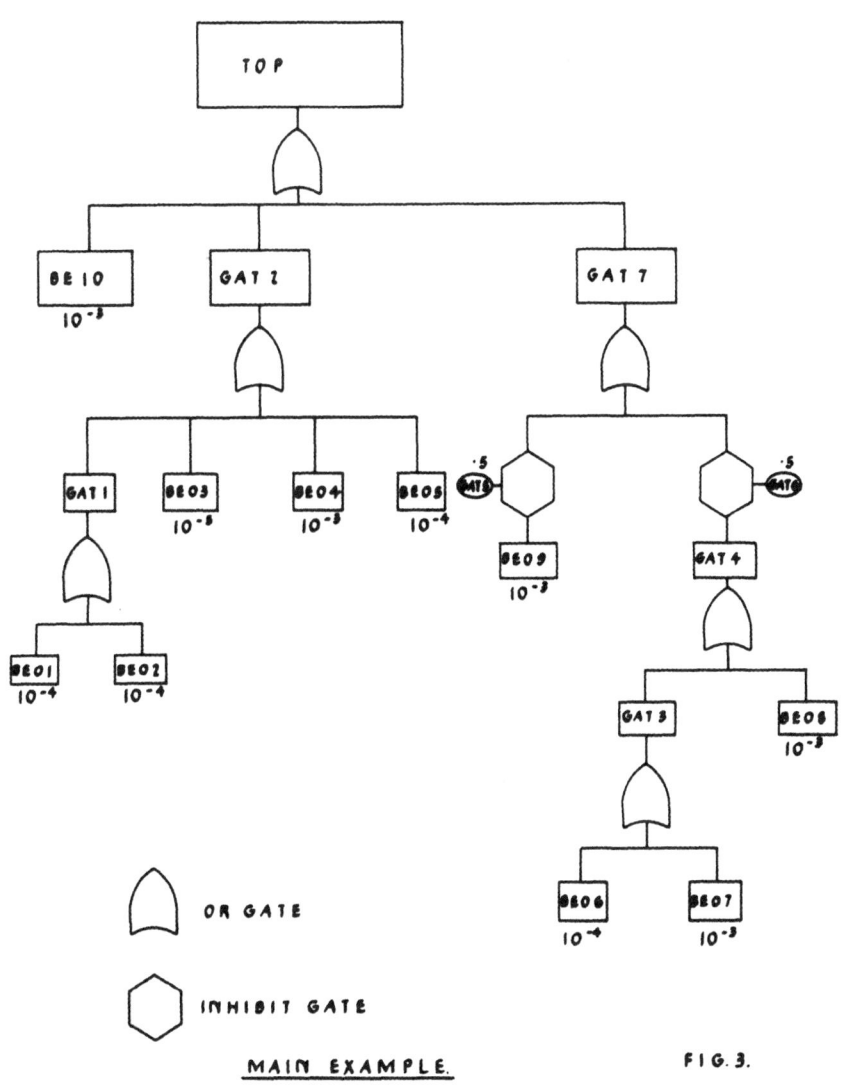

OR GATE

INHIBIT GATE

MAIN EXAMPLE.

FIG.3.

the optimism that is inherent in using the median and pessimism in using the bounds propagation methods.

A major advantage of using the Monte Carlo approach to error propagation is that the predictions obtained can be re-calculated to test for the sensitivity of the predictions to the form or size of the assumed initiating event probabilities or even changes to the tree structure. The predictions obtained using the method presented in this paper will obviously be a function of the number of simulations done and the parameter values fitted for the chosen Johnson probability distribution. The sensitivity of these predictions can be tested using an extension of the method given in Ref 6 to include all the Johnson distributions. This method produces confidence bounds on the predictions. For example, for the upper .05 probability estimate the 99% confidence interval is estimated to be .0121 to .0128. Table 1 shows how the upper .05 prediction and confidence intervals tighten as the number of simulations increase.

SIMULATIONS	PREDICTION	CONFIDENCE INTERVAL
100	.0109	.0093 , 0132
1000	.0118	.0111 , .0127
10000	.0124	.0121 , .0128

TABLE 1

The method described can be used when a large Fault Tree is disaggregated into smaller trees, and the smaller trees then used as initiating events in the reduced tree. This use of the method is illustrated by the reduction of the Fault Tree in Fig 3 to that in Fig 4.

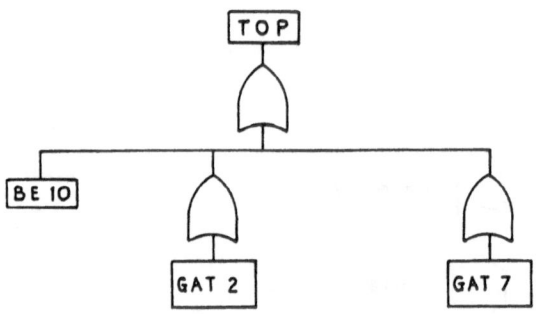

FIG 4. REDUCED FAULT TREE EXAMPLE.

The Monte Carlo simulation of the trees with top events GAT2 and GAT7 give fitted lognormal distributions with parameters (after 10000 simulations):

GAT2 : $\gamma = 7.6489$, $\eta = 1.1575$, $\lambda = 1$, $\epsilon = 0$

GAT7 : $\gamma = 9.065$, $\eta = 1.4232$, $\lambda = 1$, $\epsilon = 0$

The simulation of the reduced tree (10000 simulations) gave the upper .05 prediction of .0124 and the 99% confidence bounds of .0121 and .0128.

Further extensions of the method are in:

a) Accident Sequences where the probability of the top event (or its complement) are used in combination with other probabilities.

b) construction of Farmer curves where different initiating event probabilities are combined for specific fission product release categories.

Conclusions

The method presented gives a more statistically meaningful approach to uncertainty propagation than the traditional methods, and is more easily applied to further extensions of the analysis to Accident Sequences and Farmer curve generation. Large Fault Trees can be disaggregated into smaller trees, and the overall tree re-assembled using the probability distributions obtained from the Monte Carlo results on the smaller trees. This approach would provide a quicker solution than propagating the uncertainties through the overall Fault Tree. A typical run time for a problem of the size of Fig 3 would take around 2 mins for 10000 simulations.

Acknowledgement

This work was carried out with the support of the Procurement Executive, Ministry of Defence.

References

1. Lewis H W.
    Medians and Means in Probabilistic Risk Assessments.
    Nuclear Science and Engineering Vol 91 pp 220-222 (1985)

2. Hahn G J and Shapiro S S.
    Statistical Models in Engineering, Wiley.

3. Camarinopoulos L and Yllera J.
    Advanced Concepts in Fault Tree Modularisation.
    Nuclear Engineering and Design, Vol 91 pp 79-91 (1986)

172

4. Hill I D and Holder R L.
    Fitting Johnson Curves by Moments.
    Journal of Applied Statistics, Vol 25 pp 180-189 (1976)

5. Olson D M.
    Fitting Johnsons S_B and Su systems of curves using the Method
    of Maximum Likelihood.
    Journal of Quality Technology, Vol 11 pp 211-217 (1979)

6. Cheng R C H and Iles T C,
    Confidence bands for cumulative distribution functions of
    continuous random variables.
    Techmometrics, Vol 25 pp 77-86 (1983)

SYSTEM AVAILABILITY SYNTHESIS

A Z KELLER and Isa S QAMBER
Postgraduate School of Industrial Technology
University of Bradford,
Bradford BD7 1DP
UK

ABSTRACT

The present paper demonstrates how equations describing system behaviour can be synthesised from the Markov models for the individual sub–systems. Two methods for synthesis are presented:
(i)   Direct differentiation
(ii)   Kronecker multiplication

A number of examples are given and results checked with the two different methods. It is shown that the second method is superior in that it considerably reduces algebraic ⁻ manipulation and that it is readily generalised to systems where no uniform structure exists regarding the sub–systems. It is also demonstrated that the method is equally valid for systems described by transition probabilities as for systems described by transition rates (continuous time).

INTRODUCTION

Many systems can be considered as made up of a number of simpler components. The availability behaviour of each component is described by

$$\frac{d\underline{P}(t)}{dt} = A\underline{P}(t)$$

However, it is often difficult to derive the general overall equations for the total system from the equations describing the components. Such a knowledge is often desirable in order to obtain an understanding of the total system structure and in particular interactions between components. This is particularly desirable if one wishes to introduce interactions with regard to common mode failures.

The overall matrix, for a large system, can be built up from initial sub–system matrices by an iterative procedure which, step by step, adds to the existing one of the matrices pertinent to the other sub–systems of the system. From the knowledge of each sub–system failure and repair rate, system structure, the system matrix, and an "equivalent transition rate matrix" can be obtained.

In order to derive an overall transition rate matrix, for a large system, special techniques are required.

In the present paper, two methods are investigated: one is the use of direct differentiation and the other uses Kronecker products. Both these methods are used to obtain equivalent transition rate matrices.

STATES SUB–SYSTEMS OF FIRST–ORDER DIFFERENTIAL EQUATIONS

The problem that is studied in this section is that of discussing the availability of the system of sub–systems in Fig. 1d, where $r = 1, 2,...,n$ represents a sub–system r.

It is assumed that each sub–system $S^{(r)}$ is either in a state $S_1^{(r)}$, where the sub–system is working or dichotomously in a second state $S_2^{(r)}$, where it is failed and awaiting repair.

The probabilities of being in $S_1^{(r)}$ and $S_2^{(r)}$ are given by $P_1^{(r)}$ and $P_2^{(r)}$, respectively.

More generally one can write the basic equations governing the sub–systems to be in the form:

$$\frac{dP_1^{(r)}}{dt} = -\lambda_r P_1^{(r)} + \mu_r P_2^{(r)}$$

$$\frac{dP_2^{(r)}}{dt} = \lambda_r P_1^{(r)} - \mu_r P_2^{(r)}$$

$\underline{P}^{(r)}$ denotes the state probability vector for system r.

The overall availability of the system depicted in Fig. 1(d) can be written as:

$$A = \prod_{r=1}^{N} A_r \tag{1}$$

where $A_r = P_1^{(r)}$.

Equations for A are more conveniently obtained by embedding them within a generalised system of equations.

These equations can be obtained by one of two methods; the first method uses differentiation whilst the second method expressed these equations in a generalised form using Kronecker products of matrices (for both methods, sub–systems may have more than two states).

These two methods are now described in detail.

DIFFERENTIATION METHOD

To illustrate the method it is convenient to exemplify it first for a system with two sub–systems (Fig. 1(a)), each of which can be in one of two states, given by the equations:

$$\frac{dP_1^{(1)}}{dt} = -\lambda_1 P_1^{(1)} + \mu_1 P_2^{(1)}$$

$$\frac{dP_2^{(1)}}{dt} = \lambda_1 P_1^{(1)} - \mu_1 P_2^{(1)}$$

$$\frac{dP_1^{(2)}}{dt} = -\lambda_2 P_1^{(2)} + \mu_2 P_2^{(2)}$$

$$\frac{dP_2^{(2)}}{dt} = \lambda_2 P_1^{(2)} - \mu_2 P_2^{(2)}$$

Let:

$$
\left.
\begin{aligned}
Q_1 &= P_1^{(1)} P_1^{(2)} \\
Q_2 &= P_1^{(1)} P_2^{(2)} \\
Q_3 &= P_2^{(1)} P_1^{(2)} \\
Q_4 &= P_2^{(1)} P_2^{(2)}
\end{aligned}
\right\}
\qquad (2)
$$

It can be shown by direct differentiation and algebraic manipulation that:

$$
\begin{bmatrix}
\frac{dQ_1}{dt} \\[6pt]
\frac{dQ_2}{dt} \\[6pt]
\frac{dQ_3}{dt} \\[6pt]
\frac{dQ_4}{dt}
\end{bmatrix}
=
\begin{bmatrix}
-(\lambda_1+\lambda_2) & \mu_2 & \mu_1 & 0 \\
\lambda_2 & -(\mu_2+\lambda_1) & 0 & \mu_1 \\
\lambda_1 & 0 & -(\lambda_2+\mu_1) & \mu_2 \\
0 & \lambda_1 & \lambda_2 & -(\mu_1+\mu_2)
\end{bmatrix}
\begin{bmatrix}
Q_1 \\
Q_2 \\
Q_3 \\
Q_4
\end{bmatrix}
$$

$$(3)$$

The state space–diagram for equation (3) is given in Fig. 2.

Because each sub–system can be in either of two states, the overall system can be characterised by four states, each of which is compounded from the simple two–states for each sub–system.

The above method can readily be generalised to systems consisting of N sub–systems. Again assuming that each sub–system is in one of two states, one sees that the overall system can be described by the following vector Q where:

176

$$
\begin{aligned}
Q_1 &= P_1(1)P_1(2)P_1(3)\ \ldots\ P_1(N-2)P_1(N-1)P_1(N) \\
Q_2 &= P_1(1)P_1(2)P_1(3)\ \ldots\ P_1(N-2)P_1(N-1)P_2(N) \\
&\ \cdot \qquad\quad \cdot \qquad\qquad \cdot \\
&\ \cdot \qquad\quad \cdot \qquad\qquad \cdot \\
Q_{M-1} &= P_2(1)P_2(2)P_2(3)\ \ldots\ P_2(N-2)P_2(N-1)P_1(N) \\
Q_M &= P_2(1)P_2(2)P_2(3)\ \ldots\ P_2(N-2)P_2(N-1)P_2(N)
\end{aligned}
\tag{4}
$$

$M = 2^N$, where
M is the total number of equivalent system states
N is the number of sub-systems

Results obtained by this method for M = 3 and 4 are given below.

The case N = 3 (Fig. 1b) is now considered. Following a similar procedure as for the previous case, the overall system can be described by:

$$
\begin{bmatrix}
\frac{dQ_1}{dt} \\[4pt]
\frac{dQ_2}{dt} \\[4pt]
\frac{dQ_3}{dt} \\[4pt]
\frac{dQ_4}{dt} \\[4pt]
\frac{dQ_5}{dt} \\[4pt]
\frac{dQ_6}{dt} \\[4pt]
\frac{dQ_7}{dt} \\[4pt]
\frac{dQ_8}{dt}
\end{bmatrix}
=
\begin{bmatrix}
-A_1 & \mu_3 & \mu_2 & 0 & \mu_1 & 0 & 0 & 0 \\
\lambda_3 & -A_2 & 0 & \mu_2 & 0 & \mu_1 & 0 & 0 \\
\lambda_2 & 0 & -A_3 & \mu_3 & 0 & 0 & \mu_1 & 0 \\
0 & \lambda_2 & \lambda_3 & -A_4 & 0 & 0 & 0 & \mu_1 \\
\lambda_1 & 0 & 0 & 0 & -A_5 & \mu_3 & \mu_2 & 0 \\
0 & \lambda_1 & 0 & 0 & \lambda_3 & -A_6 & 0 & \mu_2 \\
0 & 0 & \lambda_1 & 0 & \lambda_2 & 0 & -A_7 & \mu_3 \\
0 & 0 & 0 & \lambda_1 & 0 & \lambda_2 & \lambda_3 & -A_8
\end{bmatrix}
\begin{bmatrix}
Q_1 \\ Q_2 \\ Q_3 \\ Q_4 \\ Q_5 \\ Q_6 \\ Q_7 \\ Q_8
\end{bmatrix}
$$

where:

$$
\begin{aligned}
A_1 &= \lambda_1+\lambda_2+\lambda_3 & A_2 &= \lambda_1+\lambda_2+\mu_3 \\
A_3 &= \lambda_1+\mu_2+\lambda_3 & A_4 &= \lambda_1+\mu_2+\mu_3 \\
A_5 &= \mu_1+\lambda_2+\lambda_3 & A_6 &= \mu_1+\lambda_2+\mu_3 \\
A_7 &= \mu_1+\mu_2+\lambda_3 & A_8 &= \mu_1+\mu_2+\mu_3
\end{aligned}
$$

The total system state-space diagram is given in Fig. 3.

For N=4 (Fig. 1c) the total system is described by:

i.e.

$$
\begin{bmatrix}
\dfrac{dQ_1}{dt} \\
\dfrac{dQ_2}{dt} \\
\dfrac{dQ_3}{dt} \\
\dfrac{dQ_4}{dt} \\
\dfrac{dQ_5}{dt} \\
\dfrac{dQ_6}{dt} \\
\dfrac{dQ_7}{dt} \\
\dfrac{dQ_8}{dt} \\
\dfrac{dQ_9}{dt} \\
\dfrac{dQ_{10}}{dt} \\
\dfrac{dQ_{11}}{dt} \\
\dfrac{dQ_{12}}{dt} \\
\dfrac{dQ_{13}}{dt} \\
\dfrac{dQ_{14}}{dt} \\
\dfrac{dQ_{15}}{dt} \\
\dfrac{dQ_{16}}{dt}
\end{bmatrix}
=
\begin{bmatrix}
-A_1 & \mu_4 & 0 & 0 & \mu_2 & 0 & \mu_3 & 0 & \mu_1 & 0 & 0 & 0 & 0 & 0 & 0 & 0 \\
\lambda_4 & -A_2 & \mu_3 & 0 & 0 & 0 & 0 & \mu_2 & 0 & 0 & 0 & \mu_1 & 0 & 0 & 0 & 0 \\
0 & \lambda_3 & -A_3 & \mu_2 & 0 & 0 & \lambda_4 & 0 & 0 & 0 & 0 & 0 & 0 & \mu_1 & 0 & 0 \\
0 & 0 & \lambda_2 & -A_4 & 0 & \lambda_4 & 0 & \lambda_3 & 0 & 0 & 0 & 0 & 0 & 0 & 0 & \mu_1 \\
\lambda_2 & 0 & 0 & 0 & -A_5 & \mu_3 & 0 & \mu_4 & 0 & \mu_1 & 0 & 0 & 0 & 0 & 0 & 0 \\
0 & 0 & 0 & \mu_4 & \lambda_3 & -A_6 & \lambda_2 & 0 & 0 & 0 & \mu_1 & 0 & 0 & 0 & 0 & 0 \\
\lambda_3 & 0 & \mu_4 & 0 & 0 & \mu_2 & -A_7 & 0 & 0 & 0 & 0 & \mu_1 & 0 & 0 & 0 & 0 \\
0 & \lambda_2 & 0 & \mu_3 & \lambda_4 & 0 & 0 & -A_8 & 0 & 0 & 0 & 0 & 0 & \mu_1 & 0 & 0 \\
\lambda_1 & 0 & 0 & 0 & 0 & 0 & 0 & 0 & -A_9 & \mu_2 & 0 & \mu_4 & \mu_3 & 0 & 0 & 0 \\
0 & 0 & 0 & 0 & \lambda_1 & 0 & 0 & 0 & \lambda_2 & -A_{10} & \mu_3 & 0 & 0 & 0 & \mu_4 & 0 \\
0 & 0 & 0 & 0 & 0 & \lambda_1 & 0 & 0 & 0 & \lambda_3 & -A_{11} & 0 & \lambda_2 & 0 & 0 & \mu_4 \\
0 & \lambda_1 & 0 & 0 & 0 & 0 & 0 & 0 & \lambda_4 & 0 & 0 & -A_{12} & 0 & \mu_3 & \mu_2 & 0 \\
0 & 0 & 0 & 0 & 0 & 0 & \lambda_1 & 0 & \lambda_3 & 0 & \mu_2 & 0 & -A_{13} & \mu_4 & 0 & 0 \\
0 & 0 & \lambda_1 & 0 & 0 & 0 & 0 & 0 & 0 & 0 & 0 & \lambda_3 & \lambda_4 & -A_{14} & 0 & \mu_2 \\
0 & 0 & 0 & 0 & 0 & 0 & 0 & \lambda_1 & 0 & \lambda_4 & 0 & \lambda_2 & 0 & 0 & -A_{15} & \mu_3 \\
0 & 0 & 0 & \lambda_1 & 0 & 0 & 0 & 0 & 0 & 0 & \lambda_4 & 0 & 0 & \lambda_2 & \lambda_3 & -A_{16}
\end{bmatrix}
\begin{bmatrix}
Q_1 \\ Q_2 \\ Q_3 \\ Q_4 \\ Q_5 \\ Q_6 \\ Q_7 \\ Q_8 \\ Q_9 \\ Q_{10} \\ Q_{11} \\ Q_{12} \\ Q_{13} \\ Q_{14} \\ Q_{15} \\ Q_{16}
\end{bmatrix}
$$

178

where:

$$A_1 = \lambda_1 + \lambda_2 + \lambda_3 + \lambda_4 \qquad A_2 = \lambda_1 + \lambda_2 + \lambda_3 + \mu_4$$
$$A_3 = \lambda_1 + \lambda_2 + \mu_3 + \mu_4 \qquad A_4 = \lambda_1 + \mu_2 + \mu_3 + \mu_4$$
$$A_5 = \lambda_1 + \lambda_3 + \lambda_4 + \mu_2 \qquad A_6 = \lambda_1 + \lambda_4 + \mu_2 + \mu_3$$
$$A_7 = \lambda_1 + \lambda_2 + \lambda_4 + \mu_3 \qquad A_8 = \lambda_1 + \lambda_3 + \mu_2 + \mu_4$$
$$A_9 = \lambda_2 + \lambda_3 + \lambda_4 + \mu_1 \qquad A_{10} = \lambda_3 + \lambda_4 + \mu_1 + \mu_2$$
$$A_{11} = \lambda_4 + \mu_1 + \mu_2 + \mu_3 \qquad A_{12} = \lambda_2 + \lambda_3 + \mu_1 + \mu_4$$
$$A_{13} = \lambda_2 + \lambda_4 + \mu_1 + \mu_3 \qquad A_{14} = \lambda_2 + \mu_1 + \mu_3 + \mu_4$$
$$A_{15} = \lambda_3 + \mu_1 + \mu_2 + \mu_4 \qquad A_{16} = \mu_1 + \mu_2 + \mu_3 + \mu_4$$

The total system state–space diagram is given in Fig. 4.

Obviously the method given above can be used for any value of N and can be generalised for sub–systems with more than two states. However, mathematical manipulation becomes very difficult.

KRONECKER PRODUCTS

Notation for Kronecker multiplication is now introduced and some useful results are derived and discussed.

In 1977 [1], a method for constructing the transition rate matrix using Kronecker algebra was proposed for the first time by V Amoia, as a possible method for describing compound systems consisting of sub–systems with defined transition matrices. An improved generalised method for using this is described in the present section. This generalised method is applied later to obtain results which are shown to be identical with those obtained previously using the direct–differentiation method.

Kronecker Matrix Multiplication

The general rule for Kronecker multiplication is as follows:

$$\text{Let } A = \begin{bmatrix} a_{11} & a_{12} & \cdot & \cdot & \cdot & a_{1N} \\ a_{21} & a_{22} & \cdot & \cdot & \cdot & a_{2N} \\ \cdot & \cdot & & & & \cdot \\ \cdot & \cdot & & \cdot & & \cdot \\ a_{N1} & a_{N2} & \cdot & \cdot & \cdot & a_{NN} \end{bmatrix}$$

and

$$B = \begin{bmatrix} b_{11} & b_{12} & \cdot & \cdot & \cdot & b_{1N} \\ b_{21} & b_{22} & \cdot & \cdot & \cdot & b_{2N} \\ \cdot & \cdot & \cdot & \cdot & \cdot & \cdot \\ \cdot & \cdot & \cdot & \cdot & \cdot & \cdot \\ b_{N1} & b_{N2} & \cdot & \cdot & \cdot & b_{NN} \end{bmatrix}$$

Then A⊗B is defined as:

$$
\left[
\begin{array}{cccc}
a_{11}\begin{bmatrix} b_{11} & b_{12} & \cdots & b_{1N} \\ b_{21} & b_{22} & \cdots & b_{2N} \\ \cdot & \cdot & \cdots & \cdot \\ \cdot & \cdot & & \cdot \\ \cdot & & & \\ b_{N1} & b_{N2} & \cdots & b_{NN} \end{bmatrix}
&
a_{12}\begin{bmatrix} b_{11} & b_{12} & \cdots & b_{1N} \\ b_{21} & b_{22} & \cdots & b_{2N} \\ \cdot & \cdot & \cdots & \cdot \\ \cdot & \cdot & & \cdot \\ \cdot & & & \\ b_{N1} & b_{N2} & \cdots & b_{NN} \end{bmatrix}
&
\cdots a_{1N}\begin{bmatrix} b_{11} & b_{12} & \cdots & b_{1N} \\ b_{21} & b_{22} & \cdots & b_{2N} \\ \cdot & \cdot & \cdots & \cdot \\ \cdot & \cdot & & \cdot \\ \cdot & & & \\ b_{N1} & b_{N2} & \cdots & b_{NN} \end{bmatrix}
\\[4em]
a_{21}\begin{bmatrix} b_{11} & b_{12} & \cdots & b_{1N} \\ b_{21} & b_{22} & \cdots & b_{2N} \\ \cdot & \cdot & \cdots & \cdot \\ \cdot & \cdot & & \cdot \\ \cdot & & & \\ b_{N1} & b_{N2} & \cdots & b_{NN} \end{bmatrix}
&
a_{22}\begin{bmatrix} b_{11} & b_{12} & \cdots & b_{1N} \\ b_{21} & b_{22} & \cdots & b_{2N} \\ \cdot & \cdot & \cdots & \cdot \\ \cdot & \cdot & & \cdot \\ \cdot & & & \\ b_{N1} & b_{N2} & \cdots & b_{NN} \end{bmatrix}
&
\cdots a_{2N}\begin{bmatrix} b_{21} & b_{12} & \cdots & b_{1N} \\ b_{21} & b_{22} & \cdots & b_{2N} \\ \cdot & \cdot & \cdots & \cdot \\ \cdot & \cdot & & \cdot \\ \cdot & & & \\ b_{N1} & b_{N2} & \cdots & b_{NN} \end{bmatrix}
\\[4em]
a_{N1}\begin{bmatrix} b_{11} & b_{12} & \cdots & b_{1N} \\ b_{21} & b_{22} & \cdots & b_{2N} \\ \cdot & \cdot & \cdots & \cdot \\ \cdot & \cdot & & \cdot \\ \cdot & & & \\ b_{N1} & b_{N2} & \cdots & b_{NN} \end{bmatrix}
&
a_{N2}\begin{bmatrix} b_{11} & b_{12} & \cdots & b_{1N} \\ b_{21} & b_{22} & \cdots & b_{2N} \\ \cdot & \cdot & \cdots & \cdot \\ \cdot & \cdot & & \cdot \\ \cdot & & & \\ b_{N1} & b_{N2} & \cdots & b_{NN} \end{bmatrix}
&
\cdot a_{NN}\begin{bmatrix} b_{11} & b_{12} & \cdots & b_{1N} \\ b_{21} & b_{22} & \cdots & b_{2N} \\ \cdot & \cdot & \cdots & \cdot \\ \cdot & \cdot & & \cdot \\ \cdot & & & \\ b_{N1} & b_{N2} & \cdots & b_{NN} \end{bmatrix}
\end{array}
\right]
$$

Rules and Properties for Kronecker Products

Using the above definition of multiplication, the ensuing results can be obtained, [2,3,4,5,6].

(i) Scalar Multiplication:
   If α is a scalar, then
   $$A \otimes (\alpha B) = \alpha(A \otimes B) \tag{5}$$

180

(ii) Matrix Sum and Products Association
a) $(A+B) \otimes C = A \otimes C + B \otimes C$ $\hspace{4cm}$ (6)
b) $A \otimes (B+C) = A \otimes B + A \otimes C$ $\hspace{4cm}$ (7)

(iii) Product Association
$\hspace{1cm}A \otimes (B \otimes C) = (A \otimes B) \otimes C$ $\hspace{4cm}$ (8)

(iv) Mixed Products
$\hspace{1cm}(A \otimes B).(C \otimes D) = A.C \otimes B.D$ $\hspace{3cm}$ (9)

Kronecker Sum

Consider the two matrices A(nxn) and B(mxm), then following ref. [5], the Kronecker sum can be defined by the following expression:
$\hspace{1cm}A \oplus B = A \otimes I_m + I_n \otimes B$ $\hspace{4cm}$ (10)
where: I_m is a unit matrix (mxm)
$\hspace{1.5cm}I_n$ is a unit matrix (nxn)

KRONECKER PRODUCT MODELLING

Considering the two sub–system example (Fig. 1a) previously treated, the differential equations describing the sub–systems can be re–written as:

$$\frac{dP^{(1)}}{dt} = T^{(1)}P^{(1)}$$

$$\frac{dP^{(2)}}{dt} = T^{(2)}P^{(2)}$$

Let $Q = P^{(1)} \otimes P^{(2)}$ $\hspace{5cm}$ (11)
where: Q is an equivalent system state probability vector
$\hspace{1.5cm}T^{(i)}$ is transition rate matrix for ith sub–system
$\hspace{1.5cm}P^{(i)}$ is an ith column state probability vector for the ith sub–system
It is seen that Q has components identical with that given by equation (2).

Differentiating equation (11) one obtains,

$$\frac{dQ}{dt} = \frac{dP^{(1)}}{dt} \otimes P^{(2)} + P^{(1)} \otimes \frac{dP^{(2)}}{dt}$$
$$= T^{(1)}P^{(1)} \otimes P^{(2)} + P^{(1)} \otimes T^{(2)}P^{(2)}$$
$$= T^{(1)}P^{(1)} \otimes I^{(2)}P^{(2)} + P^{(1)}I^{(1)} \otimes T^{(2)}P^{(2)}$$
$$= (T^{(1)} \otimes I^{(2)})(P^{(1)} \otimes P^{(2)}) + (I^{(1)} \otimes T^{(2)})(P^{(1)} \otimes P^{(2)})$$
$$= [T^{(1)} \otimes I^{(2)} + I^{(1)} \otimes T^{(2)}] (P^{(1)} \otimes P^{(2)})$$
$$= [T^{(1)} \otimes I^{(2)} + I^{(1)} \otimes T^{(2)}]Q \hspace{3cm} (12)$$
Therefore, the basic equations can be written as:

$$\frac{dQ}{dt} = [T^{(1)} \oplus T^{(2)}]Q \hspace{4cm} (13)$$

where: $T^{(1)} \oplus T^{(2)} = T^{(1)} \otimes I^{(2)} + I^{(1)} \otimes T^{(2)}$

Using equation (10) it is readily shown that

$$T^{(1)} \oplus T^{(2)} = \begin{bmatrix} -(\lambda_1+\lambda_2) & \mu_2 & \mu_1 & 0 \\ \lambda_2 & -(\mu_2+\lambda_1) & 0 & \mu_1 \\ \lambda_1 & 0 & -(\lambda_2+\mu_1) & \mu_2 \\ 0 & \lambda_1 & \lambda_2 & -(\mu_1+\mu_2) \end{bmatrix}$$

which is identical with the transition rate matrix contained in equation (3).

Now, considering the three sub-systems example previously treated (Fig. 1b), the differential equations describing the sub-systems can be re-written as:

$$\frac{d\underline{P}^{(1)}}{dt} = T^{(1)}\underline{P}^{(1)}$$

$$\frac{d\underline{P}^{(2)}}{dt} = T^{(2)}\underline{P}^{(2)}$$

$$\frac{d\underline{P}^{(3)}}{dt} = T^{(3)}\underline{P}^{(3)}$$

Let $\underline{Q} = \underline{P}^{(1)} \otimes \underline{P}^{(2)} \otimes \underline{P}^{(3)}$ \hfill (14)

where: $\underline{P}^{(1)}$, $\underline{P}^{(2)}$ and $\underline{P}^{(3)}$ are the first, second and third sub-systems columns probabilities vectors and $T^{(1)}$, $T^{(2)}$ and $T^{(3)}$ are their transition rate matrices.

Differentiating equation (14) one obtains:

$$\frac{d\underline{Q}}{dt} = \frac{d\underline{P}^{(1)}}{dt} \otimes \underline{P}^{(2)} \otimes \underline{P}^{(3)} + \underline{P}^{(1)} \otimes \frac{d\underline{P}^{(2)}}{dt} \otimes \underline{P}^{(3)} + \underline{P}^{(1)} \otimes \underline{P}^{(2)} \otimes \frac{d\underline{P}^{(3)}}{dt}$$

$$= T^{(1)}\underline{P}^{(1)} \otimes \underline{P}^{(2)} \otimes \underline{P}^{(3)} + \underline{P}^{(1)} \otimes T^{(2)}\underline{P}^{(2)} \otimes \underline{P}^{(3)} + \underline{P}^{(1)} \otimes \underline{P}^{(2)} \otimes T^{(3)}\underline{P}^{(3)}$$

\hfill (15)

Considering the three terms of equation (15). Therefore, the equivalent system equation can be written as:

$$\frac{d\underline{Q}}{dt} = [\{T^{(1)} \otimes I^{(2)} \otimes I^{(3)}\}\underline{Q}$$

$$+ \{I^{(1)} \otimes T^{(2)} \otimes I^{(3)}\}\underline{Q}$$
$$+ \{I^{(1)} \otimes I^{(2)} \otimes T^{(3)}\}\underline{Q}]$$
$$= [\{T^{(1)} \otimes I^{(2)} \otimes I^{(3)}\} + \{I^{(1)} \otimes T^{(2)} \otimes I^{(3)}\} + \{I^{(1)} \otimes I^{(2)} \otimes T^{(3)}\}]\underline{Q}$$
$$= [T^{(1)} \oplus T^{(2)} \oplus T^{(3)}]\underline{Q} \hfill (16)$$

Using Kronecker summation defined by equation (10), it is readily shown that

$$T^{(1)} \oplus T^{(2)} \oplus T^{(3)} = \begin{bmatrix} -A_1 & \mu_3 & \mu_2 & 0 & \mu_1 & 0 & 0 & 0 \\ \lambda_3 & -A_2 & 0 & \mu_2 & 0 & \mu_1 & 0 & 0 \\ \lambda_2 & 0 & -A_3 & \mu_3 & 0 & 0 & \mu_1 & 0 \\ 0 & \lambda_2 & \lambda_3 & -A_4 & 0 & 0 & 0 & \mu_1 \\ \lambda_1 & 0 & 0 & 0 & -A_5 & \mu_3 & \mu_2 & 0 \\ 0 & \lambda_1 & 0 & 0 & \lambda_3 & -A_6 & 0 & \mu_2 \\ 0 & 0 & \lambda_1 & 0 & \lambda_2 & 0 & -A_7 & \mu_3 \\ 0 & 0 & 0 & \lambda_1 & 0 & \lambda_2 & \lambda_3 & -A_8 \end{bmatrix}$$

where:

$$A_1 = \lambda_1 + \lambda_2 + \lambda_3 \qquad A_2 = \lambda_1 + \lambda_2 + \mu_3$$
$$A_3 = \lambda_1 + \mu_2 + \lambda_3 \qquad A_4 = \lambda_1 + \mu_2 + \mu_3$$
$$A_5 = \mu_1 + \lambda_2 + \lambda_3 \qquad A_6 = \mu_1 + \lambda_2 + \mu_3$$
$$A_7 = \mu_1 + \mu_2 + \lambda_3 \qquad A_8 = \mu_1 + \mu_2 + \mu_3$$

which is identical with the transition rate matrix obtained earlier, using direct differentiation method, by more laborious means.

General Application of Kronecker Multiplication

The results given in the previous section can be generalised to a system containing any number of sub-systems, in which the sub-systems can have any arbitrary number of states.

Assuming N sub-systems, one can define:

$$Q = \underline{P}_1^{(1)} \oplus \underline{P}^{(2)} \oplus \ldots \ldots \oplus \underline{P}^{(N)}$$

$$= \oplus \sum_{r=1}^{N} \underline{P}^{(r)} \tag{17}$$

where $\oplus \Sigma$ is a notation introduced for the general Kronecker sum.

It follows that

$$\frac{dQ}{dt} = \sum_{r=1}^{N} \underline{X}^{(r)} \tag{18}$$

with $\underline{X}^{(r)} = G(1, r-1) \otimes T^{(r)} \underline{p}^{(r)} \otimes G(r+1, N)$ (19)

where: $G(n, m) = \otimes \prod_{r=n}^{m} \underline{P}^{(r)}$

$$= \underline{P}^{(n)} \otimes \ldots \ldots \otimes \underline{P}^{(m)} \tag{20}$$

($\otimes_\Pi$ is a notation for general Kronecker products).

$\underline{X}^{(r)}$ can now be re-written as

$$\underline{X}^{(r)} = G(1,r-1) \otimes T^{(r)}\underline{p}^{(r)} \otimes I_{(N-r)} \, G(r+1,N) \qquad (21)$$

where: $I_{(N-r)}$ is the unit matrix of order (N–r). Using equation (9) successively one can proceed as follows:

$$\underline{X}^{(r)} = G(1,r-1)\otimes[T^{(r)}\otimes I_{(N-r)}.\underline{p}^{(r)}\otimes G(r+1,N)]$$
$$= G(1,r-1)\otimes[T^{(r)}\otimes I_{(N-r)}.G(r,N)]$$

$$\cdot \quad \cdot \quad \cdot \quad \cdot \quad \cdot \quad \cdot$$

$$= I_{(r-1)}G(1,r-1) \otimes[T^{(r)}\otimes I_{(N-r)}.G(r,N)]$$
$$= I_{(r-1)}\otimes T^{(r)}I_{(N-r)} \cdot \, G(1,r-1)\otimes G(r,N)$$
$$= I_{(r-1)}\otimes T^{(r)}\otimes I_{(N-r)}. \, \underline{Q} \qquad (22)$$

From equations (18) and (22) it follows that

$$\frac{d\underline{Q}}{dt} = T \, \underline{Q} \qquad (23)$$

where:

$$T = \otimes \sum_{r=1}^{N} T^{(r)}$$

$$= T^{(1)} \oplus T^{(2)} \oplus \ldots \oplus T^{(N)} \qquad (24)$$

It is seen that equation (24) contains the earlier results (equations (13) and (16)) as special cases.

DISCRETE MARKOV PROCESSES

The techniques described earlier, using the direct-differentiation and Kronecker multiplication methods, can also be applied to discrete systems. In this section one could expect that the use of an analogue to differentiation as used in direct-differentiation method, even if practicable, would be extremely cumbersome. However, it is shown in the present section that system synthesis using Kronecker products is accomplished as easily for discrete systems as for continuous time systems.

The general equation for a discrete Markov process can be written as:

$$\underline{p}^{(r)}(s+1) = A^{(r)} \, \underline{p}^{(r)}(s) \qquad (25)$$

where $\underline{P}^{(r)}(s)$ and $\underline{P}^{(r)}(s+1)$ are the state probabilities vectors at instances s and s+1 respectively, and A is the transition probability matrix for sub–system r.

As before, let:

$$\underline{Q}(s) \ - \ \otimes \prod_{r-1}^{N} \underline{P}^{(r)}(s)$$

$$- \ \underline{P}^{(1)}(s) \otimes \underline{P}^{(2)}(s) \otimes \ \dots \ \otimes \underline{P}^{(N)}(s) \qquad (26)$$

It follows that

$$\underline{Q}(s+1) \ - \ A^{(1)}\underline{P}^{(1)}(s) \otimes A^{(2)}\underline{P}^{(2)}(s) \otimes \ \dots \ \otimes A^{(N)}\underline{P}^{(N)}(s)$$

$$- \ (A^{(1)} \otimes A^{(2)})(\underline{P}^{(1)}(s) \otimes \underline{P}^{(2)}(s)) \otimes A^{(3)}\underline{P}^{(3)}(s) \otimes \dots \otimes A^{(N)}\underline{P}^{(N)}(s)$$

$$\dots\dots\dots\dots\dots\dots\dots\dots\dots\dots\dots\dots\dots\dots$$

$$\dots\dots\dots\dots\dots\dots\dots\dots\dots\dots\dots\dots\dots\dots$$

$$\dots\dots\dots\dots\dots\dots\dots\dots\dots\dots\dots\dots\dots\dots$$

$$- \ (A^{(1)}A^{(2)} \otimes \dots \otimes A^{(N)})(\underline{P}^{(1)}(s) \otimes \underline{P}^{(2)}(s) \otimes \dots \otimes \underline{P}^{(N)}(s))$$

$$- \ (\otimes \prod_{r-1}^{N} A^{(r)})(\otimes \prod_{r-1}^{N} \underline{P}^{(r)}(s))$$

$$- \ A \ \underline{Q}(s) \qquad (27)$$

where: $A \ - \ \otimes \prod_{r-1}^{N} A^{(r)}$

The total system equations can now be written as:

$$\underline{Q}(s+1) \ = \ A\underline{Q}(s) \qquad (28)$$

RELATION BETWEEN DISCRETE AND CONTINUOUS PROCESSES

As shown earlier, the equations for continuous processes are described by the system equation:

$$\frac{d\underline{Q}}{dt} \ - \ T \ \underline{Q} \qquad (29)$$

where: $T \ - \ \otimes \sum_{r-1}^{N} T^{(r)} \qquad (30)$

$\otimes \Sigma$ operation is defined earlier.

The sub–system equations are defined by:

$$\frac{dP^{(r)}}{dt} = T^{(r)} \; \underline{P}^{(r)} \tag{31}$$

Taking a sufficiently small interval of time Δt, one has the approximation:

$$\underline{P}^{(r)}(t+\Delta t) = \underline{P}^{(r)} + \Delta t \; T^{(r)} \; \underline{P}^{(r)}$$

$$= A^{(r)} \; \underline{P}^{(r)} \tag{32}$$

$$(A^{(r)} = \Delta t \; T^{(r)} + I_{(r)})$$

Using result 27 one can derive systems equations given by:

$$\underline{Q}(t+\Delta t) = A \; \underline{Q} \tag{33}$$

where:

$$A = \otimes \prod_{r=1}^{N} A^{(r)} = \otimes \prod_{r=1}^{N} (I_{(r)} + \Delta t \; T^{(r)}) \tag{34}$$

Using the notation

$$G(N,s) = \otimes \prod_{r=s}^{N} (\Delta t \; T^{(r)} + I_{(r)}) \tag{35}$$

One can re-write equation (34), neglecting terms of order $(\Delta t)^2$ as:

$$A = \Delta t \; T^{(1)} \oplus G(N,2) \tag{36}$$

Using equations (30) and (36) successively, one obtains the result

$$A = \Delta t \; (\oplus \sum_{r=1}^{N} T^{(r)}) \oplus I_{(N)} \tag{37}$$

Since it follows from equations (29) and (31) that

$$A = \Delta t \; T + I_{(N)} \tag{38}$$

Comparison of (38) with (37) gives the result

$$T = \oplus \sum_{r=1}^{N} T^{(r)}$$

previously obtained.

CONCLUSIONS

1. Kronecker multiplication can be used to derive global Markov system equations.
2. The procedure is systematic and fast and more convenient to use than direct differentiation.
3. It is readily generalisable to systems where no uniform structure exists.
4. The method is equally applicable to discrete time systems as it is for continuous time systems.

REFERENCES

1.   Amoia, V, "Computer oriented reliability analysis of large electrical systems", Summer Symposium of Circuit Theory, 1977, pp. 317-325.

2.   Bellman, R, "Introduction to Matrix Analysis", McGraw Hill, 1960.

3.   Amoia, V, De Micheli, G & Santamauro M, "Computer-oriented formulation of transition rate matrices via Kronecker algebra", IEEE Trans. Reliability, Vol. 30, No. 2, 1981.

4.   Cafaro, C, Corsi, F & Vacca, F, "Multistate Markov models and structural properties of transition-rate matrix", IEEE Trans. Reliability, Vol. 35, No. 2, 1986, pp. 192-200.

5.   Brewer, J W, "Kronecker products and matrix calculus in system theory", IEEE Trans. Circuits & Systems, Vol. 25, No. 9, 1978.

6.   Graham, A, "Kronecker Product and Calculus with Applications", John Wiley & Sons, 1981, England.

Fig. 1a   Two subsystems

Fig. 1b   Three subsystems

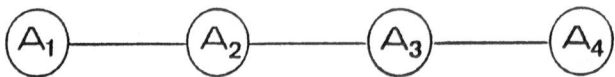

Fig. 1c   Four subsystems

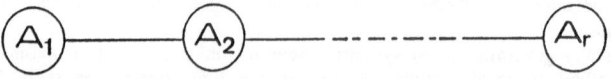

Fig. 1d   r subsystems

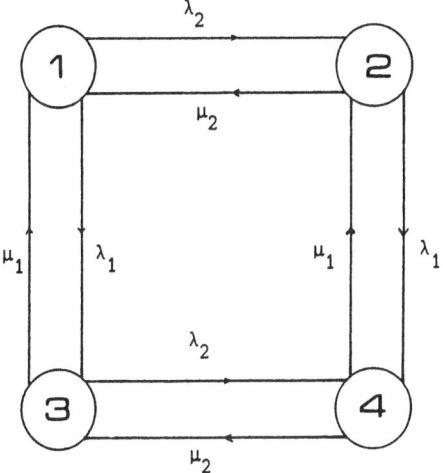

Fig. 2   The overall system state-space
         diagram of the two subsystems

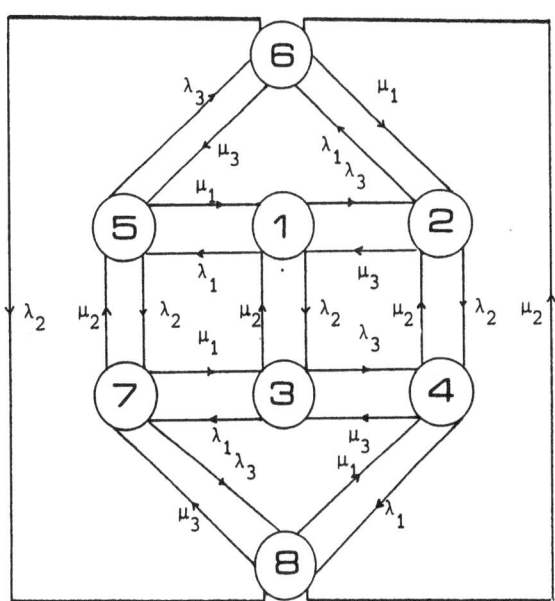

Fig. 3   The overall system state-space
         diagram of the three subsystems

188

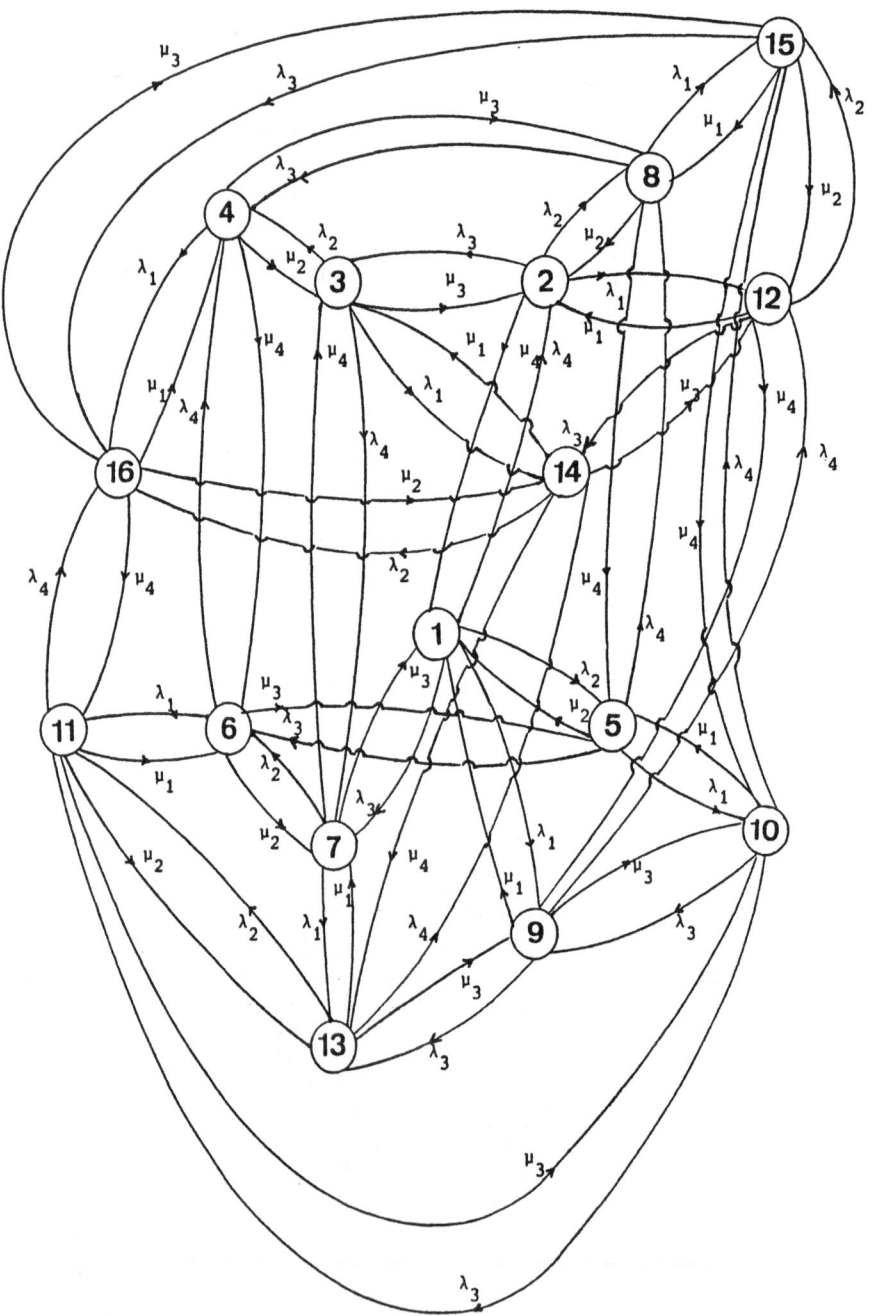

Fig. 4 The overall system state-space diagram of the four subsystems

A SOFTWARE SYSTEM
FOR ARMCO SIMULATION OF CONCEPTUAL PLANT

M P Fairclough
National Centre of Systems Reliability,
UKAEA,
Wigshaw Lane, Culcheth
Warrington, WA3 4NE

ABSTRACT

In the nuclear industry the capacity of installed or committed fuel reprocessing plant is sufficient to meet our needs until well into the next century. For future needs plant designs are only required around the turn of the century which gives time to completely review any new options that may be available. To provide a basis for assessing the relative merits of optional designs a technique of ARMCO modelling has been developed and implemented in the form of a computer software package. The package can be used for ARMCO assessments of any type of system for which reliability data is known. This paper describes the development of the package in terms of the factors influencing its specification. The current features of the package are described together with proposals for its enhancement in the future.

INTRODUCTION

The Authority and BNFL are researching and developing new techniques which could be applied in a future reprocessing plant. That is one which would be built early in the next century. Working on such a time scale allows us to step back and consider the new techniques available before the need for detailed plant design becomes pressing. The techniques are aimed at reducing costs and the wastes generated during reprocessing and some method of appraising them will be needed to choose options which need to be developed to demonstration scale. Providing safety studies do not rule out particular techniques, a method for evaluating the merits of the various options is to see what their effect is on the unit cost of reprocessing fuel in a conceptual plant design. To achieve such an assessment requires complex modelling of the availability, reliability, maintainability and cost (ARMCO) of a conceptual plant design as a change of option in one part of a plant can significantly alter the whole operation of the plant.

An NCSR study has now been started to investigate the options available for the conceptual plant designs. Included in this study is the requirement for a computer software aid to help in the evaluation of the various options and the creation of a reliability database about each option. This paper outlines the software available now to carry out plant assessments and identifies a need, which is not filled at present, for software which will evaluate many optional designs in terms of ARMCO. The development and current status of a software package for conceptual plant assessment is briefly described together with proposals for its enhancement using sophisticated sampling techniques. Modelling cannot be performed without data, hence the aspect of data collection is also considered.

COMPUTER MODELLING OF CONCEPTUAL PLANT DESIGNS

The need for the development of a conceptual plant modelling package arose because of the difficulty in assessing the relative merits of the many conceptual fuel processing plant designs which could be produced when considering the options available. The range of options being looked at is considerable and it appears that modelling software currently available cannot readily cope with this range. It is worth reviewing some of the software modelling aspects that are required.

Required Model Structure

The following points summarise the main requirements of the fuel reprocessing plant modelling package. They include comments on modelling and computer software.

1   The model needs to be a fairly simple aid to bridge the gap between the pencil and paper studies carried out in the first stages of plant design and the detailed simulation of plant processes when the design has been "frozen".

2   Primary output will be the plant availability and the unit cost of the product of interest.

3   Details of the apportionment of downtime will also be needed together with the specific requirement of the radiation doses incurred during repair and maintenance procedures.

4   Modularity of model construction is important as the system must allow options to be interchanged at will, and the option parameters must also be simple to modify.

5   A description of a plant for entry into the computer must closely resemble the physical plant to simplify construction of models.

6   To allow continuity of model development, the simulation software must be able to cope with very simple limited data at the beginning of an assessment study through to more detailed data at later stages, and give sensible results through the whole range.

7   Emphasis must be placed on data entry, editing and "housekeeping" functions. The software is specifically for conceptual plant modelling and so many options will be investigated. It is therefore

very important to be able to keep track of assessments and the values of parameters used.

8   To allow such activities as sensitivity studies, the software must be able to model the early plant designs quickly. When limited numbers of suitable plant options have been identified then this constraint can be relaxed.

9   The software package should be a general purpose modelling aid which can be used on any system for which reliability data and a logical arrangement can be established.

Current Software

To carry out the modelling required there is a range of software currently available which could be used. The range can be grouped into specific categories and these are now described in very general terms in the context of the required software.

CAD

Computer Aided Design and other similar planning software is primarily aimed at providing a draughting aid to the designer. Reliability aspects are not included in the software although quite sophisticated planning and management options are often available (Plant Design Management System and Project Engineering and Graphics System, PDMS and PEGS from the CAD Centre, Cambridge).

Process Simulation

Closely linked to CAD is process simulation software which allows the operation of chemical plant to be simulated by modelling such parameters as chemical concentrations, and temperatures and carrying out heat and mass balance calculations. Such software is again aimed at the optimisation of a system (process) which is quite well defined and detailed modelling of very specific features is needed (eg PROCESS from Sim Sci).

Reliability/Fault Tree Analysis

Perhaps the main objective of such codes is to demonstrate how a system can fail and highlight the key components of a system. The tree structure requires that a "top event" be defined and the combinations and sequences of events which may lead to this event form branches from it. As such the technique is not readily adopted to modelling plant availability in the manner required.

Visual Interactive Modelling

A very powerful operational research aid is the technique of visual interactive modelling. Plant designs can be shown graphically and the actual running of the plant illustrated. It is a statistical technique which has been used extensively for plant optimisation and other such problems as queuing. The model becomes compiled into the computer code and as such cannot be changed with ease. It is interesting to note that BNFL are using Visual Interactive Modelling with their own reliability modelling routines to simulate the new reprocessing plant THORP (eg See Why from Istel).

"Monte Carlo" Simulation of ARMCO

Perhaps the nearest type of software available to the requirements of the conceptual fuel processing plant project is the range of software which can statistically simulate a system described by events. Unlike VIM the plant description does not become compiled into the code and the system is not simulated graphically. This technique is very useful for investigating complex maintenance policies and the software available tends to concentrate on these aspects (eg MAROS from Baker, Jardine and Associates). Statistical simulation is lengthy for all but the simplest of models and generally the models are not easy to modify in the way required.

It can be seen that the above range of software overlaps in many areas but none of the range is really suitable for the conceptual plant modelling required. To fulfil the requirements, software with the features described in the next section is needed.

Proposed Software

Data input

Modularity of model construction is one of the primary requirements to be met therefore data input format and processing has high priority. Significant advantages can be gained by using a proprietary database package to handle data inputs.

a   Data input is well defined and the chance of error is reduced considerably.

b   Data can be queried to extract related sets of information.

c   "Housekeeping" is simple.

The structure is ideally suited to incorporation into a database system and in the proposed software, a database is used to hold the detailed reliability and maintainability information about individual plant items, the cost sets applicable to the plant and descriptions about how the plant is constructed. It is a simple and efficient way of filing information. For example the plant database is a computer file which is made up of separate records containing the reliability and maintainability data for individual items of plant; one record for each item in the database. A record is created by the user filling in blank fields in a form presented to him by the software. It is simple to generate a record and edit it at some later time as the software takes care of positioning of fields and data storage and retrieval. Figure 1 illustrates the structure of the database in which a particular conceptual plant design is stored as a textual record which references standard sub-system records which in turn call upon detailed records about individual plant items:

    conceptual plant design
    sub-system configurations
    plant item characteristics.

When a plant design record is used to start a simulation, the data from this record is read and used to search through the more detailed levels of the database for all the information needed to establish the plant structure and its characteristics. These descriptions are then stored and used to control the calculation of plant availability and unit cost of processing.

A feature of the future reprocessing plant study is the requirement for a database to be established recording the reliability characteristics of the various options. The use of a database for simulation input is therefore convenient and the two will be combined. How this data will be compiled is discussed in a later section. It would appear that this approach could be adopted in many other areas in which assessment studies are carried out.

Simulation method

To be able to investigate a large number of changes in both plant design and characteristics of individual items is another feature of the future reprocessing plant project. The approach used has been to allow two modes of operation of the software: An analytical method in which the time averaged output values are calculated from mathematical relationships defined in the program and a statistical method in which the model is run much in the same way as a real plant and output data is built up as items in the model fail and are then repaired.

An analytical study of a plant provides a quick and simple way of looking at the effects of changing elements of the model or altering particular parameters such as failure rates or restoration times. It does however have the drawback that the plant is not truely simulated and as such the time dependence effects of certain factors are lost. To simulate a plant we must use a statistical or Monte Carlo model and gather probability data from many computer runs which is a time consuming task particularly when sensitivity studies are carried out. It is expected that in the early stages of acquiring data to enter into a plant model, the quality of this data will be poor and so the analytical model will suffice. However as data improves in quality (and quantity) then the true plant simulation will become of value.

Sampling techniques

Initially the statistical simulation module will perform simple "next event" modelling without any attempt to reduce computer run time by sophisticated sampling techniques. However it must be appreciated that significant savings can be made by introducing improved simulation methods. Some of the techniques which may be of value are:

Russian Roulette

A technique used for simulating systems which are naturally tree structured. This reduction technique works by using the natural paths between each stage of the simulation and deciding whether to continue down the present path or not (ie if the new path to follow is interesting or uninteresting). A good example is getting the value of three from the sum of the values on two dice. If the first value is greater than two then we know not to continue.

Importance Sampling

This involves changing the shape of a distribution so that certain values which normally would not occur often occur more frequently because of their importance in the simulation. Weighting factors are later applied to restore the original form of the distribution.

Systematic Sampling

This sampling technique can be considered to be a special form of Importance sampling. In Systematic Sampling the distribution is divided into a number of equally sized intervals and for the first interval a set of random samples is taken. The pattern used is retained to take samples from each of the other intervals.

Stratified Sampling

This sampling method involves the splitting of a distribution into several sections or strata of equal probability range. When sampling for the different sections is complete then the resulting random variates are combined in a random order giving a random sample.

Latin Hypercube Sampling

Latin Hypercube Sampling is a similar type of method to Systematic and Stratified Sampling. It is used when there are a large number of input variables and involves splitting the distribution into several equal probability intervals which are then ranked by the values of random numbers assigned to each. For a simulation run samples are then taken from the same rank in each input variable.

Correlated Sampling

Correlated Sampling is a technique used for the comparison of two or more options or strategies which have a positive correlation between each other. This is achieved by using the same set of random numbers for each strategy and comparing the differences between the values created by the strategies rather than the actual values themselves.

Antithetic Variates

The idea of this method is to use two estimator values which have a negative correlation between the first and the second estimator and compensate for each others variations. If the negative correlation is very small then there can be a very high level of reduction in the sample size. Antithetic Variates is a special case of Correlated Sampling except Correlated Sampling requires a positive correlation while Antithetic Variates requires a negative correlation.

Taguchi method

Usually used for optimising product or process design this is a statistical technique which could have potential use in modelling. It can be used to identify parameters which significantly affect a plant design and those to which the plant is relatively insensitive.

At a later stage, these methods will be appraised to see which will be of value.

DATA FOR MODELLING

Simulation studies are normally carried out using well established reliability data published in standard specifications or available generally on a database. This type of data does not exist for our conceptual plant modelling as the options available are still in the experimental stage. It will be necessary to establish some form of data collection exercise to enable this data to be derived.

For our conceptual reprocessing plant, the data for running the model will be generated by analysis of basic records kept during the development of a piece of equipment or plant. It would be necessary for the R&D project teams to collect data on a regular basis to establish these records but this should not cause any problems due to the relatively small number of items under evaluation and the fact that development teams will keep records of this nature anyway.

The raw data collected can then be analysed and interpreted into the parameters needed for modelling. It has to be appreciated that initially data may be very sparse and uncertain and that some data may only become available after years of study but providing there is an awareness of the need for particular data, it can be incorporated as it becomes available.

An exercise along the lines described is being carried out at present in the PFR reprocessing plant at Dounreay and although this is at a higher stage of development than the study required, it highlights the types of problems that may be encountered.

CONCLUDING REMARKS

The paper has summarised the development of a project whose aim is to provide an ARMCO software simulation aid specifically for the assessment of the relative merits of optional design of conceptual plant. At present the software development has reached the stage where the database input and analytical module have both been designed are now being implemented. A statistical module will be added to mid 1988 followed by studies of improved statistical modelling methods.

It is expected that graphical input of plant configurations will be developed at an early stage to enhance "user friendliness". The system has been designed in a modular way and it is possible that the core module could form the basis of customised packages which have add-on routines to perform specific user requirements.

Finally it is worth restating that the software package under development is a general purpose aid for any type of ARMCO assessment which has many options to consider. Although designed for a specific project it is not restricted to that task or indeed the nuclear industry.

A case record specifies
which sub-systems will be
used and how.

A sub-system record specifies
which items make it up
and how.

An item record holds that
items reliability and
maintenance details.

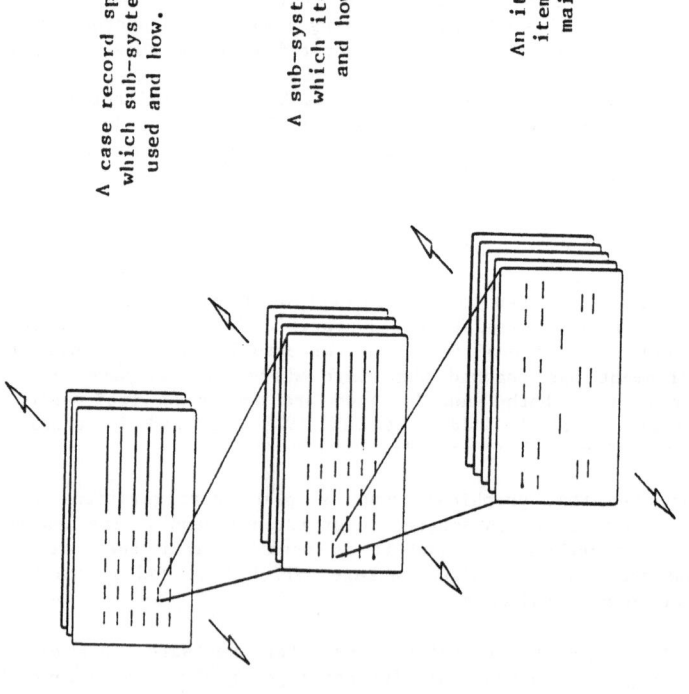

Figure 1. Databases used to enter model details

POWER GENERATION SYSTEMS AVAILABILITY
MAINTAINABILITY ASSESSMENT

Farag Ali El-Sheikhi
Ras Lanuf Oil and Gas Processing Company Inc
PO Box 296
Benghazi
Libya

and

Roy Billinton
Power System Research Group
University of Saskatchewan
Saskatoon, Saskatchewan
Canada S7N 0W0

ABSTRACT

Generating unit availability is an important component in the assessment of generation system reliability. The overall availability or unit capability is dependent upon the forced outage parameters for the unit and the planned outage and maintenance requirements. It is desirable to develop maintenance schedules that will keep the power system risk as low as possible and therefore maintenance scheduling is an important aspect in the development of an optimum generation system expansion plan. Scheduling of this capacity should permit the required work to be accomplished but not introduce excessive risk to the system. This paper presents methods for obtaining a preventive maintenance schedule using the reserve and the risk-levelizing criteria. The Gram-Charlier expansion of a distribution is used for developing the risk-levelization technique. This paper also presents a new method based on the moment cumulant method and the Tchebychev-Hermite polynomial to calculating the unit effective load carrying capability (ELCC). The paper investigates the effects of the system forecast peak load and forced outage rate of a unit on the maintenance schedule of the generation system. Detailed comparisons using the reserve and risk-levelizing methods are presented and discussed. A model of the Manitoba Hydro generating system is used to illustrate the concepts presented.

INTRODUCTION

Overall unavailability assessment of a generating unit involves an appreciation of the forced outage parameters for the unit and its maintenance requirements. The contribution towards overall unit incapability of the forced and planned contributions is shown in Figure 1. These data were taken from the 1985 Annual Report on Generation Equipment prepared by the Canadian Electrical Association. Figure 1 also shows the derating adjusted forced outage rate (DAFOR) and the available but not operating factor (ABNOF) for the general unit classes of combustion turbine, fossil, hydraulic, nuclear and internal combustion units. The utilization forced outage rate (UFOP) is shown for the turbine units.

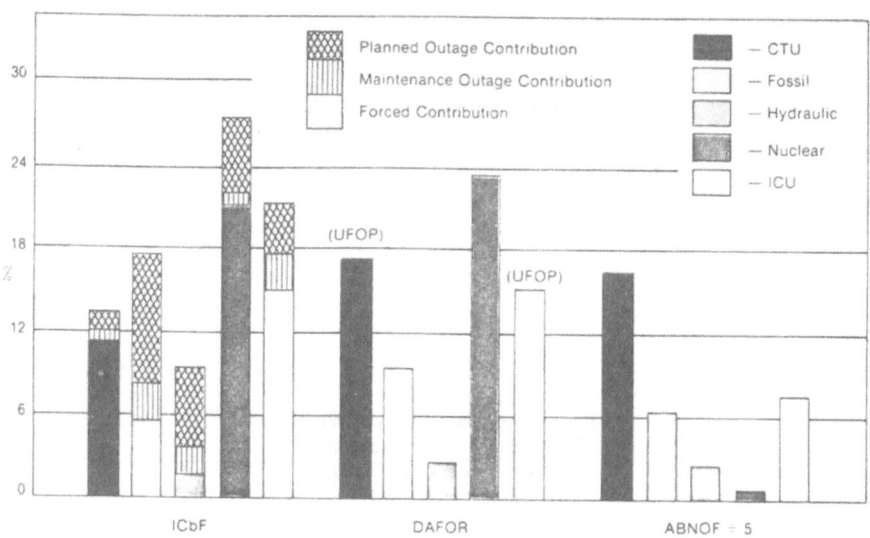

Figure 1. Comparison of ICbF, DAFOR and ABNOF for different generating
           unit types based on 1985 Canadian Electrical Association data.

It can clearly be seen from Figure 1 that generating units require a significant amount of maintenance time in order to maximize their operating reliability and to enhance the overall system adequacy. Scheduling of generating equipment for preventive maintenance plays an important role in long term generation planning. Improper generating unit preventive maintenance schedules can force a generation system expansion plan into the installation of capacity to cover maintenance at a time when economic and environmental concerns dictate that electric power utilities conserve their resources. The available installed capacity of the system is decreased during a unit maintenance period which results in an increase in the system risk. This risk should not exceed the system acceptable planning value at any time during the annual load cycle. It is important, therefore, to levelize the system adequacy throughout the year by appropriate preventive maintenance scheduling.

This levelization problem can be considered in two ways. The first approach is to levelize the system risk throughout the year and therefore this approach is designated as the risk-levelization technique. The second approach is to levelize the system reserve throughout the year and this technique is designated as the reserve-levelizing approach. Both techniques are used in this paper and a comparison between them is presented. The technique utilized in this paper for evaluating the system risk is the Loss of Load Expectation (LOLE) method [1]. Maintenance requirements can have a substantial effect on the reserve required to meet a desired LOLE level. Maintenance requirements increase the contribution of off-peak periods to the annual LOLE as most maintenance operations, especially of large units, are usually scheduled at relatively low system loads.

This paper also presents a new technique for evaluating the effective load carrying capability (ELCC) [2] of a unit based on the Edgeworth form of the Gram-Charlier series. The approach utilizes a continuous approximation of the equivalent load curve using the moments and the statistical cumulants of the system generating unit outage distribution and the hourly load distribution together with the ELCC technique to introduce a practical method for maintenance scheduling of generating equipment.

Sensitivity studies of the effect on maintenance scheduling of generating facilities with unit forced outage rate (FOR) and system hourly peak demand variations are presented using both the risk and reserve-levelization techniques. The effect of the unit FOR on its ELCC and a comparison between different maintenance plans using both reserve and risk-levelization techniques is presented and discussed. A model of the Manitoba Hydro (MH) system is used to illustrate the practical significance of the concepts proposed.

MAINTENANCE PLANNING BASED ON LEVELIZING SYSTEM RESERVE

The primary objective in this technique is the development of a practical method for automated maintenance scheduling which allows the maintenance planner to represent and resolve all the relevant and conflicting constraints on the schedule so that the computed schedules are as realistic as possible. The method is designed to schedule maintenance in weekly intervals, that is, using forecasted weekly peaks.

The method is summarized by the flow chart of Figure 2. The main aim of this method is to levelize the reserve, which is the system installed capacity minus the peak load demand and capacity out for maintenance. The system generating units can be arranged in any desired priority list. The simplest priority list for scheduling is based on generator capacity only and is formed by ordering the units on that basis with the largest first. The rationale for this method is that the largest units are the most critical and hardest to schedule. Hence, the largest units should be scheduled early in the process when the greatest flexibility exists. The generating units should be inserted in the lowest valley of the curve which consists of the weekly peak demands plus the total capacity of those units which have been previously scheduled for maintenance.

SYSTEM STUDIES

The procedure shown in Figure 1 has been applied to the MH system model. The system has 119 generating units, ranging from 127.4 MW to 2.5 MW. The installed capacity is 6454.2 MW and the system annual peak demand is 4996 MW. The detailed description of the MH generation system is given in the Appendix. Plan 1 in Table 1 shows the base case maintenance schedule.

The amount of reserve in the system depends on the forecast peak demand. The effect of the system forecast peak demand on the maintenance schedule has been tested by decreasing the original system annual peak to

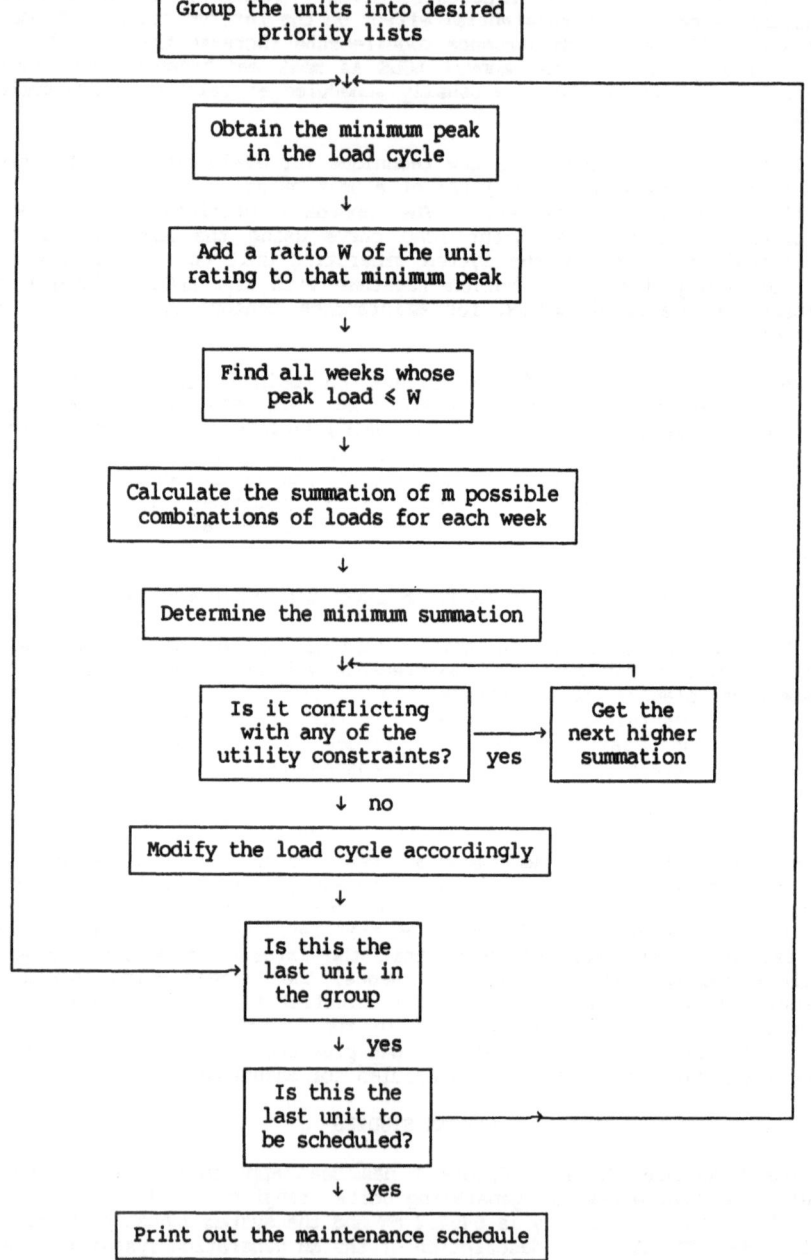

Figure 2. Generating unit preventive maintenance scheduling procedure
using the reserve-levelization concept.

Table 1

Plan 1. MH system maintenance plan using the reserve-levelization method
System annual peak = 4996 MW

Week #	Level										
	1	2	3	4	5	6	7	8	9	10	11
1-15	No maintenance										
16	117	106	66	32	22	13.67	8.5	5.5	2.5		
17	117	106	66	32	32	22	13.67	8.5	5.5	2.5	
18	117	98	66	32	22	13.67	7	3.08			
19	117	98	66	22	13.67	7	3.08				
20	106	33	25	22	13.67	7	3.08				
21	117	106	98	33	25	22	13.67	7	3.08		
22	117	106	98	33	32	21	5.5				
23	127.4	118	106	98	33	32	25	21	8.5	5.5	
24	127.4	127.4	118	117	98	98	25	8.5	7	5.5	
25	127.4	118	117	106	98	25	21	7	5.5		
26	127.4	118	117	106	98	25	21	7			
27	127.4	117	106	98	32	22	21	7	5.5		
28	106	105	66	32	22	21	5.5				
29	127.4	117	105	66	32	21	8.5	8.5	3.08		
30	127.4	127.4	117	106	105	66	32	21	8.5	8.5	3.08
31	127.4	127.4	118	106	105	66	33	5.5			
32	127.4	118	106	106	33	32	8.5	7	5.5	4.2	
33	106	106	33	33	32	22	8.5	7	4.2		
34	117	106	98	33	33	22	7	4.2			
35	127.4	117	106	98	33	21	7	4.2			
36	127.4	118	106	98	33	25	21	21	8.5	7	4.2
37	118	117	106	98	25	21	8.5	7	4.2		
38	127.4	127.4	117	117	98	33	7	3.08			
39	127.4	127.4	117	106	98	33	13.67	7	3.08		
40	117	106	33	25	13.67	5.5	2.5				
41	127.4	117	106	98	33	25	22	13.67	5.5	2.5	
42	127.4	106	98	22	13.67	8.5	5.5	2.5			
43	8.5	5.5	2.5								
44	13.67	5.5									
45	98	32	25	13.67	8.5	5	3.08				
46	98	32	25	8.5	3.08						
47-52	No maintenance										

approximately 70% of this value. The new annual system peak load 3497.2
MW and every individual weekly peak load is changed accordingly. Plan 2
in Table 2 shows the maintenance schedule of the MH system with the
revised peak loads. The weekly peak load demand cycle before and after
maintenance is shown in Figure 3. The system reserve levelizing procedure
can be seen clearly in this figure.

THE EFFECTIVE LOAD CARRYING CAPABILITY (ELCC) OF A UNIT

The ELCC of a generating unit is the increase in system load which can
be carried by the system after the addition of the unit to the system. It
can also be defined as the increase in load that the system can
accommodate in order to return the system to the reliability level that
existed prior to the addition of the unit. The ELCC is usually determined
at a specified acceptable planning expectation and depends on many
factors. Some of which are: unit size and forced outage rate (FOR), size
and FOR of the rest of the units in the system, system peak demand and the
acceptable system risk level. The number, size and FOR's of units which
are scheduled for maintenance before that unit in the same maintenance
period also has an influence on the unit ELCC.

THE UNIT ELCC CALCULATION METHOD

Description of notation:

KG_j = jth cumulant of the system generating unit outage distribution.

Table 2

Plan 2. MH system maintenance schedule using the reserve-levelization approach
System annual peak = 3497.2

Week #	Level 1	2	3	4	5	6	7	8	9	10	11
1-11	No maintenance										
12	21	8.5	7	2.5							
13	21	8.5	7	2.5							
14	No maintenance										
15	22	13.67	5.5								
16	117	106	98	25	22	13.67	5.5				
17	117	106	98	25	25	7	4.2				
18	117	106	98	25	21	7	4.2				
19	117	106	98	21							
20	117	106	32	22	8.5	5.5	3.08				
21	117	106	98	32	22	8.5	5.5	3.08			
22	118	106	98	98							
23	127.4	118	106	98	33	32	25	21	5.5		
24	127.4	127.4	117	98	33	32	25	21	8.5	5.5	2.5
25	127.4	118	117	98	98	33	8.5	2.5			
26	127.4	118	106	98	33	21	8.5	5.5			
27	127.4	117	106	105	32	21	8.5	5.5			
28	117	105	33	32	22	13.67	5.5	2.5			
29	127.4	117	105	66	33	22	13.67	5.5	2.5		
30	127.4	127.4	117	105	66	33	7	5.5			
31	127.4	127.4	117	106	66	33	22	8.5	7	5.5	
32	127.4	117	106	66	32	22	13.67	8.5	5.5	3.08	
33	118	106	98	32	13.67	5.5	3.08				
34	118	106	98	66	13.67	7	3.08				
35	127.4	117	106	66	25	22	13.67	7	3.08		
36	127.4	117	106	66	33	25	22	21	7	4.2	
37	127.4	117	106	66	33	21	7	4.2			
38	127.4	127.4	117	106	33	32	25	13.67	8.5	7	3.08
39	127.4	118	106	98	33	32	25	13.67	8.5	7	3.08
40	118	106	98	25	21	8.5	7	3.08			
41	127.4	117	106	98	25	21	8.5	7	3.08		
42	127.4	117	106	98	32	22	7				
43	33	32	32	22	8.5	7	4.2				
44	98	33	32	8.5	4.2						
45	106	98	33	13.67	5.5						
46	106	33	13.67	5.5							
47-52	No maintenance										

K_{j_i} = jth cumulant of a generating unit i outage distribution.

n = number of generating units in the system.

L_j = jth moments of the hourly load about the origin.

ℓ_i = the system load at hour i of order j.

T = number of hours in the overall maintenance period.

KL_j = jth cumulant of the system load distribution.

KE_j = j cumulant of the equivalent load curve.

C_i = capacity of unit i.

The procedure for computing the unit ELCC is based upon the Gram-Charlier expansion of a distribution. The approach proceeds as follows:

1. The system capacity model is expressed in terms of the statistical cumulants of the generating unit outage distribution as a function of the unit moment about the origin. The first six cumulants of a

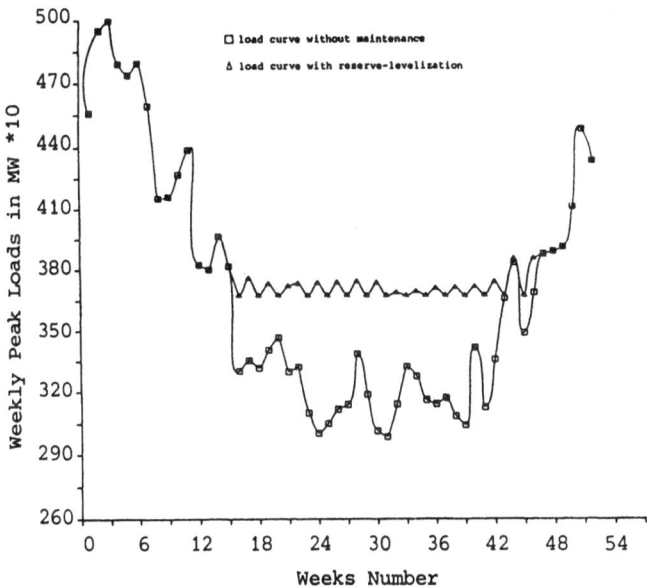

Figure 3. Manitoba Hydro system weekly peak loads with and without maintenance.

generating unit outage distribution as a function of the unit moment about the origin are calculated as follows:

$$K_{1_i} = m_{1_i}$$

$$K_{2_i} = m_{2_i} - m_{1_i}^2$$

$$K_{3_i} = m_{3_i} - 3m_{2_i} m_{1_i} + 2m_{1_i}^3$$

$$K_{4_i} = m_{4_i} - 4m_{3_i} m_{1_i} - 3m_{2_i}^2 + 12m_{2_i} m_{1_i}^2 - 6m_{1_i}^4$$

$$K_{5_i} = m_{5_i} - 5m_{4_i} m_{1_i} - 10m_{3_i} m_{2_i} + 20m_{3_i} m_{1_i}^2 + 30m_{2_i}^2 m_{1_i} - 60m_{2_i} m_{1_i}^3$$

$$+ 24m_{1_i}^5$$

$$K_{6_i} = m_{6_i} - 6m_{5_i} m_{1_i} - 15m_{4_i} m_{2_i} + 30m_{4_i} m_{1_i}^2 - 10m_{3_i}^2 + 120m_{3_i} m_{2_i} m_{1_i}$$

$$- 120m_{3_i} m_{1_i}^3 + 30m_{2_i}^3 - 270m_{2_i}^2 m_{1_i}^2 + 360m_{2_i} m_{1_i}^4 - 120m_{1_i}^6$$

2. Determine the system capacity model from the last step as follows:

$$KG_j = \sum_{i=1}^{n} K_{j_i} \qquad j = 1, 2, \ldots, 6$$

3. The system probabilistic load model is replaced by a set of moments. The load moments about the origin are calculated as follows:

$$L_j = \frac{1}{T} \sum_{i=1}^{T} \ell_i^j \qquad j = 1, 2, \ldots, 6$$

4. Cumulants of the hourly load distribution are evaluated. The first six cumulants of the system load distribution as a function of the moments about the origin are calculated as follows:

$$KL_1 = L_1$$

$$KL_2 = L_2 - L_1^2$$

$$KL_3 = L_3 - 3L_2L_1 + 2L_1^3$$

$$KL_4 = L_4 - 4L_3L_1 - 3L_2^2 + 12L_2L_1^2 - 6L_1^4$$

$$KL_5 = L_5 - 5L_4L_1 - 10L_3L_2 + 20L_3L_1^2 + 30L_2^2L_1 - 60L_2L_1^3 + 24L_1^5$$

$$KL_6 = L_6 - 6L_5L_1 - 15L_4L_2 + 30L_4L_1^2 - 10L_3^2 + 120L_3L_2L_1 - 120L_3L_1^3$$

$$+ 30L_2^3 = 270L_2^2L_1^2 + 360L_2L_1^4 - 120L_1^6$$

5. Perform the Gram-Charlier approximation of the equivalent load curve. The Edgeworth form of the Type A series is given by [3]

$$f(z) = N(z)(1 + \frac{KE_3}{6} H_3 + \frac{KE_4}{24} H_4 + \frac{KE_5}{120} H_5 + \frac{KE_6 + 10KE_3^2}{720} H_6)$$

where,

$$N(z) = \frac{1}{\sqrt{2\pi}} e^{\frac{-z^2}{2}}$$

and $H_j(z)$ is the Chebychev-Hermite polynomial of degree j in z.

The cumulants of the equivalent load curve can be obtained by convolution of the cumulants of the system generating unit outage distribution and the cumulants of the system hourly load distribution as follows:

$$KE_j = KG_j + KL_j \qquad j = 1, 2, \ldots, 6$$

6. Calculate the parameters of the equivalent load curve:

$\mu = KE_1$ mean

$\delta^2 = KE_2$ variance

7. Calculate the coefficients of the series f(z) as follows:

$$KE_j(z) = \frac{KE_{j+2}(x)}{\delta^{j+2}} \qquad j = 1, 2, \ldots$$

The standardized r.v. Z is related to the r.v. x as follows:

$$Z = \frac{x - u}{\delta}$$

8. Any point on the equivalent load curve can be found as follows [4]

$$EL(x) = u - \delta x \emptyset(z, G_j)$$

where,

$$\emptyset(z, G_j) = z + \frac{G_1}{6}(z^2 - 1) + \frac{G_2}{24}(z^3 - 3z) - \frac{G_1^2}{36}(2z^3 - 5z)$$

$$+ \frac{G_3}{120}(z^4 - 6z^2 + 3) - \frac{G_1 G_2}{24}(z^4 - 5z^2 + 2)$$

$$+ \frac{G_1^3}{324}(12z^4 - 53z^2 + 17) + \frac{G_4}{720}(z^5 - 10z^3 + 15z)$$

$$- \frac{G_2^2}{384}(3z^5 - 24z^3 + 29z) - \frac{G_1 G_3}{180}(2z^5 - 17z^3 + 21z)$$

$$+ \frac{G_1^2 G_2}{288}(14z^5 - 103z^3 + 107z) - \frac{G_1^4}{7776}(252z^5 - 1688z^3 + 1511z)$$

The function $\emptyset(z, G_j)$ has been developed using the polynomial transformation to normality method and the Inversion Theorem.

It should be noted that the standardized r.v. z can be obtained from a table lookup or by computer. For a given planning expectation, say x days/year then

$$P_x = \int_{-\infty}^{z} N(x)dx$$

9. Having developed the equivalent load curve and knowing the procedure for obtaining any value corresponding to any cumulative probability using the function $\emptyset(z, G_j)$, the ELCC of a given generating unit can be obtained as follows:

(a) Calculate $EL(X_1)$ corresponding to the specified risk level x.

(b) Subtract the cumulant of the unit from KE_j.

(c) Calculate $EL(X_2)$ for the same risk level x.

(d) The unit ELCC can now be calculated as follows:

$$ELCC = C_i - [EL(X_1) - EL(X_2)]$$

THE PROPOSED TECHNIQUE OF MAINTENANCE SCHEDULING BASED UPON LEVELIZING THE SYSTEM RISK

The technique proceeds as follows:

1. Modify the system equivalent load curve

$$KE_j = KE_j - \sum_{i=1}^{m} KG_{j_i} \qquad j = 1, 2, \ldots, 6$$

where m = number of units scheduled before the maintenance.

2. For every week in the maintenance period:

(a) Calculate $EL(X_1)$.

(b) Modify the system equivalent load curve

$$KE_j = KE_j - K_{j_i}$$

(c) Calculate $EL(X_2)$.

(d) Calculate the unit ELCC from Step 9(d).

(e) Add the unit ELCC to the peak load of the week under consideration.

(f) Repeat Steps (a)-(e) for every week in the maintenance period.

3. Repeat Steps 1 and 2 for every unit in the first priority group.

4. Repeat Steps 1-3 for every other priority group in the system.

STUDY CASES

The method described has been implemented in a computer programme and applied to the MH System model using a VAX 11/782 computer. The following sensitivity study cases were conducted.

Case 1 - Base Case

The original system data given in Appendix 1 are utilized in this case. This study is used as a base case for subsequent studies.

Plan 3 in Table 3 illustrates the maintenance plan for the MH system using the risk-levelization approach. Plan 3 indicates that the MH System has eleven levels of maintenance in some weeks as the system has 119 generating units. It can also be seen that MH has no units out for maintenance in the periods, Weeks 1-15 and Weeks 47-52, where the system

Table 3

Plan 3. MH single system maintenance schedule using the risk-levelizing criterion (base case)
System peak demand   = 4996 MW
Planning expectation = 0.1 days/year

Week #	Level										
	1	2	3	4	5	6	7	8	9	10	11
1-15	No maintenance										
16	117	106	66	32	22	13.67	8.5	5.5	3.08		
17	117	106	66	32	32	22	13.67	8.5	5.5	3.08	
18	106	105	66	32	25	13.67	8.5	5.5	3.08		
19	106	105	66	25	13.67	8.5	5.5	3.08			
20	106	105	13.67	8.5	2.5						
21	117	106	105	25	13.67	8.5	5.5	2.5			
22	117	106	33	32	25	25	21	8.5	7	5.5	3.08
23	127.4	118	106	98	33	32	25	21	8.5	7	3.08
24	127.4	127.4	118	117	98	98	33	7	4.2		
25	127.4	118	117	106	98	33	21	7	4.2		
26	127.4	118	117	106	98	21	21				
27	127.4	117	106	98	32	25	21	7	5.5		
28	106	106	32	32	25	7	7	5.5			
29	118	117	106	66	32	22	21	7	4.2		
30	127.4	127.4	118	117	98	66	22	21	7	4.2	
31	127.4	127.4	127.4	106	98	66	22	13.67	7	4.2	
32	127.4	117	106	98	66	25	22	13.67	4.2		
33	117	98	33	33	25	22	13.67	7	2.5		
34	117	106	98	33	33	22	13.67	7	5.5	2.5	
35	127.4	117	106	98	33	33	5.5				
36	127.4	118	106	98	33	33	25	22	8.5	8.5	5.5
37	118	117	106	98	25	22	8.5	8.5	5.5		
38	127.4	127.4	117	117	98	33	21	3.08			
39	127.4	127.4	117	106	98	33	21	7	3.08		
40	117	106	33	32	8.5	7	3.08				
41	127.4	117	106	98	33	32	21	8.5	7	3.08	2.5
42	127.4	106	98	21	7	5.5	2.5				
43	5.5										
44	13.67										
45	98	32	22	13.67	8.5	5.5					
46	98	32	22	8.5	5.5						
47-52	No Maintenance										

load is the highest. No constraints have been imposed on the system schedule in this plan although any constraints can be easily handled by the programme.

Case 2 - Impact of Unit FOR Values On The Maintenance Plan

The unit FOR values have a direct influence on the maintenance plan of a system as the unit ELCC depends on the equivalent load curve which depends directly on the unit FOR's. This case has been investigated for the MH system by increasing the FOR of just one of the 127.4 MW unit by a factor of ten. The planning expectation was held constant at a value of 0.1 days/year. Plan 4 in Table 4 shows this case. A comparison of this plan with Plan 3 indicates that Plan 4 has a different maintenance activity for every single week in the year except Week 44. This clearly shows the influence of the unit FOR values on a maintenance plan.

Case 3 - Effect Of The Utility Peak Loads On The Maintenance Schedule

Sensitivity of the maintenance plan to the forecast peak loads has been investigated by increasing the system peak demand by 15% of its original value. The new system peak load is 5745.5 MW. Plan 5 in Table 5 shows this case. A comparison between Plan 5 and Plan 3, i.e. the base case plan, indicates that the maintenance activities are different in every single week of the year. The system forecast peak demand, is an important factor which can influence the system maintenance schedule.

Table 4

Plan 4. MH single system maintenance schedule using the risk-levelization criterion
System peak demand = 4996 MW
FOR of the 1st 127.4 MW unit = 10 x 0.01
Planning expectation = 0.1 days/year

Week #	Level											
	1	2	3	4	5	6	7	8	9	10	11	12
1-15	No Maintenance											
16	117	106	66	32	22	13.67	8.5	5.5	2.5			
17	117	106	66	32	32	22	13.67	8.5	5.5	2.5		
18	106	105	66	32	25	13	8.5	5.5				
19	106	105	66	25	13	8.5	5.5					
20	106	105	7									
21	117	106	105	22	21	7	3.08					
22	117	106	33	32	25	22	21	3.08				
23	127.4	118	116	98	33	32	25	25	7	5.5		
24	127.4	127.4	118	117	98	33	25	21	7	7	5.5	2.5
25	127.4	118	117	106	98	33	21	7	2.5			
26	127.4	118	117	106	98	21	3.08					
27	127.4	117	106	98	32	25	21	7	3.08			
28	106	98	32	32	25	8.5	7	5.5	3.08			
29	118	117	106	66	32	22	13.67	8.5	5.5	3.08		
30	127.4	127.4	118	117	106	66	22	22	13.67	13.67	5.5	3.08
31	127.4	127.4	127.4	106	98	66	33	22	13.67	5.5	3.08	
32	127.4	117	106	98	66	33	33	8.5	7			
33	117	98	33	33	25	22	8.5	7	5.5			
34	117	106	98	33	33	25	22	21	5.5			
35	127.4	117	106	98	33	21	8.5	4.2				
36	127.4	118	106	98	98	25	8.5	8.5	4.2	4.2		
37	118	117	106	98	25	21	8.5	7	4.2			
38	127.4	127.4	117	117	98	33	21	8.5	7			
39	127.4	127.4	117	106	98	33	13.67	8.5	5.5	3.08		
40	117	106	33	32	13.67	5.5	3.08	2.5				
41	127.4	117	106	98	33	32	21	8.5	7	2.5		
42	127.4	106	98	21	8.5	7	4.2					
43	4.2											
44	13.67											
45	98	32	22	13.67	7	5.5						
46	98	32	22	7	5.5							
47-52	No Maintenance											

COMPARISON OF METHODS

The studies conducted in this paper can be used to compare the two maintenance planning methods. Comparing Plan 1 and Plan 3, the maintenance schedule is different in all the weeks of the maintenance period. This clearly indicates that the maintenance schedule of any system is a function of the method used. The method of levelizing the reserve does not recognize the differences in the availability of generating units in the system and may lead to assigning the relatively small but highly reliable units for maintenance during the higher load periods. The risk-levelization technique does recognize unit availability and therefore inherently considers system reliability while planning preventive maintenance activities of the system.

The technique utilized in this paper for evaluating the system weekly risk associated with a specified generating capacity is the LOLE method [1]. Maintenance requirements can have a substantial effect on the reserve required to meet a desired LOLE. In general, maintenance requirements increase the contribution of off-peak periods to the LOLE as most maintenance activities, especially of large units, are scheduled in off-peak periods. Extensive risk studies have been performed on the MH System using different annual peak loads [5]. These studies indicate that

Table 5

Plan 5. MH single system maintenance schedule using the risk-levelization
criterion
System peak demand = 1.15 x 4996 MW
Planning expectation = 0.1 days/year

Week #	Level												
	1	2	3	4	5	6	7	8	9	10	11	12	13
1-15	No Maintenance												
16	117	105	66	32	21	8.5	5.5	2.5					
17	117	105	66	32	25	21	8.5	5.5	2.5				
18	106	105	66	25	21	8.5	5.5	3.08					
19	106	105	66	21	8.5	5.5	3.08						
20	98	66											
21	117	106	98	66									
22	117	106	66	25	22	8.5	7	3.08					
23	127.4	118	106	106	66	25	22	8.5	7	3.08			
24	127.4	127.4	118	118	106	106	22	13.67	7				
25	127.4	118	117	106	98	32	22	13.67	7	4.2			
26	127.4	117	117	106	98	32	7	5.5	4.2				
27	127.4	117	106	98	32	25	21	7	5.5	2.5			
28	106	98	32	25	21	7	4.2	2.5					
29	118	117	106	98	22	13.67	7	4.2					
30	127.4	127.4	118	117	106	98	22	13.67	8.5	7	5.5		
31	127.4	127.4	127.4	117	98	33	32	25	13.67	8.5	7	5.5	
32	127.4	117	106	98	33	32	25	22	13.67	13.67	8.5	7	4.2
33	106	98	33	32	22	13.67	8.5	7	4.2				
34	117	106	98	33	32	32	21	8.5	5.5				
35	127.4	117	106	98	32	21	8.5	5.5					
36	127.4	118	117	106	98	25	21	8.5	7				
37	118	117	106	98	25	21	8.5	7	3.08				
38	127.4	127.4	117	106	98	98	13.67	7	3.08				
39	127.4	127.4	117	106	98	33	22	13.67	7	5.5			
40	117	98	33	32	22	5.5							
41	127.4	117	106	98	33	32	22	13.67	5.5	3.08			
42	127.4	106	33	33	33	22	13.67	5.5	3.08				
43	33	33											
44	33	33											
45	33	33	25	21	8.5	5.5	3.08	2.5					
46	25	21	8.5	5.5	3.08	2.5							
47-52	No Maintenance												

the system risk using the risk-levelization approach is lower than the
corresponding risk obtained using the reserve-levelization method over a
wide range of system peak loads.

CONCLUSIONS

Preventive maintenance of generating units is an important factor in
overall power system reliability. This paper presents a method for
scheduling the required preventive maintenance utilizing the
reserve-levelizing criterion. This paper also introduces a new technique
for evaluating the unit ELCC using the moment-cumulant method and the
Tchebychev-Hermite polynomial. A proposed algorithm for developing a
generating facilities maintenance schedule based on the Gram-Charlier
expansion of a distribution and the risk-levelizing criterion is also
presented in this paper.

The paper investigates the effect of the system forecast peak loads
and the unit forced outage rates on the generating unit preventive
maintenance schedule. It shows that the maintenance schedule is sensitive
to the system forecast peak load and that variations in the unit can have
a considerable effect on the system maintenance schedule. Weekly and
annual system risk as measured by LOLE was investigated using the two
approaches.

The generating capacity installed in a power system is a large component of the total system capital and operating cost. It is important therefore, that this capacity be adequate for the system needs without being excessive. The maintenance required for the generating capacity should be conducted such that the risk in any given period and for the entire year are acceptable values. This paper has illustrated a technique by which the factors which influence the system reliability can be incorporated into the determination of the appropriate maintenance schedule.

REFERENCES

1. R. Billinton and R.N. Allan, "Reliability Evaluation Of Power Systems," Longman, London (England)/Plenum Press, New York, 1983.

2. F.A. El-Sheikhi and R. Billinton, "Generating Unit Maintenance Scheduling For Single And Two Interconnected Systems," IEEE Trans. on PAS, Vol. PAS-103, pp. 1038-1044, May 1984.

3. F. Kendall and A. Stuart, "The Advanced Theory Of Statistics," Vol. 1, Distribution Theory, 2nd edition.

4. R. Billinton and F.A. El-Sheikhi, "Preventive Maintenance Scheduling Of Generating Units In Interconnected Systems," Proceedings 10th Annual RAM Conference, Montreal, May 1983.

5. F. El-Sheikhi, "Maintenance Scheduling Of Generating Facilities In Single And Interconnected Power Systems," Ph.D. Thesis, Department of Electrical Engineering, University of Saskatchewan, 1983.

APPENDIX

Manitoba Hydro Generation System Model

Unit Size MW	Number of Units	Unit FOR	Scheduled Maintenance
127.4	10	0.01	2
118	4	0.01	2
117	10	0.01	2
106	12	0.01	2
105	1	0.03	4
98	10	0.01	2
66	2	0.03	4
33	4	0.03	4
32	7	0.01	2
25	6	0.005	2
22	6	0.005	2
21	6	0.005	2
13.67	6	0.005	2
8.5	8	0.005	2
7	8	0.005	2
5.5	8	0.005	2
4.2	3	0.005	2
3.08	5	0.01	2
2.5	3	0.01	2

Static Analysis - A Technique for Software Verification

by

J T Webb
Rex, Thompson & Partners
'Newnhams'
West Street
Farnham
Surrey
GU9 7EQ
UK
+44 252 711414

ABSTRACT

This paper describes the use of static analysis techniques for verifying the correctness of programs and the approach taken by the MALPAS
suite of programs developed specifically for this technique. The origins
of MALPAS and the analyses it can carry out on programs will be described
In addition, some ideas for using MALPAS during the system specification
phase of a project will be presented.

INTRODUCTION

The growth of Software in systems

There is a tendency for systems to be increasingly dependent on software to fulfil the users expectations of what a system is required to do.
Other reasons put forward to justify putting more of a system's functions
in software include:

a)   it is easier to enhance a system at a later date

b)   it is easier to implement complex functions in software than
     hardware

c)   it is easier to specify a system in software terms

d)   system replication costs will be lower.

Examples of applications where such arguments have been employed
range from domestic washing machines to complex 'fly-by-wire' aircraft
control systems.. In the former case the consequences of a software
error may not be serious, a typical example being a machine which locked
itself into a cycle taking in water and immediately pumping it out!
However, systems used in applications such as 'fly-by-wire' where a
digital computer is in direct control of an aircraft are extremely criti-

cal. The possible scenario of an aircraft system failing and thus removing control from the pilot whilst approaching a busy airport can not be contemplated.

However, the history of software projects must not be forgotten. There have been many instances where complex systems have been implemented and put into service with several faults. Indeed, most users of computers will have experienced problems with compilers and operating systems. They will also have experienced problems with new releases of software which were intended to cure bugs yet which introduced a totally new set. One well known operating system was known to have several hundred bugs over a period of some years. Although successive releases were brought out to correct these bugs the number actually increased at one point. This might indeed be regarded as an interesting challenge to some users of an operating system but it can be assumed that the pilot and passengers of an aircraft will view the problem in a very different light.

It can therefore be seen that there is a conflict between the belief on one hand that an increase in the functionality of a system can be achieved by the use of software solutions to a problem and the historically proven but often overlooked fact that the generation of correct software is never achieved. All means must therefore be employed to ensure that bugs are detected and corrected.

Additionally, not only must software be error free, it must be shown to be error free. A sublime belief on the part of the software producer that his procedures are adequate to ensure its correctness is not sufficient. If software is to be used in systems where the consequences of a failure go beyond causing a temporary annoyance to a user then we need to build in formal verification and acceptance procedures to the procurement of such software.

There are various procedures which can be introduced into the procurement of software to ensure that it is correct. These include:

a) retrospective checking of the code

b) formal methods of structuring the development of the system into different levels

c) Checking each stage of the development before commencing the next.

One major difficulty with each of these procedures is the need to not only have a means of checking the code, design or software specification but also to actually know in a usable manner what the system is meant to do. There is therefore a need to ensure that all stages in the development of a software system result in a clear, unambiguous and usable definition of what is required.

The phases of software development

Basic stages of development

Over the last few years it has been recognised that the development of a software system must be performed in a highly structured and organised manner. The aim is to ensure that all aspects of the system are known, understood and fully documented. This will be achieved by proceeding from the initial need for a system to its delivery in a carefully organised sequence of controllable steps. One view details the following steps as a suitable breakdown:

a) Requirement Analysis - The specification of what is required from the system.

b) Functional Design - Initial conversion of the specified Requirements into a first level software structure.

c) Detailed Design - The refinement of the first level Functional Design into a more detailed form suitable for starting coding.

d) Coding - The production of program code.

e) Testing - The testing of code to ensure that it is logically correct and also that it meets the specifications.

f) Integration - The bringing together of various modules into an integrated working system.

g) Use - Use by the customer, the point at which it has been traditional for many errors to be found.

In practice, the above stages cannot be regarded as totally independent of each other. As a lower level phase is performed it is probable that errors will be found requiring modification at a higher level. The most common case is finding an error during testing which will require a coding change. There is also the probability that a Detailed Design step will show an inconsistency at a previous level which has to be corrected.

From the above it can be seen that there is a strong need for tools to assist all of the above stages as much as possible. One particularly important area is that of testing and finding bugs as early as possible. Considering all the stages detailed above, it is easy to see that the more levels it is necessary to iterate through in order to correct a bug the more expensive it will be. The worst case is a misunderstanding of the user's actual requirements which is not discovered until a system is in use. However, an error in design or implementation is also expensive and must be corrected as early as possible. In particular, any safety critical system cannot be allowed to enter service until shown to be correct. A reliable way of detecting such errors is therefore required.

Testing strategies - dynamic and static

Strategies for testing software can be considered as being in two main area. One is to test the software dynamically at various levels of integration by exercising it with a variety of different test cases. The alternative is to analyse the software statically to establish exactly what it is doing. It is useful at this point ot consider these two strategies in greater detail.

Dynamic testing is performed by defining a set of tests, potentially a very large set, which will exercise all of the possible combinations of inputs the program can take and ensuring that all of the outputs for each of the input cases are correct. It can easily be appreciated that there are several obvious problems with this approach as follows:

a) It is impossible to establish what all of the paths through a program of reasonable complexity are.

b) It is impossible to conceive what values and combinations of values can be expected for the different input parameters. It should be noted that this is not simply the set of expected values but also the set of all possible values which can be derived from failed devices providing input e.g. sensors. A failed altimeter could give a totally unexpected and unrealistic figure which, if the circumstance had not been envisaged, could immediately cause a program to crash.

c) It is impossible to log and check all output values for a system of any great complexity for all possible input values.

d) This form of testing in any real time application will require a large and expensive test facility to emulate all of the inputs and log all of the outputs. It will have to cope with all of the different phases of testing.

In fact, when considering the above points, the difficulty of proving the correctness of any non trivial software system can be seen. In addition, experience has shown that the cost and work load on testing facilities mitigates against conducting all of the tests known to be required.

The alternative approach is to derive information about what a system has to do in some other way. Such an alternative is static analysis, the technique of examining a program or design in depth to establish what it is doing. Work has been in progress for some years in this area and this paper describes the outcome of one line of research which has been conducted at the Royal Signals and Radar Establishment at Malvern, UK culminating in the development of the MALPAS software suite.

MALPAS

History of development

A brief mention was made in the previous section of the MLAPAS suite of software as a static analysis tool. The initial work in this area was done by Dr. C.S.E. Phillips at the Royal Signals and Radar Establishment in the mid 1970's when he developed the concept of using graphical representations of programs as a basis for analysing their structure and purpose. Initially, he developed the basic ideas and then proceeded with other organisations, notably the School of Signals at Blandford and Southampton University, to develop some programs which were able to construct the necessary graphical representation of a program and provide an analyst with details about the behaviour of that program.

Over the years, the early theoretical work has been developed into suite of programs known as MALPAS. A great deal of this later work has been done by Dr. B Bramson of the RSRE and he continues to be actively involved in further developments. However, towards the end of 1985 a Company, Defence Technology Enterprises, was formed with the specific task of taking relevant Ministry of Defence research projects and transferring them into industry where they could be more fully used and exploited. The first such project was the MALPAS suite and, after a competitive tender exercise, Rex, Thompson & Partners was granted a world wide licence to market and develop MALPAS.

The present situation is that Rex, Thompson & Partners, with the assistance of RSRE, are marketing and developing MALPAS. Many improvements have been made, a comprehensive set of user documentation has been produced and a complete support organisation set up. Work is now in progress to make MALPAS usable with as many different computer languages as possible.

Brief description of purpose

The purpose of static analysis is to validate software and verify that it meets its specification or intended function. One application of static analysis is to assist software certification once development is complete. The technique also has an important role, however, in the design and development itself. If software designs and modules are subjected to analysis as they are produced there is a potential for considerable savings by detecting faults earlier.

As has been briefly discussed, static analysis is different to dynamic testing. Static analysis does not execute the software at all, instead it makes use of mathematical techniques to reveal the structure of the software and the functional relationships within the software. The ultimate aim is to compare these functional relationships with the original software specification in order that the software may be 'proved' to perform its intended function.

It is not claimed that static analysis can totally replace conventional testing and the two approaches are to some extent complementary. However, its use during code development can considerably reduce the duration and costs of the test/fix/test approach to code production and it can also identify errors which would be difficult or extremely expensive to isolate using conventional testing techniques. A suitably planned testing strategy using both dynamic testing and static analysis techniques provides the greates probability of trapping errors with the emphasis being on the use of static analysis during the checking of individual modules.

The information provided by static analysis is diverse. Some relates to the quality of the code, providing information about such aspects as the presence of loops with multiple or no entries or exits, totally unreachable code, badly structured code etc. Other information provided includes detailing which variable are input variables, which are output variables and how output variables relate both to input variables and the flow of control through a program. Beyond this basic information, it is possible to derive information about the mathematical relationships between outputs and inputs. These relationships can be compared with the program specifications manually or, in the case of MALPAS, automatically using one of the later analysers to be developed.

The last point brings us to the fundamental requirement mentioned earlier, namely the need to have an unambiguous specification available against which the software may be checked. Indeed, if the use of static analysis does nothing else but point to the inadequacies of a specification, a great deal will have been achieved.

More detailed description of MALPAS

Overall MALPAS structure

MALPAS was developed from the basic premise that it had to be expandable to allow new or improved analysers to be incorporated as the theoretical work provided a basis on which this could be done. Another basic premise was a need to ensure that it was usable with as wide a variety of programming languages as possible Consequently MALPAS has been developed as a set of analysers all utilising a common file structure to contain the basic information about a program and its graphical representation.

The basic input to MALPAS is the program to be analysed and this is presented in an Intermediate Language developed specifically to meet the requirements of:

a) facilitating translation from as many programming languages as possible.

b) being capable of modelling the program in a manner suited to static analysis.

Currently, there are six main analysers available as follows and these are more fully described in later sections:

Control Flow Analyser.
Reveals the topographical structure of a program.
Data Use Analyser.
Checks that data is correctly used within a program, for example, that input variables are correctly read and output variables are written as intended.
Information Flow Analyser.
Identifies, for each output, the input variables on which it depends.
Semantic Analyser.
Describes the mathematical relationships between output and input variables for each path through the program.
Partial Programmer.
Partitions the program into sub-programs to allow semantic analysis to be performed for specific output variables.
Compliance Analyser.
Automatically compares the program with its specification and reveals any discrepancies.

In addition to the above analysers there are also some basic routines for reading the program, converting it into the internal graphical representation and going through various algorithms to reduce this graph to as simple a structure as possible.

Control Flow Analyser

This analyser reveals the graphical structure of the code by representing the program as a series of arcs between nodes. From the description of the nodes and their successors, simplified where possible, it is possible to construct a directed graph as illustrated below.

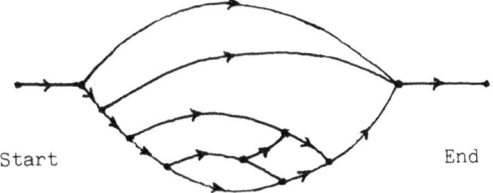

Start   End

Note that in practice the above graph would be annotated by node numbers. The graph shown indicates that there are no undesirable features sucu as unreachable code, false entry points or dynamic halts. However, the graph shown above is more complex than would be desirable, indicating that the code is not well structured.

Data Use Analyser

The Data use Analyser checks how data is used, for example, that input data is correctly read and output variables are written as intended. The use of the analyser can reveal aspects such as failing to write all output variables on all paths through a program. This would be a valuable and easily obtained indication that a program does not perform as intended.

Information Flow Analyser

This analyser identifies the input variables on which each output variable depends. It may be used to confirm that output variables are both dependent on all the input variables specified and also that no output variables have a dependency on any input variables that are not specified. For example, confirmation could be gained that an aircraft throttle opening setting is only controlled by the pilot's hand throttle and a safety override on airspeed, but that there is no dependency on cabin air temperature.

The Information Flow Analyser and the two analysers described previously can be used in a rapid, straightforward manner to provide valuable information about a program.

Path Assessor

The Path Assessor is not an independent analyser. Its function is to simply count the number of syntactically feasible paths through a program. This information assists the analyst in making fundamental decisions about the extent to which a program has to be split before it is feasible to use the two more complex analysers described later.

Semantic Analyser

This, together with the compliance analyser, is the most powerful of the static analysis techniques provided by MALPAS. The Semantic Analyser describes the mathematical relationship between program inputs and outputs for each semantically possible path through the program. Hence, for the whole range of program input variables, it reveals exactly what the program will do in all circumstances. This is important not only in itself, it also provides valuable information to aid the planning of any dynamic testing by indicating what input variables need to be varied and what corresponding output variables need to be checked.

It should be noted that the analysers which have been previously described all operate at a syntactic level. This means for example that any deductions about the program will have been based purely on finding a conditional statement. No account will have been taken of the contents of the variables which determine the branch of the conditional statement. No account will have been taken of the contents of the variables which determine the branch of the conditional to be taken. However, in some cases it is possible that values generated result in some paths being semantically impossible. The semantic analyser does consider the contents of the variables concerned and also incorporates an algebraic simplifier which has many of the basic rules of algebra built into it. In addition it can be steered to some extent by providing 'Replacement Rules' which effectively add to its knowledge about a particular situation. This allows the simplifier to reduce some expressions further than would otherwise be possible.

The output from the semantic analyser is a series of statements about what the contents of output variables will be under given input conditions. The algebraic simplifier, aided by any replacement rules supplies, will minimuse the number of possible separate conditions which will be reported.

The semantic analyser is particularly suitable for detecting subtle coding errors but can also reveal semantically possible paths through a program of which the programmer was unaware. In such cases the program may do something quite unexpected over a small range of input conditions, perhaps even outside the design range for the system. It is these errors which are particularly difficult to trace using conventional testing methods but which can occur in service, sometimes after years of operational usage.

Compliance Analyser

The Compliance Analyser may be regarded to some extent as a variant of the Semantic Analyser which automatically compares the results of the analysis with a formal specification. The hoped for result is a statement to the effect that there is no 'threat'. This may be taken as an indication that the program meets the provided specification. Naturally, it is still not a statement that the program is correct. For that to be so the specification has to be totally complete, correct and unambiguous.

The Compliance Analyser will often not be able to make the unambiguous statement that there is no threat, instead it will provide an expression which effectively states that, under a given set of circumstances a given set of output variables will take a stated set of values. It is incumbent on the analyst to:

a) Check that the statement detailing where the Compliance Analyser thinks there is a problem can actually be simplified by hand to confirm there is no danger.

b) Consider if any replacement rules could be relevant which could have the effect of aiding the algebraic simplifier such that it can deduce that there is no threat.

Carrying out the first step often leads to the second.

Although the use of the Compliance Analyser sounds quite complex, it has been found straightforward in practice. This is particularly so for code which has been written in a well structured manner from a sound, complete specification.

Analysis techniques for large programs

There are three techniques to aid the analyst in handling large programs. These are adopting a bottom up approach, making use of the Partial Programmer, and the use of node marking.

The bottom up approach to the analysis of a program requires the lowest level procedures to be analysed first. Details about what they do is taken from the MALPAS output and fed back to enable the next level to be analysed and so on. The Intermediate Language has been specifically designed in part to make such an approach particularly efficient.

The Partial Programmer provides a means whereby an analyst can perform semantic analysis on complex program in logical sections. This enables the problem to be reduced to manageable portions such that it is possible to assimilate the meaning of all the information provided by MALPAS.

The use of node marking provides an alternative means of splitting a program into sections such that the analyst need only consider one section of a program at a time. Again, IL features are provided to facilitate this approach.

Translators

All of the foregoing discussion has assumed the existence of an IL version of the program. No mention has been made of the vital step of taking the original source code in some normal high language and producing the equivalent IL program. To overcome this problem various translators are available or have been developed.

The task of translating from one of the commonly found computer languages into MALPAS IL is not too difficult. Indeed, one of the options to be considered when presented with a language for which a translator has not yet been provided is that of doing the translation by hand. If the quantity of code to be considered is not excessive then this is a perfectly viable approach. There will be the obvious problem of ensuring the correctness of the translation but it must be remembered that the function of MALPAS is to find errors. Consequently the first thing to be investigated if a problem is revealed is the correctness of the translation. However, it is intended that the number of languages supported by MALPAS will increase and should include many of the languages of interest to the producers of high integrity systems.

The languages currently supprted are as follows:

Lanugage	Availability	Comments
CORAL66	Now	Supports most of language
8080 Asm	Now	Restricted capability
8086 Asm	Now	
VAX Macro32	Now	

In addition, the following languages are either being developed or work is scheduled to start on them in the near future:

Pascal	Sept 87	Supports most of language
PLM86	End 1987	
Ada	End 1988?	

Finally, we have had considerable interest shown in the following languages and we would be very interested in finding means by which suitable translators could be resourced:

C
RTL/2
Fortran
Cobol

The need for these translators is either because the languages are in widespread use in high integrity applications or have been specifically requested by potential users. It should be stressed that high integrity applications includes many systems such as Fund Transfer Software etc. Critical systems are not just control applications where an error could result in loss of life.

It must be stressed that it is not difficult to produce a translator, the IL is well defined and there is often a compiler available which can be adapted. In some respects the problem is less than that of producing a normal compiler as there are no storage allocation problems or optimisation requirements. Indeed, the latter is positively detrimental as it could remove information of value to the analyst. On the other hand, any translators produced must provide as much support for the analyst as possible. His function is to analyse the source code given to him, not the translated IL. The IL generated must therefore be constructed to make it straightforward to relate back to the source code. Translators

221

must be designed with care and the design of the source language to IL mapping is crucial to the final efficiency of the analyst.

The use of MALPAS as a design technique and tool

I have concentrated so far on the use of MALPAS in the role of verifying software at the code level as produced by 'normal' programming languages. However, MALPAS can be applied to the other phases of a Software System development cycle. With a tendency to apply more formal methods to the requirements capture and system definition phases of the cycle and to develop techniques for generating code mechanically from the results of these formal methods, it obviously becomes relevant to apply MALPAS to these earlier parts of the cycle.

Currently, I feel there is no value in attempting to apply MALPAS to the requirements capture phases. This work is too subjective and involves a high degree of interaction with the user. Assistance in such areas is best provided by a database system which logs the requirements as they are formulated and checks that they are consistent and complete.

However, there are two ways in which MALPAS can be applied in the system specification phases. One is to produce a translator for some of the system definition methodologies, the other is to actually use MALPAS IL as the methodology itself.

Considering the application of MALPAS to existing methodologies, one finds they fall into two main classes. These are the highly formal mathematical languages such as VDM, Z AND OBJ and the standard formalisms for expressing specification concepts such as JSD, JSP and SSADM.

As MALPAS IL actually incorporates some ideas from VDM in the language one could expect the production of a translator to be perfectly feasible. Indeed, experiments have been done with hand translation from VDM and one would expect Z to also be amendable to translation. In particular, a complete project has been done where OBJ was used to specify a system. The specification was then mapped onto MALPAS Il, a fairly trivial task, and verified with the Semantic Analyser. Finally, the IL was hand translated to Coral66 which was itself verified.

Less formal techniques would also appear to be amenable to translation. Such techniques include Michael Jackson JSD and JSP and work is in hand to investigate the feasibility of using MALPAS to verify designs developed using these techniques. There is no fundamental reason why the other methodologies should not be similarly investigated.

MALPAS IL itself has been used as a specification and design language and the IL has features which support such usage. These features are:

a) The ability to define abstract data types and refine their definition at a later stage in the specification.

b) The ability to define a procedure in two stages, one which gives its specification and one which defines its body.

c) The ability to define functions and state their purpose via replacement rules which direct the actions of the algebraic simplifier as used in semantic and compliance analysis.

Using these features a system specification can be developed in a top down manner. The initial stages would develop the top level functions and data types. These would then be refined, resulting in a need to define further data types, functions, procedures etc. This process would continue until one had a complete specification of the required system. MALPAS would be used at each stage to verify the work done thus far and also to indicate what needs to be done at the next stage. Finally, one reaches the point where object code could be generated from the IL.

MALPAS IL has already been used successfully as a design language in its own right and work is going on to investigate the feasibility of integrating MALPAS and other design methodologies.

Conclusions

Although it has not been possible in the spece of this paper to describe the full pwoers of static analysis, I hope I have been able to indicate its potential, particularly when incorporated into an automatic software package like MALPAS. MALPAS has already been successfullly used in Software Systems where it has revealed significant errors which could not have been detected using traditional testing methods. Work is in hand to increase the scope of MALPAS further and to increase the range of source code to IL translators.

Mention has been made of the various possible areas of application for MALPAS, and it must be stressed that although the main use to date has been in the verification of conventionally produced code, developments taking place will encourage and support MALPAS in other phases of the software development cycle. Other areas of possible application include the checking of designs employing digital logic. Finally, although MALPAS is currently limited in its capability for investigating some aspects of real-time software, concurrency problems for example, research is being done to find a way to solve this problem. It has already been shown that timing analysis of code is possible and it is planned to include an analyser to provide this function in the near future.

NOTES ON THE SECURITY OF
PROGRAMMING LANGUAGES

B A Wichmann
Division of Information Technology and Computing
National Physical Laboratory
Teddington, Middlesex TW11 OLW
UK

Abstract
High level programming languages are being used increasingly
in systems which are safety critical. Unfortunately, the
specification of such languages does not provide the level
of security that these applications require. This paper
analyses the languages Pascal, Ada and Modula-2 and
identifies features that should be avoided when writing
critical software.

Introduction

All modern software is produced using programming languges. Low level
languages, such as assemblers, allow great freedom to the programmer which
can too easily lead to disaster - not just causing the program to produce
incorrect results but to do unpredictable things. It is therefore reasonable
to ask if high level languages can ensure that such loss of control can be
prevented. The purpose of these notes is to address this issue with
reference to three programming languages - ISO-Pascal [3], Ada [1] and
Modula-2 [2]. These three languages are chosen because of their use in
safety-critical systems. The possibility of defining new languages or
subsets of the three languages above to improve security is also considered.

What is security?

The aim in writing safety-critical software is firstly to produce an
accurate specification and then to ensure that the program meets that
specification. Unfortunately, ensuring correctness of anything more than
very small programs is currently beyond the state of the art, and hence we
must ask for something less. By 'small' is this context, we mean programs
not exceeding around 500 lines. Larger programs might be capable of analysis
given a simple structure and lack of complexity in the specification; this
issue is handled below.

The problem with asking for less than complete certainity of correctness is to characterise the security in an effective manner. Exactly what is achievable will depend upon many factors other than the programming language, for instance:

1. Are the instructions stored in ROM and hence impossible to overwrite?

2. To what extent is it possible for the software to detect malfunction and hence perform some recovery?

3. Given a recovery strategy, does the application require recovery of the 'state' (which may have been corrupted)?

4. It is possible for the software to get into a position in which it will not respond to external events?

The importance of the above questions indicates that the programming language can only perform a minor role in establishing acceptable characteristics of safety-critical software. Nevertheless, the following questions can be asked of a language (based upon experience with the issues involved):

L1: How feasible is it to show that a program cannot lose control by jumping to an arbitary store location?

L2: What language features effectively permit an arbitary store location to be overwritten, and how can potentially dangerous overwrites be avoided?

L3: To what extent is the language defined so that questions concerned with the behaviour of programs can be effectively answered?

L4: Are there procedures for validating processors for the language, and to what extent can they ensure correctness of such processors? By the term 'processor' we mean not only the compiler, but the linker, downloader, or whatever additional software is needed to run the high-level language program.

L5: Are there constructs in the language for the detection of dangerous conditions? For instance, can one check that a pointer is not NIL?

L6: Given that the software detects a malfunction, what mechanisms exist in the language to facilitate a recovery?

L7: To what extent could superior security be achieved by restriction to a subset of the language?

L8: Are there resource allocation problems which could cause a program to cease to function?

Answers for the chosen languages are given below together with some conclusions.

Complexity in programs and in their specification

The reliability of programs rapidly decreases as the complexity increases. For instance, no operating system or compiler is known to be bug-free, while many small programs and algorithms have been used for ten years or more without error. Unfortunately, the measurement of program complexity is an art rather than a science. (The author does not support the claims made by Halstead [14] which aims at measuring program complexity.)

There are particular aspects of a program and its specification which make it inherently more complex and thus potentially less reliable. Some of these factors are well known:

1. Interrelated data.
   A typical relationship might be that A and B have the same sign. If the correct functioning of the program depends upon this relationship, then this must be checked. However, such a relationship is hardly likely to be universal, since when A is changed, B cannot be changed at the same instance. A particular case with high-level languages is the relationship between local and global data. If within a subroutine, only local data is used, and the parameter mechanism is the only method used to access global data, then the relationship between the data becomes much easier to handle.

2. Interrupts.
   It is impossible to produce reliable programs using interrupts unless some discipline is followed. Serious consideration has been given to removing interrupts from much safety-critical software.

3. Loops without fixed bounds.
   If a loop is repeated a fixed number of times, then it can be 'unwound'. Hence such loops do not cause any complexity in analysis. The general form of loop can cause problems; indeed, it may not be clear that the loop terminates. (Even without loops, a worse case analysis may be needed to determine adherence to a response time requirement.)

4. Goto statements.
   In certain circumstances, goto statements cause difficulties in the analysis of programs. Labels as the destination of goto statements should also be avoided if possible, since they confuse the analysis of the flow-control in a program. (Debuggers which require the insertion of unnecessary labels are to be deprecated.)

5. Arrays.
   Arrays permit random access to their component variables. In fact, arrays are usually accessed in sequential order. In program analysis, one typically needs to verify that an array is used in a simple fashion, including showing that the indices are in range.

6. Pointers.
   Pointers come in two flavours in programming languages:

references used for parameter passing, and explicit pointers to storage which may be controlled directly (a heap in Pascal, Ada and Algol 68). Explicit pointers cause numerous problems and should be avoided if possible. References used for parameter passing are usually safer, but have dangers requiring checking.

7. Other data structures.
Apart from pointers and arrays noted above, complexity in other data structures gives rise to problems. For instance, with a record, it is usually possible to refer to a component or the entire record. Hence any checks on a component must be applied to implicit use of the component via access to the entire record.

8. Floating point.
Unlike integer arithmetic, floating point is machine dependent. To obtain the same confidence with floating point as with integer arithmetic is a research topic. Advice on the numerical stability of algorithms should be obtained, as well as checking the correctness of the floating point unit in use.

9. Recursion.
Recursion is rather like looping, but with the additional hazard that further storage is required. Program analysis can sometimes be simpler with recursion but, in general, is more complex.

10. Tasking.
Proving programs correct that use tasking (or concurrency) is a research topic.

11. Program length.
Program length in itself does not seem a concern. However, an increase in program length combined with any of the above may make program analysis infeasible.

All the above points have been listed in terms of the program rather than its specification. One can work backwards from the above list to determine the complexity of the specification. Let us say that interrupts are to be avoided. Simplicity would therefore result in a program with a single polling loop. Hence the specification should allow for such a strategy, i.e, a response time long enough for such an implementation strategy.

ISO-Pascal

The ISO-Pascal standard was agreed by ISO in 1983, several years after the language came into widespread use. In consequence, many processors implement a language based upon the previous (informal) definition of Pascal produced by Professor Niklaus Wirth [4]. Nevertheless, processors for the ISO standard are available for most computing systems.

The answers for ISO-Pascal to the questions L1-L8 are:

L1: The only construct in Pascal which potentially allows a

program to jump to an arbitary store location is the case
statement when the processor does not check the
case-expression. Most processors make this check, and
restricting oneself to those that make the check would be
adequate.

L2: An arbitary store overwrite can be performed by writing to
an 'array' element when the index-expression is unchecked,
and by writing via an incorrect pointer. These two
situations need to be considered separately. For an array,
almost all Pascal processors support bound checking, thus
ensuring the overwrite is legal. For pointers, the position
is far from satisfactory; pointers are not initialised and
even if the Pascal system attempts to initialise store,
variant records can be used to subvert any checking. Also,
the use of dispose causes all pointers pointing to the
disposed item to be illegal - effectively impossible for a
processor to enforce. The heap itself can also be corrupted
by the incorrect use of the long forms of new and dispose.
Some of these points are illustrated in Figure 1.

Figure 1

```
var
    p: ptr; { ptr is some pointer type }
    rec: record
            case tag: boolean of
               true: (x: ptr);
               false: (y: integer)
          end;

begin
p^ := ...; { p is undefined and hence may write
             to any storage! }
rec.x := ...; {some legal value}
rec.tag := false;  { rec.x is now undefined! }
rec.x^ := ...; { again perform write to anywhere! }
```

L3: The ISO standard is well defined (probably better than any
other ISO software standard). The semantics of a Pascal
program is well-defined unless an 'error' occurs, i.e, a
run-time non-conformance to the standard. Validated
processors are required to state which errors are detected.
The detection halts the execution of the program.

L4: NPL was responsible, with the University of Tasmania, for
producing the validation suite for Pascal. The validation
service is run by BSI (QAS) and is licenced by BSI to the
USA, France, Germany, Italy, China and Japan [7]. The
validation procedures attempts to ensure that illegal
programs do not compile (unlike the FORTRAN and COBOL test
suites). It is not yet feasible to prove the formal
correctness of anything as complex as a Pascal compiler.
Validated compilers do have bugs, but rarely serious ones.
The validation suite is updated annually to reduce the risk
of bugs.

L5: The Pascal type definitions, particularly enumerated types and ranges, permit simple detection of many errors, hardly possible in a language like BASIC.

L6: There is no way within the language of recovering from an error condition. This could be a significant flaw in the use of Pascal in many contexts, since one may be forced to show that a safety-critical program in Pascal is free of run-time 'error' conditions. Proving this would require tools similar to SPADE [5] or MALPAS [6].

L7: The removal of variant records from Pascal would make the language more secure (it is a simple static check to ensure they are not used). If pointers are used, then quite severe restrictions would have to be imposed. For instance, one should ensure all pointers are initialized, and that no assignment is made to a pointer involved in the with-statement.

L8: Apart from files, the two resources in Pascal are the stack and the heap for data storage. An analysis of the procedure calls within a program will allow one to estimate the stack space required unless recursion and procedure/function parameters are used. A tool can check for the absence of both of these without too much difficulty. If a heap is used, then checking that the heap space will not grow too much is very hard. The only simple restrictions to apply would be not to use new (or dispose) within a loop. Since an implementation can 'implement' dispose by means of a dummy, one would need to investigate the particular processor being used to see if it was safe to use the heap.

Conclusions

Full ISO-Pascal is not suitable for safety-critical software since store-overwrite cannot be controlled in practice. A safe subset can be defined, but this still has the severe drawback that recovery from a run-time error is not possible. Proving that a Pascal program, written in a safe subset is free of run-time errors is only feasible for small programs using such tools as Spade.

Ada

The Ada programming language was designed for the Department of Defense for 'embedded' systems and is therefore targetted on an application area similar to safety-critical systems. The design aims of Ada are explained in a rationale written by the design team and published recently [15]. The language has been standardized by ISO recently and is accepted by NATO and the EEC.

Answers to the questions are as follows:

L1: There is no construct in Ada which allows a jump to an arbitary store location. For instance, the case statement is defined to be checked, raising CONSTRAINT_ERROR if the check fails. (Contrast with Pascal where the check is not required.) Another situation would be to call an entry to a task of a task variable that did not exist. This is also checked, since the task variable is always initialized.

L2: There is no construct in Ada which gives rise to an arbitary store overwrite. Array subscripts are checked and pointers are initialized to NIL (which is checked). The use of the low-level facilities of the language could give rise to security problems, see L7 below. Optimizing Ada compilers will omit unecessary checks and hence an obvious concern is whether any such optimization is correct, see Figure 2.

Figure 2

```
type Person is record
                Age:...
                end record;

type PersonId is access Person;

Peter: PersonId := new(...);
Class: array( 1..10 ) of PersonId;

begin
if Peter.Age < 21 then
            -- compiler need not check for nil
    ...;
end if;

for I in 1..10 loop
  Class(I) := ...;
        -- compiler need not check I is in range
end loop;
-- pragma SUPPRESS could be used 'safely' here
```

L3: The Ada standard is written (very carefully) in English. The complexity of the language is nevertheless such that ambiguities do arise in the more complex features. Around 500 queries have been posed to DoD on Ada, compared with only 20 to ISO for Pascal. Very few of these problems are of practical significance, and none seem to be relevant to safety-critical software. The main problem area arises with 'erroneous' programs: these are illegal programs which do not have to be rejected by processors. The language manual makes the status of these programs clear; the problem is to avoid writing them. The initialization order of library packages is one such issue, to which no effective solution is available.

L4: The Ada validation suite has been developed for DoD by SofTech and the procedures developed by IDA for DoD. The tests are very similar to those for Pascal, but the increased size of the language means that the comprehensiveness is not as good as for Pascal. Validated Ada compilers are currently less reliable than validated Pascal compilers. It will be 2 or 3 years before the Ada procedures/suite catches up that currently available for Pascal.

L5: The Ada strong type rules are broadly similar to Pascal and hence provide good checking. The handling of variant records in Ada is much more secure than Pascal. This means that various tricks to subvert to type system which work for Pascal are not possible with Ada. Any subversion of the type system in Ada has to use the low-level features, see L7.

L6: All run-time malfunctions in Ada result in the raising of an exception. Recovery is possible by means of exception handlers.

L7: The answers given to the rest of the questions here assume that the low-level features have not been used. In a sense, Ada already has a subset defined - the language without address clauses, UNCHECKED_CONVERSION, UNCHECKED_DEALLOCATION and without calling routines written in other languages. For safety-critical applications, this is the only sensible way to use the language. The checks that an Ada system is required to perform, such as ensuring that on dereferencing a pointer it is not NIL, can be suppressed by means of a pragma. We assume here that this pragma is not used. All these low-level facilities in Ada are highly visible in the source text - a deliberate choice in the language design. Hence management of these features should be simple, indeed compilers may require explicit permission for them to be invoked.

L8: The resources required by Ada are the stack and heap space for each task. If the storage for any task is exhausted, the exception STORAGE_ERROR is raised. Calculating the storage used by an Ada task is not easy; tool support is needed here. Currently, some Ada compilers contain errors which cause STORAGE_ERROR to be raised at unpredictable times.

There is some confusion concerning Ada subsets. The US Department of Defense copyrighted the name Ada to refer solely to the full language. This policy has prevented dialects of the language developing, something which for other other languages has been a pitfall to portability. Validated Ada compilers must accept the whole language - as least in the mode in which they are validated. However, an Ada development environment could call the compiler in a special mode so that the unsafe features listed above are inaccessible. This facility could then be used by management to control the use of the unsafe low-level features. Tools need not process the entire language. Program proving tools would impose severe restrictions, and in some cases may require the addition of specification information in the form of comments (see the language Anna [12]).

Conclusions
The language features of Ada make it a much more attractive language than ISO-Pascal. It is easy to avoid using the low-level features of the language which could give rise to security problems. Currently, the reliability of compilers would be a serious concern. The size of the language is such that tool development is much more expensive. Ada looks like being a good choice for safety-critical applications in the future - assuming some of the problems noted here are overcome. (The recently announced Ada evaluation service [16] should contribute to the reliable use of Ada.)

Modula-2

The Modula-2 programming language was designed by Professor Niklaus Wirth as a successor to Pascal. The important additions to Pascal are modules, information hiding via opaque type, and a simple form of concurrency. The language is currently being standardized by BSI for ISO. A draft standard is expected in 1988.

Answers to the questions are as follows:

L1: Modula-2 contains procedure variables. Hence the call of a 'procedure' via an unset procedure variable would be unpredictable. Otherwise, the position is the same as for Pascal. (Case-statements could be unchecked; it depends upon the processor in use.)

L2: The situation here is the same as for Pascal but with one small improvement; tagless variants may not be allowed in ISO-Modula-2.

L3: BSI have agreed to specify the Standard in VDM. This will give a firm basis for program proofs and for ensuring that there are no semantic gaps in the language. This work is part of an Alvey project (with BSI, NPL and the University of Leicester [8]). It is hoped that the experience with both Pascal and Ada can be built upon to give a superior position in determining the semantics of any program. For instance, the VDM will be used to derive a model of the semantics of a program, which will be used to check the validation suite and could be used to check individual programs (but it will be much less efficient than a true compiler).

L4: A validation suite of a high standard already exists for the language. It will be necessary to ensure this conforms to the ISO standard (when agreed) and to base a validation service upon this. BSI (QAS) is committed to do this.

L5: The type system for Modula-2 is essentially the same as Pascal.

L6: There is no facility defined in the language to allow the user to specify the recovery action on detecting an 'error' (i.e, run-time malfunction). It this sense, the language is the same as Pascal. However, the intention is that the standard will allow an extension to permit such recovery and that at least some processors will implement some form of recovery (since Modula-2 is designed for 'embedded' systems for which such recovery is important.)

L7: A safe subset should ensure initialization of procedure variables and opaque variables, as well as pointers. The exclusion of variant records is also essential. There is a feature corresponding to the UNCHECKED_CONVERSION of Ada to be avoided as well.

L8: The resource allocation problem is easier to control than with either Pascal or Ada. The space allocated for the heap and tasks is under programmer control. The price paid for this is that low-level features of Modula-2 have to be used for allocation, and these are essentially dangerous. Hence to provide a secure Modula-2 system, it would be necessary to validate the space allocation routines.

Conclusions

It would be difficult to use the safe subset of Modula-2 for a real- time system. A more reasonable approach would be to make limited and very careful use of the dangerous features. Security can sometimes be increased by use of such facilities - for instance, by inspecting the stack pointer, one can ensure that the storage space resource is adequate. Modula-2 is a very much simpler language to compile than Ada and hence it is feasible to consider trying to produce a formally verified compiler. Leicester University would be interested to produce such a compiler after the current project to define the language is completed.

Other Languages

The programming research group at Toronto has introduced the concept of 'faithful' to imply that every program accepted by a compiler is legal and has defined semantics if it executes to completion. A faithful language would be nearly ideal for safety-critical systems, but there are drawbacks. Firstly, the execution must be performed 'with checks' which could place unacceptable overheads on the execution speed - for example, the need to test every variable access to ensure it has been assigned a value. Secondly, to gain confidence in the program, one needs to execute the code very comprehensively; we know that this is hard from current practice. Nevertheless, the ideal of faithfulness is a target against which any language can be judged - see the answers to L3 above.

The Toronto group has designed a language which is faithful; it is called Turing [9] and has a comparable expressive power to Pascal. An alternative approach is to use a subset of an existing language to try to obtain the same effect. An example is Spade Pascal [5], although other goals are present in the language design.

There is an essential conflict between safety-critical software and ordinary programming-in-the-large principles. In safety-critical systems, nothing should be hidden, while information hiding is important in conventional large systems. The way round this dilemma is to provide tools for proving parts of a program so that its internal characteristics can be ignored. For instance, given a sort routine, a tool should be able to verify the routine from its specification and store the proof and conditions in a data-base. Then every call of the sort routine would invoke this proof including checking any additional pre-conditions required of a call.

The programming language "C" cannot be recommended for safety-critical software. There are several reasons for this. Firstly, there is no proper definition - the only effective current test is to try it on the Portable "C" compiler which is part of UNIX. (This will change when the standardization process is complete and compilers have been changed to accept the new language - a process that will take some years.) Secondly, the language does not have strong typing and hence is open to obscure tricks. These tricks are not apparent from the source text, unlike in Ada. Thirdly, the form of the language does not aid readability, which is vital for independent validation and verification.

The programming language NewSpeak, designed by I Currie [10], is an attempt to provide a system for safety-critical software. The language aims at being faithful, provides even stronger type structures than Ada, and helps program proof methods. The semantics of the language are defined mathematically, like the future ISO Modula-2 standard.

The problem with any new language is not just the language itself, but all the support that it requires: compilers for all reasonable machines, textbooks, support environments and tools. In the future, tools are likely to be the critical factor. Program proof tools are language specific, and very expensive to produce, yet they are going to be essential for highly critical software.

The size of a language and its degree of acceptance are very important characteristics. If a language is large, then tool development (including compilers) is expensive. On the other hand, if a large language like Ada has widespread acceptance, then the cost of the tools can be borne by the larger market. Hence all the three main languages considered here could succeed. The current most cost-effective option is probably the use of subsets of Pascal. In the future, it could be the use of quite large subsets of Ada.

Reliability of compilers

One can produce highly reliable software with compilers which are much less reliable. This is fortunate, since all commercially available compilers known to the author contain bugs. It is not easy to judge the reliability of a compiler. Some compilers are 'mature' and have been in widespread use but are certainly unreliable. I use at home the Pascal compiler for the BBC "B" computer. I have the first release which has a few small bugs but I would recommend it wholeheartedly. One reason for my confidence in the compiler is that it is very simple and contains no optimization.

Existing validation services can be regarded as an accelerated wear test (with maintenance!). Such methods cannot explore adequately the total capabilities of a compiler. To try to improve this situation, I have developed a test program generator for Pascal compilers. The generator revealed significant defects in 3 of 4 validated compilers exercised to date with the tool. Fortunately, the bugs discovered are unlikely to arise in safety-critical software.

One long-term solution with compilers is to produce formal proofs. This would be just about possible for Pascal or Modula-2 (I believe), but is not currently feasible with Ada.

Given this essential uncertainity with compilers, the only option that should be used now with highly critical software is to de-compile the assembler/binary produced by the compiler and check that it corresponds to the source text. Some tools can aid this process.

Given the uncertainity introduced by compilers, the traditional use of assemblers for small items of software is likely to continue in this area. In fact, with very high discipline, good quality software whose properties can be guaranteed, can be produced using assemblers. Uncertainities can still arise in the processor design; indeed, one compiler supplier traced a bug reported to him to a microcode fault. One possible development to overcome these problems is that of special processor chips, see [13].

Reliability of development environments

Apart from compilers, numerous tools are used to develop software. Fortunately, many are aids whose reliability does not impact upon the final product. An example would be an editor. There are tools which are critical to the final product. Examples are: link editors, proof tools and analysers. Some of these can be quite complex and hence, given the current state of the software development process, are likely to contain significant bugs. Some validation tools are small enough for them to be applied to themselves - this is a very useful step which increases the quality of the product but is often not visible to the user. This is a general issue with software; how can a user judge its quality? No universal solution is to hand, by the developers could reasonably be registered under BS5750 as a step in ensuring that quality procedures were used to construct the product [11].

Future Developments

Interest in these notes has been shown by RSRE which is expected to result in the inclusion of this material within a Defence Standard. It is also hoped that these notes can contribute to national and international standards suitable for the civil sector. Comments on the issues raised here are therefore requested.

Acknowledgements

Dr D Schofield of NPL, Mr N Nash of Admiral Computing and Professor B Randell of Newcastle University commented on earlier drafts which have contributed to this paper.

References

[1] J D Ichbiah et al, Reference Manual for the Ada Programming Language. ANSI/MIL-STD-1815A, 1983.

[2] N Wirth, Programming in Modula-2. Springer-Verlag. 1983.

[3] ISO 7185 (also BS 6192). The specification for the computer programming language Pascal. British Standards Institution, 1982.

[4] K Jensen and N Wirth, Pascal user manual and report. Springer-Verlag, 1974.

[5] SPADE Pascal, Program Validation Limited. Southampton. 1987.

[6] MALPAS. Rex, Thompson and Partners, Farnham, 1987.

[7] Pascal Compiler Validation Service. British Standards Institution - Quality Assurance Services. Milton Keynes. 1987.

[8] D J Andrews. The formal definition of Modula-2 and an associated Interpreter. Alvey Project SE.063. Alvey Posters. 1986

[9] R C Holt and J N P Hume. Introduction to Computer Science using the Turing programming language. Reston Publishing Co. Virginia. 1984.

[10] I F Currie. NewSpeak - an unexceptional language. RSRE 1986.

[11] BS 5750, Part1. Specification for design, manufacture and installation. British Standards Institution. 1979. (Similar to ISO 9000).

[12] D C Luckham, F W von Henke, B Krieg-Breuckner and O Owe. Anna: A Language for annotating Ada Programs. Stanford University. 1984.

[13] W J Cullyer and C H Pygott. Application of formal methods to the VIPER microprocessor. IEE Proceedings, Vol 134, May 1987, pp133-141.

[14] M H Halstead. Elements of Software Science. Elsevier, 1977.

[15] J D Ichbiah, et al. Rationale for the design of the Ada programming language. Department of Defense. 1987.

[16] BSI Ada evaluation service. Preliminary information sheet. October 1987.

SOFTWARE - CORRECT BEYOND REASONABLE DOUBT?

R A Humphreys
Rolls-Royce and Associates Limited
P O BOX 31
DERBY   DE2 8BJ

Abstract

Two microprocessor modules are incorporated in a safety critical design.
The safety case for the two systems involved, makes the claim that design
error will not significantly affect the performance of the system; that
the software is correct beyond reasonable doubt. This paper discusses the
basis of this claim and presents the reliability arguments which were used
to support the evidence submitted.

Introduction

The objectives of this paper are somewhat obscured by the title itself.
The subject of software reliability although of justifiable concern, is
only a part of a general problem created by the introduction of general
purpose electronic components which are configured by the user to suit a
particular purpose. Such user configured systems whether configured by
software or alternative hardware design techniques are vulnerable to
design error. This paper is written in the context of a genuine
application, but the real issue is not the safety of that particular
design, but whether it is possible to provide adequate evidence that any
such application is safe. This paper presents an approach to the
assessment of a user configured system in which software is only one
element. Two computation units are then considered in the context of that
approach.

Two safety critical, microprocessor modules have been produced, in which a
small dedicated microprocessor system is used to compensate for
shortcomings in the output of a sensor. It is intended to present the
reliability arguments that attempt to justify the use of such units.

Scope

Within the scope of this discussion are all kinds of instrumentation and
protection systems, for it has to be acknowledged that very few sensors
are ideal. The ideal sensor would respond only to changes in the
parameter of interest and would have a linear response throughout its
working range. Some sensors depart so far from the ideal that it has been
necessary in the past, to resort to frequent recalibration and complicated
systems of manual compensation. Such systems are vulnerable to abuse and
error and are not favoured by engineers or safety assessors. By using a
microprocessor with associated software it becomes possible to compensate
for the non-linearity of sensors automatically. The incentive to use such
devices is consequently very high, and they should, in theory, result in
safer and more accurate systems.

The possibility of errors in software cause great concern to those
involved with the assessment of safety. Such errors are common mode in
their affects, destroying the benefits of hardware redundancy, and the
effect of an unknown error is almost impossible to predict. In a safety

application, it is not sufficient to provide evidence that the software is free from error, for errors in software are only one form of design error. There is an increased probability of design error in complex digital systems, in part it arises from the unknown circuits within the chips themselves. To be convincing the safety case must also address the possibility of such errors in hardware.

This paper is concerned only with the possibility of design error in a particular type of application. Hardware faults are not to be discussed. Thus the argument - while it may not be possible to prove a program correct it is possible to prove that it will not "hang-up" - is only made in the context of design error. Of course, there are hardware faults which can stop a microprocessor.

Design Errors in Software

Only those who have been engaged in the writing or analysis of software really appreciate the enormous variety of errors that are possible. Nevertheless, as it was pointed out many years ago, these errors can be divided into just two basic categories:

1 - Absolute errors, which can result in a program which is not viable. That is to say, coding errors which cause the program to abort. or perhaps to disappear into the black hole of an endless loop.
This is a program that no one would want.
It is possible to detect errors of this type by comparison with absolute standards, without even referring to the specification.

2- Relative Errors, which result in a viable program which is not correct. Errors in the specification or interpretation of the specification, which produce subtle and occasional deviations from correct behaviour.
This is a program that someone may want, and correctness can only be established by comparison with the true needs of the system.

The significance of this distinction is that category 2 errors will not stop a program from running and it is now becoming increasingly possible to guarantee that category 1 errors are not present in programs of some size. Such claims are made possible by the development of well constructed languages and the development of formal methods of examination.

Some discussions of software reliability have an underlying assumption of catastrophic failure, such failure will not occur in this application and it is more appropriate to regard failure from the viewpoint of accuracy. Calculations are performed and if the output differs from the correct value by more than the quoted accuracy the system is in a state of failure.

Note: It is possible to disappear in an endless loop in a correct program. e.g. Find the first prime number that starts with a certain combination of digits. - Such a number may not exist and the search would never cease. It follows, that to guarantee exit from each module of a program it is necessary to prohibit certain constructs in coding.

Objectives of Assessment

In the design of a computing system, it is a long and twisted path that links "that which is needed" with "that which is done", but if we are to be assured of safety we must at least consider the whole of this path.

Needs
↓
Stated Requirements
↓
Specifications
↓
Implementation
↓
Action

a - Assurance is necessary that the specification for both software and hardware properly reflects that which is actually needed, not just that which is said to be required.

b - Does the design implemented in terms of both hardware and software, meet the specifications.

c - Does the theoretically correct design perform the actions predicted.

In the real world very few things are found to be perfect, and it would be preferable if deviations in each of the above categories were quantified. Failing this it is necessary to convince an assessor that each will not be significant in terms of safety.

Software Reliability and Performance

When considering the actual performance of a computer that is defective in some respect, it is surprisingly difficult to define what is meant by failure. Various definitions are used or simply implied, and indeed different definitions are perhaps appropriate in a given context. For the purposes of this discussion we will define failure in terms of overall function. Given that a hardware fault or software error is permanently present, the output of the computer may or may not be correct. In a digital system, only the smallest fraction of components are in use at any given instant of time. The effect of a given defect will depend on when that part of the code or circuit is invoked. Both hardware faults and software errors result in unpredictable and temporary deviations of the output. Whilst there are obvious intrinsic differences in the nature of hardware fault and software error, one should not be misled into thinking that they produce a different effect. Indeed, it is possible to show an identical effect in many cases:

e.g. Consider that part of a computer memory which is occupied by code. If a single binary bit is set incorrectly it can be due to either hardware fault or software error. If is is a hardware fault it can be argued that the fault has produced a software error. Either way, the effect on the computer is identical, and the cause cannot be determined by observation alone.

Much work has been done under the heading of "software reliability" which seeks to determine the number of errors which remain in a program. Such work is clearly of great value to the producers of software, but it is not very useful to the safety assessor who wishes to know the effect of any remaining errors.

A large program will have many functions and thus many sections of code which are only used occasionally. By this means the affect of software error is constrained, often being limited to the maloperation of a single function. For such programs, the idea of degraded performance is understood. A matter of common experience. Large programs are never perfect, and numerical estimates of Reliability or Availability are meaningful in terms of everyday operation. Also, in the case of a large program, where the worst effect of a single error is limited, it can be seen that the degradation in performance will be proportional to the total number of errors remaining. On the other hand, in the case of small programs where all the code is in more or less continuous use, the idea of degraded operation has little meaning. In such a program a single error can have an effect that ranges from the insignificant to the catastrophic. In the extreme case a specification error could mean that the output was incorrect at all times. Performance is dominated, not by the number of errors, but by the nature of a single error.

If a program is less than perfect in use we may discuss its performance in numerical terms. If we tell someone that a program has no major problems, and that it is, say 99% correct, the statement would be meaningful if we were talking about a Word Processor. However, it would not make much sense if we were referring to the "addition" function on a calculator. Arithmetic functions that work most of the time are not tolerable. They are among the class of small programs which must be correct.

Broadly speaking there are two ways in which to describe the effects of failure. In terms of frequency (Reliability, MTBF, failure rate, etc.), or in terms of down time (Fractional Down Time, Unavailability, Probability of failure on demand, etc.). In a particular case, one of these measures will be relevant in terms of safety. In those circumstances where a single output deviation can cause a disastrous result we are obviously concerned with frequency. Most obvious in this category, are control applications, where an error closes a valve, or moves an aileron. Alternatively, in those circumstances where the system must respond to a demand which may come at any time, the duration of the failure state, is all important.

In general computers are very unreliable due to imperfections in software design. It is possible to identify frequent failures in any ordinary application, in so far that it is always possible to see small defects and inconsistencies obviously not intended by the designer. On the other hand computers have an extremely good intrinsic unavailability. Provided that a crash can be avoided, the down time due to deviation from desired operation is very short. It is this combination of frequent failure with a very short duration of the failure state that explains why children are able to write programs that are quite satisfactory.

Prediction or Measurement

There is considerable difficulty experienced in measuring or predicting
the actual performance of small programs, and this situation may not
improve in the future.   In all safety applications there is a balance of
both deterministic and probabilistic techniques.   For large numbers of
components, statistical methods predict the total number of failures in a
given period of time.   With small numbers of components statistical
methods are of less value.   If safety depends on a single component,
success or failure cannot be regarded as a random event,   it becomes
necessary to examine the particular component in its actual environment.
As computer programs become smaller, similar considerations may apply;
deterministic methods become more appropriate.

Prediction of Reliability Performance

After only a very short experience of computer programs it becomes
apparent to the user that there are inherent difficulties in trying to
predict the performance of a new program.   For small programs in
particular, even words like "typical" or "on average" have little meaning,
let alone the precise language of mathematics.   Without exaggerating in
any way it can be said that a small program may be correct, performing
without problems until the day the computer is thrown out.   On the other
hand, it may not work at all!   This may seem to be stating the obvious but
it is important to understand that the idea of intrinsic failure rate is
not appropriate for software.   The failure rate of electronic components
is not greatly influenced by the configuration of the circuit in which the
are installed.   The performance of software is entirely determined by the
configuration of individual program instructions.   Errors in computer
programs are a class of design errors. Apparently random departures from
the perfect implementation of a design concept result from human error in
all its many forms.   Given sufficient data it would be possible to quote
the mean failures per line of code.   This figure would be accurate and
indisputably correct.   The problem is that the confidence limits that
applied to the figure would be so wide as to make it unusable.

It therefore follows that any technique which predicts program performance
must take into account the actual program configuration.   By limiting the
effects of the many different sources of program variation, such as
language used, type of application, and by applying strict control to the
methods of production, some measure of consistency may be introduced.   If
at this stage the program is examined in detail, decisions can be made on
the relative complexity of the programming task and predictions made on
the potential number of errors.   Some success can be claimed for such
techniques and with better software metrics they will undoubtedly improve
in the future.

Measurement of Performance

One technique for estimating software performance is to measure that
performance during the design of software and predict by extrapolation the
performance of the final product.   In practice the application of this
technique is limited to recording the detection of programming errors and
predicting the number of errors remaining.   The effect of errors on the
program output is rarely discussed.   Variations on this process are
classed as "Reliability Growth Models".   It can be seen, that if
predictions are to be accurate, it is necessary that errors are detected

at a rate which is proportional to the total number of errors remaining. If our earlier assertion is correct – that in small programs, performance is dominated by the effect of errors rather than the number of errors – it follows that growth models are most suited to large programs with many different functions.

It is possible to prove that a small program is correct, by showing that the final code is equivalent to a formal mathematical specification of the program. In practice it is not feasible to extend such a proof to programs with thousands of lines, this limitation arising quite simply from the amount of labour involved. The maximum size for which formal proof can be undertaken is being extended by the increased use of formal specification languages and tools like those of Static Analysis.

On a very few applications it is possible to carry out 100% testing. That is, to test all possible combinations of inputs, and check that the output is correct. Unfortunately, on the vast majority of applications this is impractical and the following limitations apply:

a. The maximum number of tests that can be carried out represents an insignificant fraction of all possible input combinations.

b. Software errors are not distributed evenly with respect to the values of inputs. This means that any sample may not be representative of the whole population.

c. The best that can be done to establish correct output is to compare the output with that of another version of the same program. In high reliability applications there would be concern about common errors which could allow the occasional deviation to escape detection.

d. Testing may not be not done in real time, but at a fraction of normal speed. Any statistical calculation based on the equivalent of a few hours real time running would therefore represent a considerable extrapolation.

Any claims to evaluate reliability numerically on the basis of such testing can be challenged on the grounds that the test data is not representative of real time operation. Acceptable proof of reliability on the basis of testing would require speed greater than normal and high confidence in the method used to detect incorrect outputs.

Summary of Reliability Performance

The situation is complex, for while it can be stated that those applications which require good Availability are more tolerant of software error than those demanding high Reliability, it is also true to say that in a small program the effect of a single error is quite unpredictable. There is no present technique by which the Availability of a small program can be predicted or measured; neither are there good prospects for such techniques in the near future. In these circumstances the responsible designer cannot tolerate the possibility of a single error remaining in a program, where the effect of that error is unknown. The idea of degraded performance is quite invalid for such programs. There is simply no place for second class software in this context, where a single specification error can have devastating effect. To suggest that a program of unknown pedigree might be allocated an unavailability of 0.01 or 0.001 would be a gamble not a conservative estimate. Deviations of a wild and unpredictable nature cannot be allowed to dominate the performance of a safety critical system. Programs written for this purpose should be correct; if they cannot be made correct they should not be used.

It can be argued that it is not possible to be certain that any program is correct; even with formal proof of equivalence there will remain some chance of error within the proof or possible affects associated with the implementation of the language. To allow for such reservations in any real application it is necessary to moderate extreme claims. Correctness should be established beyond reasonable doubt; both designer and assessor must be confident that program deficiencies will not be significant in the performance of the complete system.

The Application

Two computation units were regarded as safety critical, they can be briefly described as follows:

1 - Neutron Power measurement

2.5K (each byte 12 bits) of machine code complete with look-up tables Several auxiliary inputs (some 2 state) enable compensation for temperature etc.

2 - Level measurement

1.25K (each byte 12 bits) machine code complete with look-up tables Takes the average of the last 32 consecutive input values

It is necessary to be convinced that both programs are satisfactory over the complete working range. Trips are preset at a number of different positions, but it should be possible to change the settings in the future without rework on the safety assessment.

(Note: Both units were programmed in assembler code, however this should not be regarded as highly significant. The project took a number of years and the decision which strongly influenced hardware design was taken long ago. A case can be put forward for using a high level language with a well proven compiler, particularly in association with the process of Dynamic Testing described in the next section.)

The Strategy Adopted

An obvious objective was to aim for software of the very highest quality, a topic which must lie outside the scope of this paper. It was also decided to adopt the following techniques:

1 - Limitations

As already mentioned it is inevitable that restrictions must be imposed on coding and design if analysis of the results is to be practical. Some of the most notable limitations imposed were:

Single processor system

Interrupts not used

Comprehensive coding standards were written to control the instructions
and structures permitted.

2 - Static Analysis

In the process of Static Analysis, a program (or more accurately a
suite of programs), is used to examine the software under test. It is
important to understand that the program which is being examined is not
executed. Static Analysis in its simplest form can be regarded as an
extension of those facilities which announce syntax errors during
program development. Such facilities can be extended so that, for
example, all the possible paths through the software can be mapped
together with the conditions for entering each section of code. The
MALPAS package was chosen for this project.

For any program Static Analysis can be regarded as a most useful tool
with which to examine code. It presents both decisions and
calculations in an alternative format, thus casting a different light
on that which is done. If coding limitations are accepted, in terms of
instruction type, constructs permitted, and overall structure, the
power of the tools are greater. It can then be claimed that all
absolute errors would be detected. These are those errors which would
cause execution to halt.

In some circumstances the process of Static Analysis can come close to
providing a formal proof of equivalence between specification and
code. However such a claim could only be made when a perfect and
independent specification is provided at the same level as each section
of code. It would not be sufficient to compare, that which the
programmer intended, with that which the programmer did!

3 - Dynamic Testing

It was decided that after formal acceptance of the complete computing
module (i.e. the hardware with embedded software), extended testing
would be carried out. In order that automatic testing could take
place, a separate version of the computation unit was implemented, but
this time entirely in software. The test rig consisted of a
minicomputer programmed in CORAL66. Test data which had been
previously prepared, was presented to the minicomputer which then
provided that data via an interface to the module under test. The
results of this parallel computation were then compared by the
minicomputer. No margin was allowed for comparative accuracy, a bit
for bit match was required.

This completely diverse form of testing was seen to be a practical
defense against hardware design error, an example of which would be
errors in the micro code of microprocessor itself.

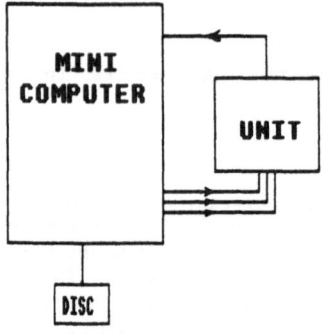

Figure 1 - Dynamic Test Rig

Referring back to the Objectives of Assessment, the role of Static Analysis and Dynamic Testing are seen to be as follows:

Dynamic Testing provides a good practical demonstration of correct operation and is unique in that it also can be seen to give assurance against hardware design errors. Static Analysis on the other hand was used in this particular case as an independent tool for assessment during the development of the software, and its role is to show the equivalence of specification and code.

A good deal of effort was devoted at the specification stage, which was recognised as a potential source of common error, although a formal specification language was not used on this project. Separate engineering requirements specifications were drawn up for unit and monitor programs. Whilst this was obviously desirable to reduce common error, it was in any case necessary in that each monitor program was required to model both hardware and software of the production unit. The monitor programs were written and Dynamic Testing carried out by one software house. A second software house provided the computation unit programs which were subjected to user acceptance tests before formal acceptance. Only then were they submitted for Dynamic testing.

The gap that remains between the true "needs" of the application and the "engineering requirements" remains. It is necessary that the assessor should be convinced that no significant error should result from an incorrect perception of what was really necessary. In this context it is only possible to suggest appropriate criteria.

Within the range of the possible applications of a computing system there
are requirements which can be stated with certainty. As in the
requirement to add two numbers together. Or, at the other extreme, those
circumstances in which development is inevitable. As in the requirements
for the first "word processor" programs. It is noticeable that there are
less errors in the design of those programs which are intended to carry
out calculations. Probably not because such programs are less complex
than others, but because people are able to use ordinary mathematical
notation in thinking about what is intended.

There is increased confidence if the requirements can be stated in
mathematical terms. The sight of charts, graphs and equations would
please any system analyst, who dreads most of all the round table
discussion as to what might be a good idea.

In most conventional electronics systems, design error has not been found
to represent a major problem. In those circumstances where a particular
design has previously been implemented using analogue techniques there is
reason to be confident that errors will not be worse in this part of the
design process.

Results of Testing and Analysis

The program was subjected to Static Analysis during development and in its
final form, using the MALPAS tools. Static Analysis is used in the early
stages of program development to confirm that the coding is of good
quality with particular regard to complexity and structure. At this stage
it is confirmed that the program can be analysed in detail by the MALPAS
tools. At the final stage the Semantic Analyser was used for detailed
comparison of specification with code.

The reports of Static Analysis are interesting and somewhat different in
character to those of other forms of testing and analysis, comments often
concern the maintainability of the program, with details of the effects of
changing the values of constants, etc.

One error was found in the "self test routine" using Static Analysis. The
program failed to test all of RAM as was specified, omitting to test one
location. This was in fact a RAM location used for the checksum
operation! It is of interest to note that this error could not have been
detected by any external test.

The sets of test data for Dynamic testing were generated separately, prior
to the tests themselves. Various types of random input sets were used
some with inputs outside range. For the majority of input sets, the
random factors used were constrained to allow each set of inputs to
represent a set that might occur in practice.

In the case of Neutron Power measurement over one million input sets were
applied without detecting any error in the hardware module or in the high
level monitor program. An exact bit for bit correspondence was obtained,
when the outputs were compared.

Over 500,000 sets of test data were applied to the Level measurement unit,
each set consisting of 32 correlated inputs. No error was detected, an
exact bit for bit correspondence was obtained, in the outputs of the
hardware module and monitor program.

Conclusions

This paper has attempted to outline the overall reliability thinking that
links a number of techniques used to support the installation of a user
configured system in a safety critical application. It was decided that
for an application of this type it was not appropriate to quote figures
for the affect of design error on performance, or proper to derive such
figures on the basis of testing. Rather, the claim is made that on the
basis of evidence submitted with regard to analysis, testing, and software
quality, design error will not significantly affect plant safety. The
software is claimed to be correct beyond reasonable doubt.

(Note: This project has taken some time, and it should not be assumed
that various techniques not mentioned were considered and rejected. For
example, subsequent analysis of specification errors has revealed that 75%
of all such errors would have been avoided by the use of VDM.)

Reference

1.   Achieving Safety and Reliability with Computer Systems - SARSS'87
     B. K. Daniels - ELSEVIER APPLIED SCIENCE

COMPARATIVE RELIABILITY
OF
SOME COMMON DESIGN AND SPECIFICATION METHODS

DECEMBER 1987

CLIFF LEACH
MANAGER, SOFTWARE PRODUCTION ENGINEERING
DIGITAL EQUIPMENT CORPORATION

INTRODUCTION

In my work in the field of software engineering, I have been
concerned with the application of methods and supporting
tools to the task of commercial (industrial) software
development. In this situation, the common theme of many
development problems centers around cost and time to market
coupled with the cost of support and maintenance. It has
been possible for me to gather a considerable amount of data
on a wide number of projects and it is from this source that
I draw the material for this paper. As the information has
been provided to me in confidence, I have stripped all
confidentiality from it and I also wish to point out that
none of the information in this document relates to Digital
Equipment Corporation. The reader will I am sure appreciate
the commercial sensitivity involved in this area.

I am, in a small way trying to answer the question "so just
HOW effective is this new Method/tool/etc" in a quantitative
way. Structured design methods are, like God, Motherhood and
Reforestation, good for the soul. However this is as near as
we often come to looking at a quantitative answer. My view
of reliability and maintainability is essentially that it
needs only to be sufficient for the intended purpose and
therefore, in the case of a sizable software system, it may
be that the requirements for reliability and maintainability
are heterogeneous, depending on the criticality of the
function or component to the end user or the system as a
whole. I have begun in this paper to look at the differing
reliability of components in large software systems and
between the systems themselves. I hope the work points the
way for further research into this most crucial area.

CRITERIA FOR COMPARISON
It is arguable that many sophisticated measures exist for
the comparative assessment of reliability of both products
and processes. However many of these measures lack a
significant amount of either empirical validation or
industrial acceptance. The job of importing methods and
supporting tools into an industrial context is therefore not
aided by the use of such measures at the moment. I do feel
that more complex and representative indicators are vital in
the long term for studies such as this but this study is
drawn from my own industrial experience and therefore
reflects the need to justify given recommendations and
courses of action to the senior levels of management of
commercial enterprises.

Given this, I have used the following simple metrics for
comparison:

a)   SOFTWARE TYPE
It is important to try to group the software by type
(OLTP/DB/OS etc) if for no other reason than to allow ones
prospective client audience to identify with the software
under question. It is also reasonable to suppose that
certain applications have certain things in common i.e. all
OLTP systems use some realtime applications, most batch
payroll systems use few or no realtime applications etc.

b)   SOFTWARE SIZE
There is no assured way to compare the complexity of large
scale software systems. However in commercial software it is
apparent that there exists some sort of correlation between
the size of the software and its complexity.

c)   LANGUAGE

The implementation language(s) of a given software system
has several implications for the project and the software,
firstly some languages (Pascal,Ada etc) are intrinsically
more highly structured than others (C, Assemblers etc). This
does not assure structured software implementation but does
considerably increase the probability that such will be the
correlation. Secondly, certain languages (Ada for example)
may come with powerful tools or compilers which in
themselves are well validated and therefore less error prone
than others. Lastly there is the "mind set" approach, the
users of highly structured languages are often people who
design in a highly structured way, the converse of this,
that C or assembler programmers tend not to approach the
problem in so structured a fashion is also true (note here
we are talking about trends)

d)   MTBF

Good old Mean Time Between Failure is the simplest
meaningful metric for software reliability. Having said this
I am aware of the very great shortcomings of such. However
even if we can't use MTBF to predict the reliability of a
specific system under a range of conditions ,we can do one
thing, predict with a reasonable level of accuracy, the
expected support traffic (and therefore cost) of a given
software system. In short, if the MTBF is falling (ie
getting smaller) then the traffic is increasing and so are
the costs of support. If on the other hand the MTBF is
rising (ie getting larger) then the traffic is decreasing as
are the support costs.

e)   MTTR

Mean Time To Repair. In software failure measurement, MTBF
tells us a lot about the likely costs of support.
Maintenance on the other hand requires us to predict not
only how many faults require fixing but how expensive each
one is to fix. (less therefore does not imply cheaper to
maintain). MTTR is the measure of how our average repair
time (and therefore cost, since my measure is in Man Weeks)
looks over a given period.

THE DATA
In the following section I include data from several
projects based on the preceding metrics.

Project A
This consisted of: 1.2 Million Lines Of Deliverable Source
Code (LODSC), written in Fortran 77 and Macro. The code was
entirely hacked together by a reasonably experienced group,
was unstructured, undocumented and barely commented and
little or no buddy checking. Essentially this was a
centralized DB system.

Project B
This consisted of: 1.8 Million LODSC written mainly in C but
around 20% in Fortran 77. The code was much more structured
and well documented and commented but little buddy checking.
It was built by a similar group to A and was a centralized
DB system with some TP capability.

Project C
This consisted of: 970K LODSC written entirely in Pascal,
designed using Yourdon SAD. The code was highly structured,
well documented and buddy checking of the source was
extensive. This was a TP system with extensive external
comms to other heterogeneous systems and was fully
distributed.

Project D
This consisted of: 1.3 Million LODSC Written entirely in C
and designed using Yourdon SAD. Full Fagin Inspections were
used on all code and design and the code was highly
structured except where performance optimization required
otherwise. DEC's Language Sensitive Editor (LSE) was used to
ensure structure in the code. The system was a distributed
realtime OLTP system.

Project E.1
This consisted of:1.1 Million LODSC Written in C and
including some AI applications and developed exactly as D,
by the same group as their next project. Since there were
very significant reliability implications for this system,
the core which handled not only the transactions themselves
but full journal and security facilities, was developed as a
separate project (E.2)

Project E.2
This core system was entirely designed using Yourdon SAD but
specified formally using Z. It consisted of 102K LODSC and
full Fagin inspections were used. The software was written
using C mainly by new graduates.

THE RESULTS

MTBF HOURS	CPU HOURS X1000							
	0.5	5	10	15	20	30	40	50
A	9.2	8	10.1	11	13.2	11.8	19	39
B	11.2	14	18	31.2	40.7	40.9	53	90
C	21	60.7	87	87	123.2	212	253	371
D	27	98.6	109	200.1	221.3	286	391	404
E.1	57.9	111	111	246	272	398	437	591
E.2	$1*10^3$	$2*10^3$	$2*10^3$	$2*10^3$	$2*10^3$	$2.5*10^3$		

TABLE 1 MTBF BY SYSTEM

MTTR (man weeks) AND TEST	DIAGNOSE	REPAIR	INTEGRATE
A	6.9	18.2	7.3
B	6.0	9.4	7.1
C	5.2	8.3	4.1
D	3.1	3.1	2.9
E.1	3.2	3.2	3.1
E.2	2.6	2.5	3.1

TABLE 2 MTTR BY SYSTEM AND FUNCTION

Note, Integration and Test in Project E.1 and E.2 are higher
since configuration management on the projects was very
crude leading to much delay in this phase of repair (and
indeed in the development of the original product)

A FRAMEWORK FOR DEPENDENT FAILURE ANALYSIS

R P Hughes
Central Electricity Generating Board
Berkeley Nuclear Laboratories
Berkeley, Gloucestershire GL13 9PB
United Kingdom

ABSTRACT

The Distributed Failure Probability (DFP) approach to the problem of dependent failures in systems is presented. The basis of the approach is that the failure probability of a component is a variable. The source of this variability is the change in the "environment" of the component, where the term "environment" is used to mean not only obvious environmental factors such as temperature etc., but also such factors as the quality of maintenance and manufacture. The failure probability is distributed among these various "environments" giving rise to the Distributed Failure Probability method. Within the framework which this method represents, modelling assumptions can be made, based both on engineering judgment and on the data directly. As such, this DFP approach provides a soundly based and scrutable technique by which dependent failures can be quantitatively assessed.

INTRODUCTION

One of the principal design features used to achieve reliable systems is that of redundancy. If an important function is required of a component within a system, then several such components might be used in parallel in order that the function should still be performed even if one or more of those components should fail. This is sound engineering practice, which is widely followed.

In order to assess the benefits of such redundancy, it is necessary to be able to quantify the reliability of systems which rely on the working of several similar components. Existing techniques, such as fault tree analysis, are based upon the assumption that the components within the system being analysed are independent of one another. This implies that the reliability of a system could be improved arbitrarily by increasing the level of redundancy. This is not achieved in practice, because there are dependencies between components which lead to a level of reliability which cannot be improved upon. In effect, the reliability of the system has a cut-off value as the level of redundancy is increased. This cut-off behaviour is not a characteristic which arises out of standard fault tree methods, unless a cut-off is introduced into the tree as a separate element. It is this cut-off approach to dependent failures in systems which the CEGB has used for some years. The value of this cut-off cannot be determined by standard fault tree methods and is based upon engineering judgment. The CEGB, wishing to improve on the techniques available to it for reliability studies, identified a requirement, therefore, for a method of treating dependent failures between redundant components, which does allow them to be included in fault tree analysis in a consistent way, and which can properly evaluate their effects on system reliability.

Since the recognition of the significance of dependent failures for redundant systems, many approaches have been put forward with a view to evaluating their effects. Some seek to evaluate the effect of dependent failures on the individual cut-sets of the system fault tree (e.g. the Beta Factor method[1] , the Multiple Greek Letter (MGL) method[2], the Binomial Failure Rate (BFR) method[3], the C-factor method[4], while others attempt to form a judgment about the system in a structured way (e.g. the Partial Beta Factor method and its extension[5]).

The choice available to the systems analyst seeking to quantify the effect of dependent failures on a system is wide. This wide choice is not a benefit to the analyst, but only adds to his problems - he is faced with the question of which method to chose and why. In practice, his choice will be motivated by factors such as familiarity, data availability, ease of application etc.. Having made his choice, it will be difficult to defend over the alternatives, because the interrelationship between all the different approaches is not generally understood. Indeed, it is precisely this lack of a coherent framework in which modelling can be considered which has led to the existing diversity of models.

Our purpose here is to create that framework. The important benefits are that this framework represents a new approach to dependent failure analysis in which

- certain other methods can be seen as special cases

- other more appropriate assumptions can be justified directly from the data or using engineering judgment

We will call this new approach the Distributed Failure Probability (DFP) method, for reasons which will emerge below. The details of this approach have already been published[6,7]. The intention here is to give a fuller description of the basic concepts.

THEORETICAL BASIS OF THE DFP APPROACH

In standard fault tree analysis, the basic elements are generally the components which comprise the system being analysed. The structure of the system is expressed in terms of a fault tree which is solved to produce minimal cut-sets of these components, and the reliability of the system is estimated by evaluating these cut-sets in terms of the individual failure probabilities of the components. As discussed in the introduction, this process does not account for the presence of dependent failures among the components.

The underlying reason for this, as we shall see below, is the fact that the values which we use for the failure probabilities of the components are average values. This averaging has been performed over two sources of variability, which can be illustrated as follows. Suppose we are interested in a particular type of component, a pump say. This type of pump may be manufactured by several manufacturers. One manufacturer may produce pumps which are "more reliable" than most others because of the quality of his machines, the detail of his design etc.. It is natural, therefore, to expect a variability of pump performance depending on manufacturer. Superimposed on this kind of variability will be the usual random differences in pump reliability arising out of the random variations in the manufacturing process. The difference between these two types of variation in reliability is that if several pumps are chosen from the same manufacturer, their reliabilities will be correlated as far as the first kind of variation is concerned, but uncorrelated as far as the random variations of the manufacturing process are concerned.

It is with this correlated type of variation that we are concerned here, because it has a marked effect on the reliability characteristics of systems of several similar components. As well as the example given above of the influence of a common manufacturer, the reliability of a component can be influenced by environmental factors such as temperature and humidity. It can also be affected by the quality of maintenance. We shall refer to all these correlated variabilities as due to the "environment" of the component, where "environment" means all those factors which influence the failure probability of the component as discussed above. Some of these "environmental" factors will change with time, and in doing so will change the failure probability of the component. If this happens, we say that the "environment" is now different. Clearly, if one of the factors has no effect on the component, then any change in it is irrelevant and does not change the "environment".

Thus the failure probability of a component varies with time as its "environment" varies, but within a given "environment" the failure probability is fixed. If we denote this failure probability within a particular "environment" E_j by $p(F|E_j)$ and the probability that the component will be in that particular "environment" by $p(E_j)$, then the average failure probability is given by

$$p(F) = \sum_j p(F|E_j) \times p(E_j) \tag{1}$$

with the sum running over all possible different "environments". Equation (1) shows clearly how the component failure probability is distributed over various possible values associated with each of the various "environments".

If we now consider several such components, n say, together in the same "environment", then the probability that they all fail together is given by

$$p(F^n) = \sum_j p(F|E_j)^n \times p(E_j) \tag{2}$$

This is because, given a particular "environment" E_j's in which they all exist together, their failure probabilities are each $p(F|E_j)$. The components fail independently of one another (we are not considering cascade failure here for simplicity, although they could be included very straightforwardly) but the failure probabilities of each are directly correlated by the common "environment" in which they find themselves.

Generally, $p(F)^n \neq p(F^n)$, since in the first expression, the averaging over "environments" is performed before raising to the power n. The correct value for the failure of several similar components together is given by equation (2). Both equations (1) and (2) make plain the fact that the failure probability of a component is a variable, varying from "environment" to "environment", and that the proper distribution of failure probability over these "environments" is essential in order to evaluate correctly the failure probability of several components together. This key concept forms the basis of the Distributed Failure Probability (DFP) approach.

The way in which the failure probability of the component is distributed as implied by Equation (1) can be shown graphically as in Figure 1. This represents a discrete probability distribution since the failure probability within each "environment" $p(F|E_j)$ is assumed to be precisely known. In reality, there will be some uncertainty associated with each of these probabilities, and the discrete values will become blurred as shown in Figure 2. The result of this blurring is to produce a continuous distribution for the failure probability (see Hughes[8] for the details). In this situation, the generalisation of equation (1) is

$$p(F) = \sum_j \int_0^1 x \times f_j(x) \ dx$$
$$= \int_0^1 x \times f(x) \ dx$$

(3)

where the failure probability of the component is represented by x (corresponding to $p(F|E_j)$), the probability of a given value of x is given by the probability density function $f_j(x)$ (corresponding to $p(E_j)$) and the sum over the different "environments" has been replaced by the integral over x. Equation (2), representing the failure probability of n such components together, becomes

$$p(F^n) = \int_0^1 x^n \times f(x) \ dx$$

(4)

Notice that the system structure in terms of its components is embodied in the term x^n appearing in (4) (i.e. the system fails if all of its components fail). Systems with more complicated structures would have more complicated algebraic structure functions. For example, a 2 out of 4 system would have the algebraic structure function $4x^3(1-x) + x^4$. All the information on component behaviour in the different circumstances is included in the function $f(x)$.

We have now established the framework which we were seeking. It is one in which the modelling of the system is explicit and factored out of the modelling of the component behaviour. We shall turn now to the modelling of this behaviour, as represented by $f(x)$.

MODELLING

In order to evaluate the failure probability of our system from Equation (4), we need to model the function f(x). From the way in which f(x) arose, it is clear that we should be able to do this from an understanding of the circumstances or "environments" in which the components of the system will exist, and the behaviour of the components within each "environment".

One possible model is represented in Figure 3, where we see that there are three different "environments", one of which is lethal (i.e. has a failure probability of 1 associated with it), one is severe but not lethal in that the failure probability of the component in it is p, and one which is fairly benign in that the failure probability in it is small (λ). These "environments" occur with probabilities ω, μ , and $(1 - \omega - \mu)$ respectively. Assuming that all the probabilities are precisely known, the f_j's in this example are then $\delta(1 - x)$, $\delta(x - p)$ and $\delta(x - \lambda)$ with the p_j 's being ω, μ, and $(1 - \omega - \mu)$ repectively. Equations (1) and (2) give

$$p(F) = (1 - \omega - \mu) \times \lambda + \mu \times p + \omega$$
$$\sim \lambda + \mu \times p + \omega$$

and

$$p(F^n) = (1 - \omega - \mu) \times \lambda^n + \mu \times p^n + \omega$$
$$\sim \mu \times p^n + \omega$$

since all the parameters are small, and λ is assumed to be much smaller than p.

The modelling of f(x) described above represents exactly the Binomial Failure Rate (BFR) model[3]. Notice how as n increases, the limiting value of $p(F^n)$ is ω, the effective cut-off value for our simple system. If we had chosen only two "environments" by putting $\mu = 0$, (i.e. taking all shocks as being lethal) then this would represent the assumption of the Beta Factor and the C-Factor methods - if one component fails due to a common cause of failure, then all the components of the system fail.

Thus, these three familiar models of dependent failure are seen to be special cases within the general DFP framework. This fulfils the first of our objectives.

Such modelling assumptions as give rise to these models are very specific. Data may not be able to be fitted adequately to them[3] . Different component types may be characterised by quite different distributions f(x) and any attempt to 'model' the form of f(x) a priori as specifically as this, is suspect. The assumption of two hazardous "environments" one of which is lethal, for example, is not supported by experience. A far better procedure is to see what the data on a particular component can tell us about the distribution f(x) directly. This is what we shall consider in the following section.

DIRECT APPROACH

We saw in the previous section how the Distributed Failure Probability approach could be used as a framework within which certain modelling assumptions could be used to produce various existing models of dependent failure as special cases. More general assumptions, say on the number and "harshness" of the "environments", could easily be made on the basis of the analyst's engineering knowledge of the system being assessed. We shall not persue this here, but concentrate on an approach which uses data directly to construct the distribution f(x).

This direct approach is best explained by way of an example. Here, we shall use data given by Apostolakis and Moieni[9]. (Other examples are given in reference 7). The system consists of three similar components, and the data are gathered from 500 system tests in which there were 13 single failures, 2 double failures and 1 single failure as shown in Table 1.

	system events	no. of comps failing	component events successes	failures
	484	0	1452	0
	13	1	26	13
	2	2	2	4
	1	3	0	3
totals	500		1480	20

Table 1 - System Failure Information

No information as to the causes of the component failures is given, so we must make assumptions about the possible different "environments" which the component experiences. If further engineering knowledge was available, this would clearly guide and influence these assumptions. We shall assume throughout that the 500 tests are representative of the "environments" in which the component exists. If we knew this was not the case, the test data would be weighted accordingly.

The details of how the function f(x) is estimated from the data are given in full elsewhere (Hughes[7]) so here we shall give only a brief indication. We shall consider first the possibility that there are 2 different "environments" affecting our system, and that all the multiple failures were caused by one "environment". For the remaining tests, we assume that the system existed in another "environment". We shall label these "environments" as (2,3) and (0,1) respectively.

The (0,1) "environment" occurred (from Table 1) 497 times out of 500 with (2,3) occurring 3 times. Crude estimates of the x_j 's ($x_j = P(F|E_j)$) are therefore 497/500 and 3/500 respectively.

In the (0,1) "environment" there were 497×3 component tests and 13 failures were observed, giving a crude estimate of the failure probability of the components in the (0,1) "environment" of $13/(497 \times 3) = 8.73 \times 10^{-3}$. Similarly for the (2,3) case, there were 3×3 component tests with 7 failures giving a failure probability of 7.78×10^{-1} .

For a single component, then, the failure probability is

$$p(F) = (497 \times x_1 + 3 \times x_2)/500$$
$$= 1.33 \times 10^{-2}$$

Similarly, the failure probability of all three components failing together is

$$p(F^3) = (497 \times x_1^3 + 3 \times x_2^3)/500$$
$$= 2.82 \times 10^{-3}$$

We have stressed above the "crudeness" of the estimates of the probabilities we have used. The more sophisticated Bayesian technique given in reference 7 gives slightly different results, and these are given in detail in Table 2, not just for the assumption of 2 "environments" (0,1) and (2,3) considered above, but for several splittings of the data among different "environments".

Data Split	p(F) (s)	$p(F^3)$ (s)
(0,1)(2,3)	1.51×10^{-2} (3.94×10^{-3})	3.34×10^{-3} (2.41×10^{-3})
(0,1)(2)(3)	1.62×10^{-2} (4.16×10^{-3})	4.01×10^{-3} (2.66×10^{-3})
(0,1,2)(3)	1.52×10^{-2} (3.65×10^{-3})	2.28×10^{-3} (2.08×10^{-3})
(0)(1)(2)(3)	1.70×10^{-2} (4.71×10^{-3})	5.26×10^{-3} (2.79×10^{-3})
(0,1,2,3)	1.40×10^{-2} (3.03×10^{-3})	3.13×10^{-6} (2.12×10^{-6})

Table 2 - Failure Probabilities for various Data Splits

The figures in brackets are the standard deviations of the probabilities given, and give some indication of the confidence we can place in them given the limited amount of data.

Notice how the probability of a single component failing is very insensitive to our assumptions of how the data might be split into the different "environments". This is generally true for the probability that three components will fail together, except for the last case in Table 2 where only one "environment" is assumed. This corresponds directly to the usual process of pooling all the data on a component to form a single number and then cubing it to get the failure probability of three components together. (In fact, cubing p(F) in this case does not produce exactly $p(F^3)$ since in our calculation of Table 2, we have made proper allowance for the uncertainty in p(F) arising from the data.)

In order to help us choose which of the data splits is most appropriate, we can take the calculations one step further, and estimate the probability that, given each of our assumptions about the "environments" in Table 2, we would have actually observed the failures as detailed in

Table 1 in a series of 500 tests. The results are as shown in Table 3 below.

Data Split	Probability of Observing the Observations
(0,1)(2,3)	.87
(0,1)(2)(3)	.72
(0,1,2)(3)	.28
(0)(1)(2)(3)	.049
(0,1,2,3)	.00013

Table 3 - Likelihoods that Data Splits are Realistic

Clearly, the assumptions about the data splits are in decreasing order of likelihood. The last choice of a single environment is obviously not supported by the data, since if it were true, the chance of actually observing the results of Table 1 is so small. Of all our assumptions, the first two are the most likely to be correct. Since the results in Table 1 are so similar for these two cases, there is no significant difference between them.

If we use the first assumption (0,1)(2,3) for definiteness, we can now use the distribution f(x), on which the first line of Table 2 is based, to calculate the failure probabilities of other systems by an appropriate choice of the algebraic structure function in the integral of Equation (4). All the component and "environmental" information is embodied in the distribution f(x). Table 4 shows the failure probabilities of 1 out of n systems for several values of n.

n	$p(F^n) = \int x^n \times f(x)dx$
1	1.51×10^{-2}
2	4.44×10^{-3}
3	3.34×10^{-3}
4	2.63×10^{-3}
5	2.10×10^{-3}

Table 4 - System Failure Probabilities for Different Redundancies

The results for n=1 and 3 are already in Table 2, but we can see from Table 4 how the reliability of the system increases as the degree of redundancy is increased. Clearly, after n=2, the benefits are limited in terms of reliability. Also, the degradation of the system caused by taking one component out of service for maintenance, for example, is also small, provided at least two components remain.

DISCUSSION

Our objective here has been to present the basic concept of the Distributed Failure Probability (DFP) method for dependent failures, and this has been illustrated throughout with very simple systems of several similar redundant components all in the same "environment". We have not considered more complicated systems where, for example, some components might be in the same "environment" with respect to one factor (high temperature, say) but in different "environments" with respect to another (diverse maintenance, say). The extension to such cases is straightforward and will form the subject of later papers. Suffice it to say here that the technique is applicable to these more general types of situation. The CEGB, recognising the benefits and practicallity of the DFP approach, will use it as a basis of their treatment of dependent failures for the final safety assessment of the Sizewell B Nuclear Power Station currently under construction.

CONCLUSIONS

A framework for dependent failure analysis has been presented. This framework is based upon the concept of a Distributed Failure Probability (DFP) for a component, the distribution being over the various different "environments" in which the component is required to operate. Certain existing models for dependent failure can be seen to be special cases within this framework. The main benefit is that more realistic and representative assumptions as to the component's response in different circumstances can be readily accommodated. These assumptions can be guided by engineering judgment, by the data directly as illustrated above, or by a combination of both. As such, the DFP method provides a credible, soundly based, scrutable and flexible tool with which to address the problem of dependent failures.

REFERENCES

1. Fleming, K. N., 1975, GA-A13284

2. Fleming, K. N., et al, Nuclear Eng. and Design 93 (1986) 245

3. Atwood, C. L., 1983, NUREG/CR-3289, EGG-EA-2258

4. Evans, M. G. K., et al, Reliability Engineering 9 (1984) 107

5. Johnson, B. D., 1987, Structured Procedure for Dependent Failure Analysis, Proceedings of 'Reliability 87', Birmingham, UK

6. Hughes, R. P., 1986, CEGB report TPRD/B/0813/R86

7. Hughes, R. P., Reliability Engineering 17 (1987) 211

8. Hughes, R. P., to be published

9. Apostolakis, G. and Moieni, P., Reliability Engineering 18 (1987) 177

AKNOWLEDGEMENT

This paper is published with the permission of the Central Electricity Generating Board.

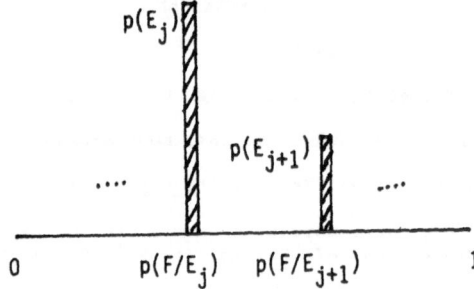

Figure 1 - The Discrete Distribution of Failure Probability

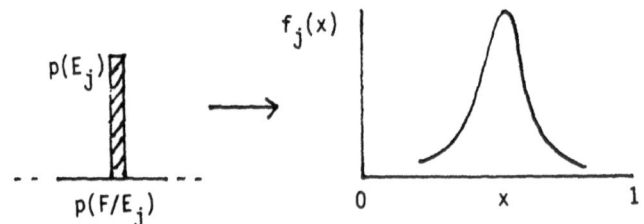

Figure 2 - The Blurring of a Discrete Probability into
a Continuous Probability Distribution

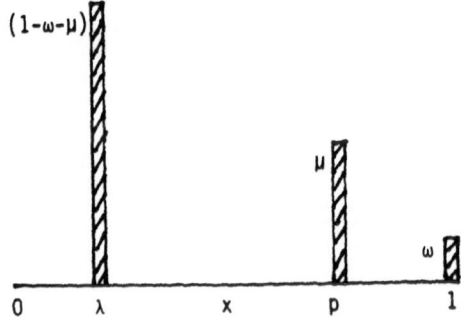

Figure 3 - The Discrete Distribution of the BFR Model

ASSESSING ENVIRONMENTAL EFFECTS

J. I. Ansell
Department of Management Systems and Science,
School of Management,
University of Hull,
Hull,
HU6 7RX.

and

M. J. Phillips,
Department of Mathematics,
University of Leicester,
Leicester,
LE1 7RH.

ABSTRACT

Data collected from working systems are often the most useful in
assessing the performance of the components in situ. Such field data
often suffers from environmental effects which are jointly experienced by
a number of components. A model is proposed for estimating such effects.
The effects may be of interest in themselves or one may purely wish to
remove them. The model may also be used in the context of common
stresses or dependency of components.

There are several models for dependency of components based
primarily on bivariate or multivariate models, such as Downton's
bivariate exponential [1] and the beta factor model [2]. These models do
not usually consider sequences of components in situ. The model proposed
in this paper does, assuming that there are a number of components
suffering the same stresses at any time, that the components are replaced
on failure and that the time of failure for the components are noted. The
model follows a suggestion by Lindley and Singpurwalla [3] for such
situations using a formulation similar to proportional hazards. The
formulation is similar though not identical to reliability growth models
as described by Jewell [4].

The paper considered estimation and fitting of the model, using an
example. Comparisons are made to other methods. Some extensions of the
model are contemplated.

INTRODUCTION.

The analysis of field data is usually most rewarding when estimating the 'real' reliability of components. Unfortunately field data is rarely clean data. There may be numerous problems associated with such data. One of the factors which may effect the data is varying environmental conditions. In this paper we consider a model for estimating these effects.

Several models/methodologies have already been proposed for similar problems such as common cause or dependent failure models. Downton's model [1] for statistical dependency of components followed a series of similar statistical models. Other approaches have attracted considerable interest in the reliability literature such as the Beta Factor Method, Flemming [2]. Both of these models will be briefly reviewed as typical examples of their type in the next section.

Both of these models may be viewed as concerned with single components in situ in a static environment. This paper explores a more general model in which the failed components are replaced and the environmental conditions may change. Figure 1 illustrates such a process, which statistically may be referred to as a superimposition of renewal processes.

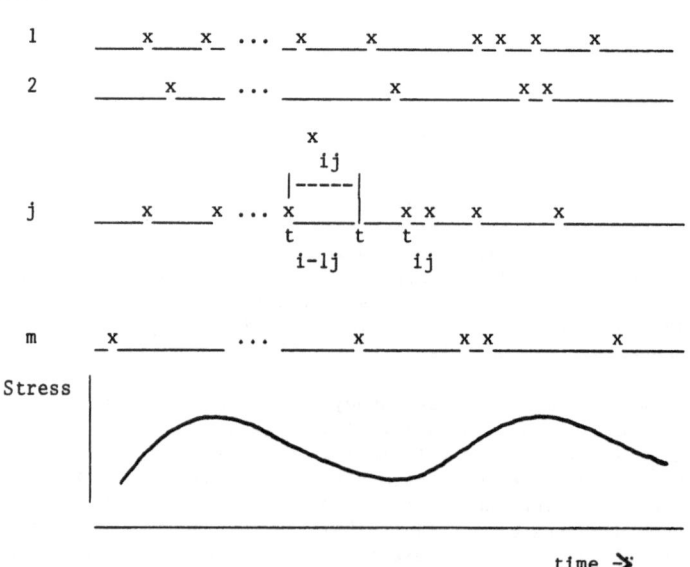

x denotes event. x_{ij}, t_{ij}, t_{i-1j} and t are defined in the text.

Figure 1. Failure times for sequences of components with Common Environmental Stress

Initially it will be assumed that only information on component failure times are collected and that the environmental conditions cannot be directly measured. The environmental conditions will therefore be estimated from the joint data on component performance. A possible model for such common environmental effects is suggested in the paper. The model is a development of a suggestion in Lindley and Singpurwalla [3] and is similar in formulation to Cox's regression model, or Proportional Hazards Model, [5,6].

Having estimated the common environmental effect it is then possible to adjust the estimates of the components performance accordingly. It may also be possible using the estimates of common environmental effects to produce 'better' estimates of Beta Factors for future modelling.

The practicalities of using the model are considered in subsequent sections. Firstly statistical aspects of the model such as estimation and inference are explored through the use of a data set. Then we compare the model to other similar models in Reliability such as those reviewed by Jewell [4]. Extension to the model proposed are considered both from a Reliability and Statistical Standpoint.

MODELS FOR DEPENDENCY

There exists a considerable literature on dependency both in the Statistical and Reliability literature. The form of the dependency being defined from a physical or theoretical base. It is not our desire to review all such models but we choose to select a model from each area, both illustrative approaches to dependency.

DOWNTON'S BIVARIATE EXPONENTIAL MODEL

Modelling dependency statistically means the construction of bivariate or multivariate distribution models. Typical of such is Downton's Bivariate Exponential Model [1]. Although the model was known prior to Downton's paper, he indicated how it might arise naturally in reliability as a shock model. His paper presents the model as a joint distribution density function for two random variables X and Y;

$$f(x,y) = \{ab/1-p\} \, \exp\{-(ax+by)/1-p\} \; I_0\{2 \, \sqrt{(pabxy)} \; /1-p\} \quad (1)$$

where I_0 is the modified Bessel function of the first kind of zero order. The parameters a and b represent the components' independent performance characteristics, for X and Y respectively, the parameter p representing their dependency. It is noticeable that p has to be a positive correlation coefficient in the model, interpreted as damage to one component may also damage the other component. The model appeared following a series of other bivariate models such as Marshall-Olkins Bivariate Exponential [7], which Downton presented as a distribution function as follows;

$$F(x,y) = 1 - \exp\{-(\lambda_1+\lambda_{12})x\} - \exp\{-(\lambda_2+\lambda_{12})y\}$$

$$+ \exp\left\{ -\lambda_1 x - \lambda_2 y - \lambda_{12} \max(x,y) \right\} \quad (2)$$

The main interest for Downton was to study various properties of the model rather than its application to reliability directly. The paper therefore concentrated on the conditional distribution and other properties rather than exploring practical aspects of the model. It did illustrate differences between the two models because as the dependency between the variables changed, p increased.

Downton suggested that his model was superior to Marshall-Olkins model since it was mathematically simpler, and hence the properties of the distribution may be obtained more easily. The Marshall-Olkin model suffered in Downton's view from a singularity which leads to complications for mathematical tractability. The singularity results from a non-zero probability of a joint failure at a specified time. This non-zero probability however is a central theme to common cause failure models. In the practical context, therefore, Marshall-Olkin's model seems more appropriate.

Downton's model may be generalised to include other underlying distributional forms, as is suggested in his original paper. Examples such as Marshall-Olkins indicate that other forms of dependency can be considered. Extension though to cover the current concern of this paper would not be easy. Downton's model relates a single component to another single component not sequences of components.

The model in some senses illustrates the statistical concern which, having produced a probabilistic model, is to derive its characteristics. The characteristics may lead secondly to a comprehension of the practical context.

BETA FACTOR MODEL

A more familiar model in the reliability literature is the Beta Factor Model, Flemming [2]. The model has strong similarities to the Marshall-Olkin model, though no distributional form is usually presented with the Beta Factor model and no mathematical model is posed for the dependency.

The model is usually described in terms of failure rates, with a components failure rate being composed of an independent part and a 'common cause failure' (CCF) part, as

$$\lambda = \lambda_i + \lambda_c . \quad (3)$$

It is referred to as the Beta Factor Model since   is usually obtained by multiplying the components overall assessed failure rate by a beta factor.

$$\lambda_c = \beta\lambda . \quad (4)$$

The selection of a beta factor depends on several features of the system being studied. Assessment is often based partially on historic data and partially on subjective assessment. Given the nature of the problems usually tackled the latter plays a central role.

Producing adequate and sound methodologies for the subjective assessment has become the prime role for the Reliability Engineer. Various strategies and methodologies have been suggested in the literature [8,9,10,11]. The range of factors which may be taken into account is large, and may for example include:

1) Systems Structure;

2) Degree and Form of Redundancy;

3) Degree of Diversity;

4) Environmental;

5) Defense Strategies Employed.

Given the subjective nature of the estimation of the beta factor one would expect Bayesian methodologies to be more widely employed. There have been some applications [12]. Lindley and Singpurwalla [13] suggests a Bayesian approach to Fault Trees when expert opinion is used which also might be helpful in such studies.

Although no distributional assumption are made about the components' lifetimes, there seems to be an acceptance throughout of the exponential distribution for the time to failure distribution. The model proposed for the dependency is not stated in a mathematical form, though it is explored through Engineering Judgements.

The estimates produced by the approach may be crude since usually aggregate values for components' performance are used. These are then refined by possibly subjectively assessed beta factors. The model proposed in the next section uses historic data to estimate both the components performance and environmental factors. In modelling the environmental conditions it may shed light on the types of values which might be used in the Beta Factor method.

Whilst the reliability models attempt to address the practical problems they sometimes make assumptions which are not supported necessarily by a proven model or data, [14].

PROPOSED MODEL

Again we use the failure rate to describe the model illustrated in Figure 1. At time t the ith component in sequence j has a rate failure which depends both on the time since last replacement in the jth sequence $x_{ij}, \lambda_j(x_{ij})$, and a factor dependent on the common environmental conditions, $S(t)$. The environmental conditions are assumed to vary with time and hence the dependency on total time t for $S(t)$.

It is not assumed in the model that the different sequences of components have the same distributions for failure times. It seems mostly that they will have differing failure rates but also different failure distributions.

The model can be expressed as

$$\lambda_{ij}(t) = g\{\lambda_j(x_{ij}), S(t)\} \qquad (5)$$

where g is an arbitrary function. Two simple possibilities for g are the additive and multiplicative models;

$$\lambda_{ij}(t) = \lambda_j(x_{ij}) + S(t) \qquad (6)$$

$$\lambda_{ij}(t) = \lambda_j(x_{ij}) \cdot S(t) \qquad (7)$$

Following Cox's Regression model we develop the multiplicative model, though it would be quite plausible to use the additive model. (As in the Extended Hazard Model of Etezadi-Amoli and Ciampi [15], an extension of the Proportional Hazards Model, it may be plausible to allow for both additive and multiplicative models in the same model.)

Having chosen the form of g we must specify the form of S(t). Cox's suggested for Proportional Hazards the simplest function would be the exponential function which would give in this case

$$s(t) = \exp\{\beta t\} \qquad (8)$$

For the present application this would imply an increasing, or decreasing, effect for the environmental effect throughout the period. Since this may be unrealistic in most applications we develop a model based on a polynomial form for S(t).

$$S(t) = \sum_{s=0}^{r} a_s \, t^s \qquad (9)$$

It must be noted though that S(t) must be positive over the range considered, so that

$$\sum_{s=0}^{r} a_s \, t^s > 0 \quad \text{for all t in } (0,T],$$

where T is the end of the observation period. This restricts the choice of parameter values and hence the form of polynomials which can be fitted to the problem. It may also cause some problems when estimating the parameter values. Similar problems were encountered with the precursors to Proportional Hazards Model, for example Feigl and Zelen [16] proposed the Exponential Regression Models with polynomial or inverse polynomial multiplicative factors on the hazard.

There is no reason why other models could not be taken for S(t). For example harmonic models could be explored which also could circumvent the problem of negative hazard.

Again unlike Cox's regression model, we will develop a model which is parametric. We perceive though no difficulty in producing a semi-parametric model, similar to the Proportional Hazards Model. Therefore we assume it is possible to specify the components' failure distribution, which is frequently the case in practice. Hence we will assume that the rate of failure for the ith component in sequence j after a period w can be described as

$$\lambda_j(w) = h_j(w) \qquad (10)$$

where $h_j(w)$ will be the hazard function from a specific distribution.

Obviously one could select the function $h_j(w)$ to reflect Engineering Judgements. Previous knowledge of the performance of the component or the defense strategies employed in the system could be incorporated into the model.

STATISTICAL ASPECTS

In this section we consider both the estimation of the environmental effect and components lifetime distribution, and the inferential approach which might be employed. We have chosen to follow a traditional Maximum Likelihood approach, though a Bayesian methodology could very useful be employed.

If t represents the time at which the ith component of the jth sequence fails then the log likelihood may be written as

$$\sum_{j=1}^{m} \sum_{i=1}^{k_j} \left\{ \log[h_j(x_{ij})] + \log[S(t_{ij})] - \int_0^{t_{ij} - t_{i-1j}} h_j(u) S(u + t_{i-1j}) du \right\} \qquad (11)$$

If the underlying distributions were assumed to be exponential with parameter then the form would be

$$\sum_{j=1}^{m} [k \log(\lambda_j) - \lambda_j \int_0^T G(u) du] + \sum_{j=1}^{m} \sum_{i=1}^{k} \log[S(t_{ij})] \qquad (12)$$

The effect of this assumption is to produce a set of NHPPs.

In this case the estimates of the component failure rates would be

$$\hat{\lambda}_j = k_j / \int_0^T S(u) \, du \qquad (13)$$

and the maximum likelihood for the parameters of S(t) would be given by the following equations;

$$\sum_{j=1}^{m} \sum_{i=1}^{k} \frac{d}{da} \log[S(t_{ij})] = \sum_{j=1}^{m} k_j \frac{d}{da} \log [\int_0^T S(u) \, du] \qquad (14)$$

where a are the parameters of S(t).

As indicated above we will develop the model for the polynomial form for S(t). In this case (13) would become

$$\hat{\lambda}_j = k_j / \sum_{s=0}^{r} (\hat{a}_s T^{s+1} / s+1) \qquad (15)$$

and (14) would be

$$\sum_{j=1}^{m} \sum_{i=1}^{k} t_{ij} / \sum_{s=0}^{m} (\hat{a}_s t^s_{ij}) = (\sum_{j=1}^{m} k_j)[T^{r+1}/r+1]/[\sum_{s=0}^{r} (\hat{a}_s T^{s+1}/s+1)] \qquad (16)$$

The estimators cannot be explicitly obtained and hence derivation of their distribution would be difficult. For this reason, in common with Proportional Hazards, we use standard asymptotic results to derive inferential procedures.

The confidence intervals for the parameters can be constructed based on the the result that the maximum likelihood estimators are asymptotically normal with mean the true parameter values and variance given by the expectation of the derivative of the log likelihood function. Tests for hierarchies of parameters can also be derived based on the usual Chi square test on the log likelihood. The details are given in any standard book of Statistical Inference, such as Silvey [17].

EXAMPLE

The model has been fitted to the data presented in Table 1. Plots of the data illustrate that the failure rate is neither increasing or decreasing consistently throughout the interval. It therefore seems wiser to apply the polynomial model rather than the exponential. If a polynomial is to be fitted then it would appear it required to be at least of order three to provide an adequate fit. The underlying distribution for simplicity was chosen to be exponential.

Table 1. Failure times for the four systems

System	Failure Times
1	55 62 64 66 68 101 102 104 105 108 112 115 124 125 126 127 130 135 139 141 142 143 157 165 167
2	31 103 122 171 173
3	30 34 117 118 131 132 150 152 170 180
4	47 129 136 137 166

Using a Newton-Raphson procedure it is possible to estimate the parameter values. Given that at the boundary of the admissible region the log likelihood decreases rapidly towards minus infinity then great care is needed in using the approach. Finding a suitable starting point needs attention as well as the monitoring of changes in the log likelihood value to ensure convergence towards a maximum.

The results of fitting polynomials up to order 5 are presented in Table 2. It is noticeable that the log likelihood value does not dramatically change from fitting a linear form to quintic. The change between fitting the mean and linear is also not significant.

Table 2. Results of the Analysis

Model	Polynomial Coefficients (Variance)					Likelihood
	a1	a2	a3	a4	a5	
Mean						45.21
Linear	4.34 (4.48)					50.85
Quadratic	3.93 (6.72)	0.18 (2.38)				50.86
Cubic	-3.52 (2.20)	9.63 (6.26)	-4.04 (3.04)			52.42
Quartic	2.20 (10.47)	-9.42 (30.48)	16.12 (32.62)	-6.32 (10.52)		52.82
Quintic	-3.94 (12.54)	21.39 (59.30)	-35.79 (97.97)	28.50 (67.01)	-8.04 (16.21)	52.97

Interpretation of these results would suggests that the there is no common environmental effect present. In itself an interesting result since runs counter to other information subsequently gathered. Hence according to the present results there would be no reason to incorporate an environmental factor into a future analysis.

COMPARISON

The model described in the previous section has similarities to other reliability models, ie reliability growth models. A partial review of reliability growth models was made by Jewell [4] in special edition of the Operations Research Journal on Reliability. The article was principally based on work by Crow [18]. In the paper the two differing growth models were presented Type I and Type II. The first of these was referred to as the 'environmental model' whilst the latter was described as the 'product' model. Generally the hazard of the ith component at time t with lifetime x, replacing the (i-1)th component at time t, can be described as

$$\lambda(x,t;t_{i-1}) \quad = \quad \theta g(t;t_{i-1}) + w\, h_c(x)$$

where the $h_c(x)$ is the hazard of a component and $g(t,t_{i-1})$ is the learning curve, $x=t-t_{i-1}$ and $t > t_{i-1} > 0$. The two models are;

Type I (environmental) : $g(t,t_{i-1}) = g(t)$

Type II (product)   : $g(t,t_{i-1}) = g(t_{i-1})$

For the hazards to be positive throughout

$$\theta\, g(t;t_{i-1}) + w\, h(x) > 0.$$

Our model is similar to the Type I formulation but we have assumed a more general form than the additive model suggested in Jewell's paper. It should also be stated the Type II model is assuming a single sequence of components where we are assuming several sequences of components.

Typically in these models users have been misled into using models which predetermine an improvement in the performance, see for example LLoyd [19].   This occurs by choosing for $g(t;t_{i-1})$ either an exponential or power function of t. If such a choice is inappropriate it will obviously lead to misleading results. Several authors have attempted to correct this mistake, see Crow [18] who employs a Bayesian approach to Reliability Growth Models. If the model proposed employs a polynomial form it will not suffer from this drawback.

EXTENSIONS AND DEVELOPMENTS

We have already made several suggestions on how the model might be developed both from a statistical and reliability viewpoint. In this section we bring together those suggestions and add others which may also be applicable to reliability.

The proposed model provides a model to estimate both the components performance and the environmental effects. As presently formulated it does not allow specifically for the use of data other than failure times. There are several ways around this shortcoming.

Allowance for external factors could be built into the current model either through the underlying distribution or alternatively the common effect S(t). Hence we could select the underlying distribution to reflect some known information. This may be through the choice of distribution or parameterization of the distribution. For example we could model the failure rate exactly as the Beta Factor model does, whilst still allowing some environmental conditions to vary with time, as S(t). Other strategies, such as defensive strategies, could obviously be built into the model this way.

By generalizing the common effect model we may also achieve a similar aim. Suppose we assume that the common effect may be written as $S(t,z)$ where z represents a vector of known conditions which may or may not vary with time. In this way the known factors incorporated through z and t would allow for the effect of other unspecified factors which vary with time. Choice of suitable models for $S(t,z)$ would depend on the context. Some choices will be easier to employ than others. As with Proportional Hazards Modelling an exploratory approach may be very revealing.

Another approach to the problem of incorporation of other information besides the failure times would be via Bayesian Methodologies. Subjective information could be characterized by the prior distribution. There would be no real problem envisaged in producing a Bayesian parallel to the above Maximum Likelihood Estimation method.

The model arises from a statistical viewpoint. It requires the refinement that methods such as the Beta Factor Method have already received before becoming a tool of the Reliability Engineer. Given these refinements then there is no reason why the model should not achieved 'better' estimates under the right conditions.

DISCUSSION

The model proposed estimates common environmental effects to which a number of components may be subject, based only on component failure times. The estimates obtained of the environmental conditions could be compared with other information to find the possible the significant factors. Also the estimates may well be usefully in future modelling, giving appropriate Beta Factors.

Using the estimates will allow an appropriate adjustment to the estimates of the components performance. The proposed model should produce 'better' estimates than simple aggregates when environmental conditions are suspected of affecting the data.

The model proposed could be described as an extension of the proportional hazards model, as suggested by Lindley and Singpurwalla [3]. Other close relatives are the Reliability Growth Models [4].

The current model is relatively unrefined and plausibly requires further work before becoming a standard tool for Reliability Engineers. The method of estimation suggested in the paper is very sensitive to the choice of starting point. Other forms of estimation could be considered and might be more fruitful. Also the present model does not posses some of the refinement which have been developed for the Beta Factor Model in a practical reliability context. Some of the possible extensions and development of the model which may be helpful in exploring these problems have been suggested in the last section of the paper.

REFERENCES

1. Downton, F, [1970], 'Bivariate Exponential Distributions in Reliability Theory', JRSS,B,32, 408-417.

2. Flemming, K N, [1974], 'A reliability model for common mode failures in redundant safety systems', General Atomic Report, GA-13284.

3. Lindley, D V, and Singpurwalla, N D, [1986], 'Multivariate distributions for the life lengths of components of a system sharing a common environment', J App Prob, 23,418-431.

4. Jewell, W S, [1984], 'A general framework for learning curve reliability growth models', Opn Res, 32,547-558.

5. Cox, D R, [1972], 'Regression models and life tables (with discussion)', JRSS, B,34, 187-220.

6. Cox, D R, [1975], 'Partial likelihoods', Biometrika, 62, 269-276.

7. Marshall, A W, and Olkin, I, [1967], 'A multivariate exponential distribution', JASA,62,30-44.

8. Johnston, B D, [1987], 'A structured procedure for dependant failure analysis (DFA)', Reliability '87.

9. Martin, B R, and Wright, R I, [1987], 'A practical method of common cause failure modelling', Reliability '87.

10. Humphreys, P, Games, A M, and Smith, A M, [1987], 'Progress towards a better understanding of dependent failure by data collection, classification and improved modelling techniques', Reliability '87.

11. Humphreys, R A, [1987], 'Assigning a numerical value to the beta factor common cause evaluation', Reliability '87.

12. Haim, M, and Apostolakis, G, [1987], 'Statistical dependence in risk and reliability', Reliability '87.

13. Lindley, D V, and Singpurwalla, N, [1986], 'Reliability (and fault tree) analysis using expert opinion', JASA, 81,393,87-90.

14. Ansell, J I, and Bendell, A, 'Practical aspects of fault tree analysis and the use of Markov reliability models', Reliability '85.

15. Etezadi-Amoli, J, and Ciampi, A, [1987], 'Extended hazard regression for censored survival data with covariates : a spline approximation for the baseline hazard function', Biometrics, 43,181-192.

16. Fiegel, P, and Zelen, M, [1965], 'Estimation of exponential survival probabilities with concomitant information',

Biometrics,21,826-838.

17. Silvey, S D, [1975], Statistical Inference, Chapman and Hall, London.

18. Crow, L H, [1982], 'Confidence interval procedures for the Weibull process with applications to reliability growth', Technometrics, 24,67-72.

19. Lloyd, D K, [1986], 'Forecasting Reliability Growth', ARTS 86.

THE HARDNESS OF SEMICONDUCTOR DEVICES EXPOSED TO
SIMULATED ELECTROMAGNETIC PULSES

D.M. Barry
M. Meniconi
H. Lee

Department of Electrical Engineering
Lakehead University
Thunder Bay, Ontario
Canada
P7B 5E1

ABSTRACT

Reliability models of the survivability of selected semiconductor devices tested to failure using simulated EMP are presented. An EMP simulator providing variable pulse widths and power outputs was designed and built. Failure thresholds of devices such as transistors and low power Schottky gates were determined experimentally. Failure models were developed as functions of pulse width, local models being generated using such distributions as the lognormal, the inverse Gaussian, the Weibull and the exponential power distribution. Empirical equations relating failure thresholds for pulse width and power were also devised and global models generated.

INTRODUCTION

An electromagnetic pulse (EMP) can be defined as a large impulsive type of electromagnetic wave generated by a nuclear explosion. The EMP can couple transients into a system containing semiconductor integrated circuit components. These transients are functions of the type of nuclear weapon used in the explosion, and, of course, the system component configuration.

The EMP can reach a peak field of 50 kV/m in about 10ns and has a pulse width or duration in the range between 10 nanoseconds and 10µs. Large metal conductors such as aircraft act like antennae and couple an EMP pulse and large currents and voltages into an electronic circuit.

The hardness of any system containing electronic components can be defined in terms of the survivability of the system when exposed to EMP. This hardness is a direct function of the survivability of the components themselves, as semiconductor devices are the weakest elements with respect to the EMP. In the past many studies of damage to semiconductor devices produced by an EMP have been performed. Experimentally, the mode of interaction of the EMP and the device-under-test (DUT), has taken the form of simulating an EMP as it is coupled into input lines interfacing with the device (1). In this manner a large magnitude induced voltage pulse is produced across the semiconductor junction of interest or a large magnitude current pulse is injected into the junction.

In the study reported here, failure by simulated EMP was achieved using a step stress method. In order to be able to predict EMP damage to semiconductor devices, it is necessary to establish a failure threshold for a device. Successive pulses of varying pulse widths and hence pulse power carrying capacities were applied to DUTs, and empirical equations relating the ensueing failure thresholds to pulse width were generated. This method of producing empirical and semi-empirical models for damage prediction was introduced by Wunsch and Bell (2). Two sets of equations may be used to determine failure threshold. One can be obtained by the empirical method just referred to, and also a theoretical predictive method (3) may be employed. This theoretical model uses classical thermal analysis to predict the breakdown conditions for semiconductor junctions. Further statistical methods for applying empirical and theoretical models specifically to integrated circuits (ICs) exposed to EMP pulse damamge can also be employed (4).

This paper presents, therefore, a method of producing failure threshold data as a function of pulse width for semiconductor devices exposed to simulated EMP. The data thus produced are then modelled using such distributions as the lognormal, the inverse Gaussian, the Gamma, the Weibull and the exponential power distribution. Empirical and theoretical models are then discussed and parameters for them are estimated for the particular devices used in this study. Global models are then generated and the results are compared to the classic Wunsch-Bell approach (2).

RELIABILITY MODELS

The reliability models used for local modelling were chosen from those with whom successful modelling was achieved in the past, (5,6) for semiconductor IC and VLSI devices. A broad choice of local models was decided upon and those chosen are listed below in terms of their cumulative distribution functions (c.d.f) and their failure densities.

$$NORMAL: \quad F(x) = \phi\left(\frac{x - \alpha}{\beta}\right), \quad -\infty < x < \infty$$

where standard normal distribution $\phi(x)$ is:

$$\phi(x) = \frac{1}{\sqrt{2\pi}} \int\limits_{-\infty}^{x} \exp\left(-\frac{1}{2}z^2\right) dz$$

$$f(x) = \frac{1}{\sqrt{2\pi}\beta} \exp\left(-\frac{1}{2}\left(\frac{x - \alpha}{\beta}\right)^2\right)$$

α is the mean
β^2 is the standard deviation

LOGNORMAL: $F(x) = \phi\left(\dfrac{log(x) - \log(\alpha)}{\beta}\right), \quad x > 0$

$f(x) = \dfrac{1}{\sqrt{2\pi}} \dfrac{1}{\beta x} \exp\left[\dfrac{-1}{2}\left(\log\left(\dfrac{x/\alpha}{\beta}\right)\right)^2\right]$

α is the scale parameter
β is the shape parameter

TWO–PARAMETER EXPONENTIAL: $F(X) = 1 - \exp\left(-\dfrac{\left[x - \alpha\right]}{\beta}\right), \quad x > \alpha$

$f(x) = \dfrac{1}{\beta} \exp\left(-\dfrac{\left[x - \alpha\right]}{\beta}\right)$

α is the location parameter
β is the scale parameter

WEIBULL: $F(x) = 1 - \exp\left(-\left(x/\alpha\right)^\beta\right), \quad x > 0$

$f(x) = \dfrac{\beta\, x^{\beta-1}}{\alpha^\beta} \exp\left(\left(-x/\alpha\right)^\beta\right)$

α is the scale parameter
β is the shape parameter

GAMMA: $F(x) = 1 - \exp\left(-x/\alpha\right) \displaystyle\sum_{j=0}^{\beta-1}\left(\dfrac{x}{\alpha}\right)^j \dfrac{1}{\Gamma(j+1)}$

$f(x) = \dfrac{x^{\beta-1}}{\alpha^\beta \Gamma(\beta)} \exp\left(-x/\alpha\right)$

α is the shape parameter
β is the scale parameter

INVERSE GAUSSIAN: $F(x) = \phi\left[\sqrt{\beta/x}\left(x/\alpha - 1\right)\right] + \exp\left(\alpha\alpha/\beta\right)\phi\left[\sqrt{\beta/x}\left(x/\alpha + 1\right)\right]$

279

$$f(x) = \sqrt{\beta / \left[2\pi x^3\right]} \, \exp\left(\frac{-\beta \left[x - \alpha\right]^2}{2\alpha^2 x}\right)$$

α is the mean
β is the shape parameter

POWER FUNCTION: $F(x) = \left(\dfrac{x}{\alpha}\right)^{\beta}, \quad 0 < x < \alpha$

$$f(x) = \frac{\beta \, x^{\beta-1}}{\alpha^{\beta}}$$

α is the scale parameter
β is the shape parameter

EXPONENTIAL POWER: $1 - \exp\left[1 - \left(\exp\left(\dfrac{x - \alpha}{\beta}\right)\right)\right], \quad x > \alpha$

$$f(x) = \frac{1}{\beta} \exp\left[1 + \frac{x - \alpha}{\beta} - \exp\left(\frac{x - \alpha}{\beta}\right)\right]$$

α is the location parameter
β is the shape parameter

UNIFORM: $F(x) = \dfrac{x - \alpha}{\beta - \alpha}, \quad \alpha < x < \beta$

$$f(x) = \frac{1}{\beta - \alpha}$$

α is the location parameter
β is the shape parameter

A method of matching moments was used to estimate the various parameters, α and β , and to ensure validity Kolmogorov-Smirnov goodness-of-fit tests were performed on the local models.

Empirical and theoretical models of device failure mechanism were applied to the local models thus generating global models. Wunsch and Bell (2) introduced a semi-empirical model for predicting EMP damage to semiconductor devices as a function of electrical overstress applied to the devices. According to this model, the power required to cause failure in a device is given by:

$$P = A\,\tau_p^{-0.5}$$

Here P is the power; A is a constant which is a function of junction area, thermal conductivity, density and specific heat and is known as the Wunsch damage factor; τ_p is the duration of the pulse. Lasca (3) used classical thermal analysis to expand the Wunsch equation which, ignoring steady state heating can be written:

$$P = P_0\left(\sqrt{t_0 t_1/t} + \sqrt{t_1/t}\right)$$

where t_0 is a constant in seconds describing adiabatic heating and t_1 is the quasi-adiabatic heating term. For a typical semiconductor, the transition from adiabatic to quasi-adiabatic heating is around 10ns to 100ns and the quasi-adiabatic region spans about three decades. Another empirical model of the Wunsch form was proposed by Jenkins and Durgin (4), for integrated circuits. It takes the form:

$$P = A\,t^{-b}$$

where P is failure power and t is the time to failure. The coefficients A and b may be found by least squares for a particular set of data.

In this particular study, the pulse-power width equivalent form of the theoretical and empirical models respectively take the form of:

$$x = x_0\left(\sqrt{t_0 t_1/t} + \sqrt{t_1/t}\right) \qquad \text{and} \qquad x = x_0 t^{-b}$$

When these are imposed on the local models, the distributions become functions of pulse width t. F(x) becomes F(x,t). Also, the α - and β - parameters become functions of pulse width.

$$\alpha = \alpha_0\left(\sqrt{t_0 t_1/t} + \sqrt{t_1/t}\right), \quad \beta = \beta_0\left(\sqrt{t_0 t_1/t} + \sqrt{t_1/t}\right)$$

$$\alpha = \alpha_0 t^{-b}, \quad b = \beta_0 t^{-b}$$

For a number of local models whose data was generated by performing tests at different pulse widths, the parameters may be estimated using linear regression techniques. As before, a Kolmogorov-Smirnov goodness-of-fit test was applied to each estimated global model and to the

failure densities.   This determines the worst deviation and thus model
validity.

DEVICE TESTING

Although the empirical and theoretical models referred to above have
been developed for square-and sine-pulses, they may be used to approximate
the total energy delivered to a device by a simulated EMP pulse which takes
the approximate form shown in Figure 1 (7).

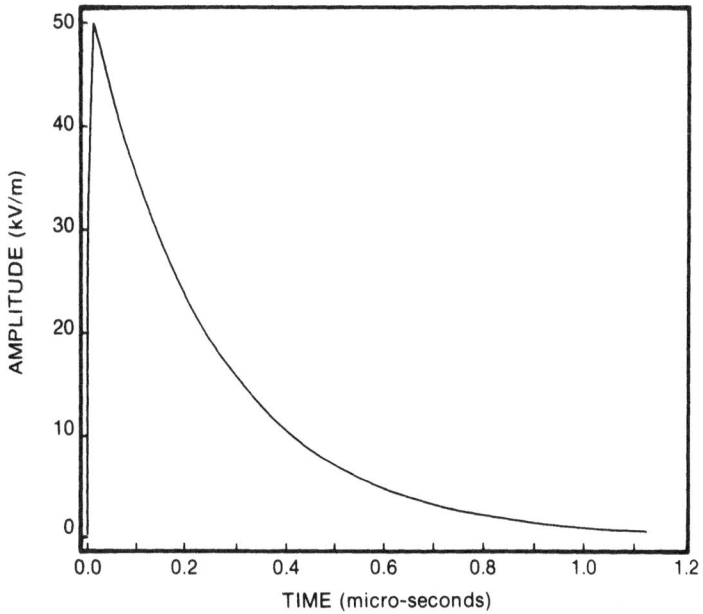

FIGURE 1: Typical EMP Waveform

The devices used in test were 2N3904 bipolar transistors, and low
power Skottky quad OR-gates. The test circuit (EMP simulator) is shown in
Figure 2.   The simulator consisted of two main blocks, a pulse generator
and a power amplifier.   The pulse generator consisted of two 555 timers,
the first was an astable multivibrator which trigged the pulses on its
falling edges.   The period of the output of the time was varied using a
variable resistor, thus controlling the duty cycle of the pulse train.   The
second timer is a monostable multistarter which was triggered on the
falling edges of the output of the first timer. The width of these pulses
being controlled by another resistor.   The power amplifier consisted of a

single NPN power MOSFEST in the common emitter mode. This was used because of its fast switching ability and very low turn-on resistance. This particular MOSFET was capable of amplifying pulses down to about 2 to 3 µs.

Using the equipment described above, high energy pulses were applied to the device under test (DUT). The voltage drop across the device was measured as was the current through the DUT. DUT resistance was quite low so was ignored.

The pulse amplitude was initially set to zero and after each test the amplitude was measured until a failure was detected. The power-to-failure in the preceding pulse was then noted. The DUT's were tested in reverse bias mode and failure criteria were defined to be step charges in the reverse breakdown voltage. Usually a single pulse was applied to the DUT at a time. Trains of pulses may also be used with a very low duly cycle of the waveform, so that the actual power of the pulse train is greatly reduced. In order to observe changes due to trains of pulses, transistors were subjected to a series of pulses and impedence was monitored. No changes in impedence were observed.

DATA GENERATION

Four batches of statistical samples of the 2N3904 bipolar transistors were tested at 3.4, 5.0, 7.5 and 10.0 µs respectively. The transistors underwent a pre-test impedance check and the test terminals used were the emitter-base in reverse bias mode with the collector open-circuited. The reverse breakdown voltage was 6V and the voltage needed to induce catastrophic failure in the transistors was about 12-15V -2 to 3 times the reverse breakdown voltage. Post-test impedance checks showed shorts across the base-emitter junction in the range of 1 - 7Ω.

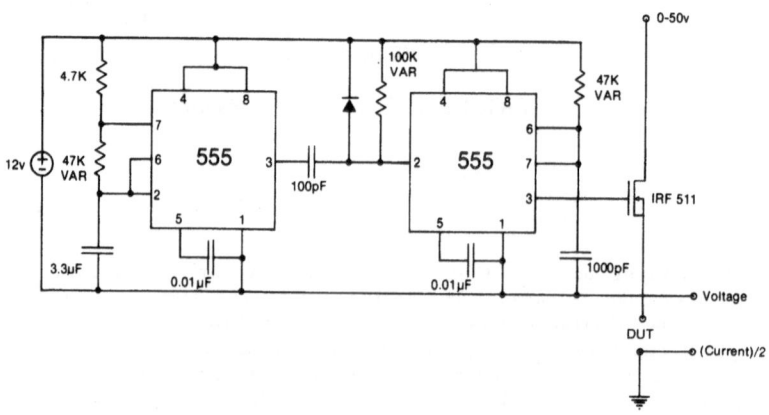

FIGURE 2: EMP Simulator Test Circuit

Three batches of statistical samples of low-power Schottky quad OR gates were tested at 5.0, 10.0 and 20 µs respectively. The positive terminal on the output and the negative on one of the inputs were the test terminals. In this manner all four gates on the device was capable of being tested individually. Hence four tests could be performed on each device. the reverse breakdown voltage and the failure voltages were noted. Failure criteria were set by a pre-test logic operation check. After the tests logic operation was found to be nonexistant and an impedence check showed a short across the tested terminals.

RELIABILITY MODELLING

For each distribution introduced above, the failure data was modelled thus producing ten local models. A method of matching moments was used to estimate the distribution parameters. For each estimation, the Kolmogorov-Smirnov goodness-of-fit test was performed and the parameter that gave the best goodness-of-fit was used for the local model. Probability-of-failure, hazard rates and reliability curves plotted with respect to absorbed power were generated for each distribution used.

As examples Figures 3 and 4 show the relability of the devices for the transistors and for the OR-gates respectively, plotted with respect to absorbed power, each for two of the distributions which produced good fits for the local models. Both Figures show results for pulse widths of 5µs.

The best fit distributions were the Weibull and power distribution for the transistors and the lognormal and Weibull for the OR-gates. Estimated parameters and maximum deviations for each of these distributions are shown in Table 1.

TABLE 1

2N3904 TRANSISTORS

DISTRIBUTION	α	β	Dmax
Weibull	149.3646	1.4481	0.1663
Power	1.0915	243.914	0.1356

LOGIC GATES

DISTRIBUTION	α	β	Dmax
lognormal	4.8859	0.5430	0.1155
Weibull	6.2160	2.1939	0.1255

284

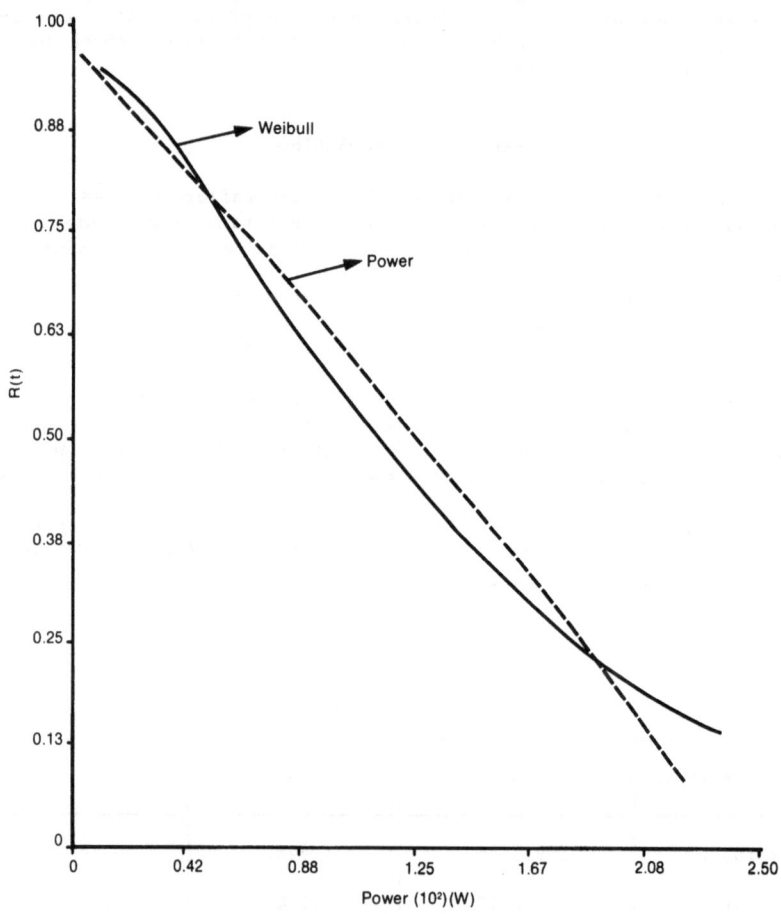

FIGURE 3: Reliability vs Absorbed Power for Local Distributions Weibull
and Power for 2N3904 Bipolar Transistors.

285

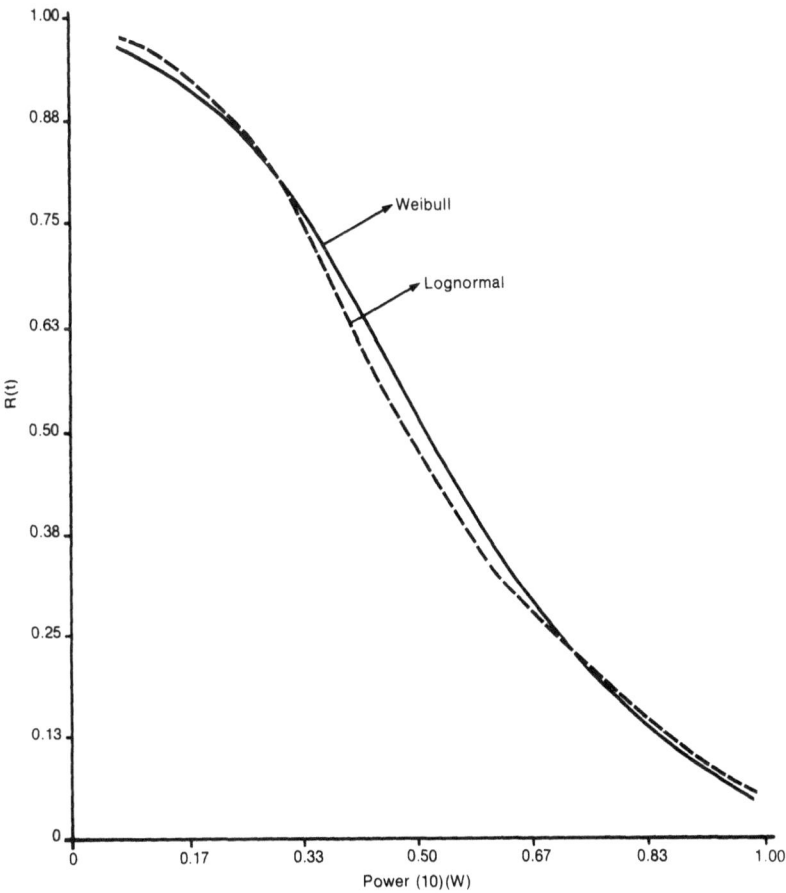

FIGURE 4: Reliability vs Absorbed Power for Local Distributions Weibull
and Lognormal for the Logic Gates.

As explained previously, the empirical and theoretical models were then imposed upon the local models in order to generate global models for the data. Table 2 shows relavent values for parameters α and β and the A and B estimations from the global models for the two best-fit global models - power function for the transistors and lognormal for the logic gates.

TABLE 2

POWER FUNCTION GLOBAL MODEL FOR 2N3904 TRANSISTORS				
α	β	A	B	Dmax
$639t^{(-0.587)}$	1.09	639	0.597	0.1540
LOGNORMAL GLOBAL MODEL FOR LOGIC GATES				
α	β	A	B	Dmax
$10.12t^{(-0.459)}$	0.2853	10.12	0.459	0.2608

The Dmax values refer to a 5% D-value used for a maximum deviation of 0.27 for a sample size of 24.

DISCUSSION

All the local models passed the goodness-of-fit test for the transistors. Two distributions, the two-parameter exponential and the inverse-Gaussian local models did not pass for the logic gates. Difficulty was encountered using the theoretical global model as none of the models passed the goodness-of-fit test. As can be seen from the previous section, the empirical model was perservered with and good fit results were obtained.

The results showed that for the transistors data the empirical model imposed upon the Weibull and Power Function local distributions provided the best fit with with the Power Function being slightly better. The cumulative failure density function for population below x at pulse width t follows the empirical power function global model:

$$F(x) = \left(\frac{x}{639\,t^{-0.597}} \right)^{1.09} , \quad x < 639\,t^{-0.597}$$

For the logic gates the empirical model imposed on the Weibull and the lognormal provided the best fits, with the lognormal slightly the better. The cumulative density function for the population below x at pulse width t follows the empirical lognormal global model:

287

$$F(x) = \phi \left[\frac{\log \left(\dfrac{x}{10.12\, t^{-0.459}} \right)}{0.2853} \right]$$

Both models follow the Bell-Wunsch equation closely and both indicate an increasing hazard rate in the test range. An increasing hazard rate can imply that a first strike can be devestating to an entire electronics system.

ACKNOWLEDGMENTS

The authors gratefully wish to acknowledge the financial support of the Natural Sciences and Engineering Research Council of Canada through the award of NSERC Grant No. A6039. The authors also wish to acknowledge the technical help of Mr. D. Roussy.

REFERENCES

1. Dickhaut, R.H., "Electromagnetic Pulse Damage to Bipolar Devices", Circuits and Systems, April 1976, pp. 8-21.

2. Wunsch, D.C. and Bell, R.R., "Determination of Threshold Failure Levels for Semiconductor Voltages", IEEE Transactions on Nuclear Science, Vol. NS-15, No. 6, December 1968, pp. 244-259.

3. Tasca, D.M., "Pulse Power Failure Models in Semiconductors", IEEE Transactions on Nuclear Science, Vol. NS-17, No. 6, December 1970, pp. 364-372.

4. Jenkins, C.R. and Durgin, D.L., "An Evaluation of IC EMP Failure Statistics", IEEE Transactions on Nuclear Science, Vol. NS-24, No. 6, Decmeber 1977, pp. 2361-2364.

5. Barry, D.M., Meniconi, M and Weir, N.A. "Reliability Modelling of Dynamic Random Access Memory Devices Subjected to High Temperature Soak Tests", Reliability Engineering, 17, 1987, pp. 255-266.

6. Bain, L.J., Smith, R.M., "An Exponential Power Life-Testing Distribution", Communications in Statistics, 4(5), 1975, pp. 469-481.

7. Jenkins, C.R. and Durgin, D.L., "EMP Susceptibility of Integrated Circuits", IEEE Transactions on Nuclear Science, Vol. NS-22, No. 6, December 1975, pp. 2494-2499.

THE USE OF POSSIBILITY CONCEPTS IN FUZZY LOGIC RELIABILITY APPLICATIONS

C. Kara-Zaitri
A. Z. Keller
Postgraduate School of Studies in Industrial Technology.
University of Bradford.

The concept of possibility as a quantity which is capable of logical expression but is not quantifiable in a probabilistic sense is developed with the use of fuzzy membership functions. It is demonstrated that because of its close relationship with every day semantics, it is of potential value in supplying information and interpreting results of reliability and safety assessments. It is shown that the technique is particularly valuable for "fusing" subjective information with other types of information and offers an alternative to Bayesian techniques. The utility of the concept and the associated methodology is demonstrated by applications to selected examples.

INTRODUCTION

A common practice of reliability engineering and safety assessment in particular is the use of "soft" data which is neither exact nor lends itself to exact analysis. This soft data can be vague and imprecise in many ways. Firstly, it may or may not be readily available and secondly, when it is available, it may or may not be appropriate. Although, probability theory has been utilized in such situations, it still does not delineate rigorously between what is probable and what is possible. For instance, in order to provide maximum system safety [1,2], it would seem that one would wish to protect against the possible as well as the probable. A theory that fuses the two features together is possibility theory [3].

One of the first to suggest the use of possibility rather than probability when handling soft data was the economist, Shackle [4] who defined possibility as a measure of potential surprise and presented axioms for his proposals. Shackle associated perfect possibility with a zero degree of surprise and impossibility with a maximum degree of surprise. Zadeh [5] also suggested the notion of a possibility interpretation of rational judgement and stated that, if not forced to, people think more of what is possible than what is probable. Later, Ellen Hisdal [6,7] extensively compared the relations between probability independance and possibility non-interraction of two fuzzy events.

The purpose of the present paper is to focus on the application of possibility theory to risk and reliability studies. Keller and Kara-Zaitri [8] have reviewed relevent tools of fuzzy set theory needed for understanding the proposed methodology.

PROBABILITY AND POSSIBILITY

One of the basic aspects of reliability engineering is that of a repeatable experiment. In well controlled laboratory conditions, it is often easy to perform an experiment several times. However, the conditions are similar and similar results are obtained. The notion of probability stems from that idea of repeated tests and the necessity of values to sum to one. It is based on Boolean theory and is therefore more useful for Natural Sciences but most inapropriate whenever subjective or rational judgement is required. In the real world, experiments are not always repeatable or sometimes even possible. In such circumstances, practitioners very rarely use the term probability which is a very well defined word. They are forced to get away from the restrictions of using one single and "crisp" value and use a whole spectrum of values each assigned a different degree of possibility. This may be extended further to the idea of probability of a possibility or even possibility of a probability [3].

It is important to note that the concepts of probability and possibility are not equivalent. The fundamental distinction between the two concepts is that events with a high possibility do not necessarily have a high probability. As stated by Zadeh [3], "What is possible may not be probable and what is improbable need not be impossible".

The difference between probability (Prob) and possibility (Poss) is best illustrated by an example. Let A be the event that represents the failure of the braking system of a car. Imagine that failure is caused by oil leakage (expressed in %), i.e., the ratio of lost and total oil capacity. Subjective values for possibilities and probabilities of A may be:

Leakage	0%	10%	20%	30%	40%	50%
Poss(A)	0	.3	.6	1	1	1
Prob(A)	0	.1	.5	.1	0	0

Thus it can be said that:

1. It is impossible as well as improbable that the brakes will fail if the leakage is 0%.

2. It is highly possible that the brakes will fail when the leakage exceeds 30%, but it is very improbable.

Imagine that the possibility of brakes failure is enquired when the leakage is small or large. In general, if:

N = Reference set
x = Element of reference set
$A \subset N$ and $Poss(A) = \left\{ u_A(x_i)/x_i \right\}$

$B \subset N$ and $\quad B = \left\{ \dfrac{u_{B}(x_i)/x_i}{} \right\}$

then,

$\quad$ Poss(A occuring with B) = Max $\left\{ \min [u_B(x_i), u_A(x_i)]/x_i \right\}$

or,

$\quad$ Poss(A occuring with B) = B o A

This operation is usually refered to as the min-max composition.

"Small" and "Large" are fuzzy words and can be defined by the following membership functions;

Leakage	0%	10%	20%	30%	40%	50%
Small leakage	1	.8	.2	.1	0	0
Large leakage	0	0	.1	.6	.8	.1

Note that "Large" does not represent "Not small" which is expressed as:

$$u_{Not\ small} = \left\{ 0/0,\ .2/10,\ .8/20,\ .9/30,\ 1/40,\ 1/50 \right\}$$

Clearly, the possibility of system failure caused by a small leakage should be low. By contrast, the possibility of system failure caused by a large leakage should be high; this can be verified as follows:

Poss(Failure when leakage is small) = Max$\left\{ \min[Poss(x_i), u_{Small}(x_i)]/x_i \right\}$

This is equal to :

Max$\left\{ \min(0,1),\ \min(.3,.8),\ \min(.6,.2),\ \min(1,.1),\ \min(1,0),\ \min(1,0) \right\}$
and therefore;

Poss(Failure when leakage is small) = .3

Similarly;

Poss(Failure when leakage is large) = Max$\left\{ \min[Poss(x_i), u_{Large}(x_i)]/x_i \right\}$

Poss(Failure when leakage is large) = 1

CONSISTENCY PRINCIPLE

One should note that although the notions of probability and possibility are not equivalent, they still have some common aspects. This consistency [3] can be summarised as follows:

1. An event which is highly impossible is bound to be improbable

2. A high level of possibility does not necessarily imply a high level of probability

3. The less the possibility of an event is, the less its probability of occuring, but not vice versa.

4. The degree of possibility of an event is always greater than its degree of probability.

APPLICATION TO HAZARD ASSESSMENT

Increasing the safety factor of a hazardous material containment system by increasing the burst pressure and working pressure ratio of a tank introduces new dangers in the event of a rupture.

Imagine that R (Rupture) is the fuzzy relation that links the burst and working pressures together. R can be thought of as the possibility of rupture and can be expressed as:

Values of R

		40	50	60	70	Burst pressure in units of pressure
Working pressure in units of pressure	10	0	0	0	0	
	20	.1	.1	0	0	
	30	.4	.3	.1	.1	
	40	.9	.9	.3	.2	
	50	1	.9	.9	.9	

To calculate the possibility of rupture when the working pressure is either low or high, assume that the fuzzy words high and low are defined by the following:

Working pressure	10	20	30	40	50
Low	1	.3	.2	.1	0
High	0	.1	.2	.9	1

Therefore,
Poss(Rupture when working pressure is low) = Low o R
This can be rewritten as:

$$= \begin{bmatrix} 1 & .3 & .2 & .1 & 0 \end{bmatrix} \circ \begin{bmatrix} 0 & 0 & 0 & 0 \\ .1 & .1 & 0 & 0 \\ .4 & .3 & .1 & .1 \\ .9 & .9 & .3 & .2 \\ 1 & .9 & .9 & .9 \end{bmatrix}$$

$$= \{Max(0,.1,.2,.1,0), Max(0,.1,.2,.1,0), Max(0,0,.1,.1,0), Max(0,0,.1,.1,0)\}$$

$$= \{.2/40, .2/50, .1/60, .1/70\}$$

This result makes sense because if the working pressure is low (approximately equal to 10 units of pressure) then the possibility of rupture for any of the previous burst pressures is very low.

Similarly,
Poss(Rupture when working pressure is high) = High o R

$$= \{1/40, .9/50, .9/60, .9/70\}$$

Note the high possibility of rupture for every burst pressure and particularly for that of 40.

APPLICATION TO HUMAN RELIABILITY

Imagine that human acts can be divided into two categories: Deliberate and non-deliberate acts. To illustrate the next example consider the first category only and suppose that these acts may be:

- Blind eye   (B)
- Sharp practice (S)
- Theft   (T)
- Fraud   (F)

It is emphasised that these four types of acts are fuzzy and overlap considerably but are hopefully useful for illustration. Let the sample space S be.

$$S = \{B, S, T, F\}$$

Consider the following fuzzy relation R expressed as:

	B	S	T	F
B	.50	.25	.13	.10
S	.80	.60	.14	.11
T	.90	.80	.90	.60
F	1	1	1	.90

The entries of the first row can be thought of as the possibility of the various kinds of acts following an act type B (Blind eye), and those of the second, third and fourth row represent possibility values following a sharp practice, fraud and theft respectively. Note that whilst the possibility of act B following B is .5, the possibility of F following B is .1.

Suppose that the human being under consideration has committed an act type B, what is the possibility of a moderate or grave act respectively ? Assume that moderate and grave can be defined by the following fuzzy sets.

	B	S	T	F
Moderate	1	.6	.2	.1
Grave	.1	.3	.7	1

Thus,

Poss(Moderate act after B)   = Moderate o B

$$= [1 .6 .2 .1] \quad o \quad \begin{bmatrix} .50 \\ .25 \\ .13 \\ .10 \end{bmatrix} = .5$$

Poss(Grave error after B)   = Grave o B

$$= [.1 .3 .7 1] \quad o \quad \begin{bmatrix} .50 \\ .25 \\ .13 \\ .10 \end{bmatrix} = .25$$

and therefore,
 Poss(Moderate act after B) > Poss(Grave act after B)

This result was expected because if one has turned a blind eye, one is more likely to commit a moderate act than a grave act.

APPLICATION TO DECISION TAKING

Possibility theory can also be applied to decision taking and provides a ranking of truth of different alternatives constrained by various states of the system.

Suppose that one is asked to buy both a cheap and reliable system. Cheap and reliable are fuzzy words and can perhaps be expressed in terms of available amount of money and possibility of failure respectively.

Suppose that a decision is to be made on whether it is better or worse to choose a very cheap or quite cheap system and a very reliable or not entirely reliable system. Assume that subjective assessments led to the following membership functions:

Possibility of failure	$\frac{-10}{10}$	$\frac{-8}{10}$	$\frac{-6}{10}$	$\frac{-4}{10}$
Reliable	1	.9	.3	.1
Not entirely reliable	0	.2	.9	.1
Very reliable	1	.8	.1	0

Available amount of money	¥100	¥300	¥600
Cheap	1	.6	.1
Very cheap	1	.4	0
Quite cheap	.2	1	.6

Since the goal is cheap (C) and reliable (R), the fuzzy relation T is defined in the Cartesian product (CxR) by the following transition matrix:

T

	$\frac{-10}{10}$	$\frac{-8}{10}$	$\frac{-6}{10}$	$\frac{-4}{10}$
¥100	1	.9	.3	.1
¥300	.6	.6	.3	.1
¥600	.1	.1	.1	.1

The elements of T(i,j) [3] are simply equal to min{cheap(i),reliable(j)}

Consequently, the truth [9,10] of each of the four alternatives can be computed in the following manner;

Alternative 1 as <u>Very cheap</u> AND <u>Very reliable</u>

Truth(Alternative 1) = Very cheap o T o Very reliable

$$= [1 \quad .4 \quad 0] \circ \begin{bmatrix} 1 & .9 & .3 & .1 \\ .6 & .6 & .3 & .1 \\ .1 & .1 & .1 & .1 \end{bmatrix} \circ \begin{bmatrix} 1 \\ .8 \\ .1 \\ 0 \end{bmatrix} = 1$$

Therefore, the four alternatives can be ranked as follows:

Alternative			Truth
1. Very cheap	AND	Very reliable	1
2. Quite cheap	AND	Very reliable	.6
3. Quite cheap	AND	Not entirely reliable	.3
4. Very cheap	AND	Not entirely reliable	.1

FUTURE WORK

It is intended to extend the use of fuzzy set theory to many other areas in the field of reliability studies. One such area is fuzzy function minimization which has found applications in optimization, artificial languages and pattern recognition. It is hoped to apply this minimization process to fault tree analysis for the determination of essential fuzzy prime implicants. Even for moderate size trees and without the fuziness of events, the manipulation is complex. An algorithm, based on bit manipulation techniques, has been developed and will be presented elsewhere.

CONCLUSIONS

- The concepts of possibility theory are appropriate for describing complex systems with a large uncertainty content such as human behaviour.

- Since the fundamental operation in possibility theory (min-max composition) is relatively simple, it can readily be implemented on computers for the studies of complex systems.

- The proposed methodology can be utilized in reliability and safety expert systems to determine the possibility or likelihood of a fuzzy event.

REFERENCES

1. A. Kandel. Fuzzy statistics and system security. Proceedings of the 1980 International Conference. Security through science and engineering.

2. L. J. Hoffman. Security and privacy in computer systems. Los Angeles. Melville Publishing Co. 1973

3. L. A. Zadeh. Fuzzy sets as a basis for a theory of possibility. Fuzzy sets and systems, 1, pp. 3-28, 1978.

4. Shackle. ......

5. L. A. Zadeh. Probability measures of fuzzy events. J. Math. Anal. Appl., 23, pp. 421-427.

6. E. Hisdal. Possibilically dependent variables and a general theory of fuzzy sets. Advances in fuzzy set theory and applications. pp.215-234, 1979.

7. E. Hisdal. Conditional possibilities. Independance and non-interractivity. Int. J. Fuzzy sets and systems, 1, pp. 283-297, 1978.

8. A. Z. Keller and C. Kara-Zaitri. Application of fuzzy logic to reliablity assessment. Proceedings of Reliability '87. Volume 1, 3A/3.

9. R. Bellman and L. A. Zadeh. Decision making in a fuzzy environment. Management science, 17, pp. 141-164, 1970.

10. P. P. Wang. Advances in fuzzy sets, possibility theory and applications. Plenum Press. New York. 1983.

PROBABILISTIC RISK ASSESSMENT ASSOCIATED WITH

FIRE SPREAD IN SEGREGATED STRUCTURES

A. Veevers and T.B. Boffey
Department of Statistics and Computational Mathematics

and

D.F. Yates
Department of Computer Science
University of Liverpool
P O Box 147, LIVERPOOL L69 3BX
UK

ABSTRACT

A segregated structure is regarded as a collection of volumes joined together by barriers. A fire, starting in one of the volumes, may spread through its bounding barriers to other volumes. The safety assessment of a vital piece of equipment contained in a particular volume, or passing through several volumes, needs to contain a contribution associated with the risk of damage by fire which starts in or spreads to these crucial volumes. A technique is described for doing this using the directed graph of a reduced network with 'fire transition' probabilities assigned to its arcs. This time-independent approach has been implemented in a computer program. Examples illustrating its use are given. Extensions which include the temporal aspect, and involve the stochastic modelling of the time taken for a fire to breach a barrier, are discussed.

INTRODUCTION

Fire is recognized as a major hazard in all walks of life. Wherever there is combustible material and a supportive atmosphere the threat of fire is present. Emphasis in the construction industry is on the prevention and the containment of fire; the risk to human life being the dominating consideration. It is often the case when loss of life is involved that death is caused by the inhalation of smoke or toxic fumes rather than by direct contact with fire itself. Consequently, a great deal of effort has been and continues to be, put into modelling the spread of smoke and fumes in buildings and other structures (1). If, however, interest is directed instead towards a vital piece of equipment (a component) which would be compromised by fire but not necessarily by smoke or fumes, then a different emphasis is necessary. For example, the component may be a safety device monitoring a nuclear reactor or a computer controlling processes in a chemical factory. One form of quantitative risk assessment for this situation treats fire in the same way as earthquakes and other external hazards; an occurrence probability is estimated and incorporated into the overall safety assessment. Techniques for arriving at the estimated figure are based on such things as past experience, total volume at risk, type of combustibles present, and are discussed, for example, in (2-4). The approach adopted here is to model quantitatively the spread of fire through the structure towards the vital component. Because of the uncertainties involved, a probabilistic model is required.

PROBABILISTIC NETWORK MODEL

Network models for postflashover fire spread from room to room in a building have been discussed by Ling and Williamson (see for example (5) and the references therein). The proposals made here are in the spirit of these developments.

Any structure which can be regarded as a collection of segregated volumes, for example, a building or a ship, can be represented by a network with volumes as nodes and barriers as arcs joining the appropriate nodes. The probability of a barrier being breached by fire, given that a fire exists on one side, is assigned to the corresponding arc of the network. Applying to this some standard and some novel techniques of graph theory, (6), enables the spread of fire through the network to be modelled.

To construct the model it is necessary to assign to each volume i an ignition probability p_i representing the probability of fire starting in that volume. These probabilities can be based on any realistic unit of time, such as one year. In addition, breach probabilities need to be assigned to each barrier separating adjoining volumes. These probabilities represent the likelihood of fire breaching the barrier given that fire exists on one side of it. Each barrier may have two such probabilities attached, one for each direction of burn-through. The severity of a fire in a volume clearly affects its ability to breach a barrier, so another probability can be assigned to each volume representing the likelihood of the fire becoming significant, e.g., reaching flashover, given that it begins in or enters the volume. These probabilities can, however, be subsumed into the barrier breach figures which may now be interpreted as follows. For adjacent volumes i and j, p_{ij} is the probability of fire reaching a significant intensity in volume i and breaching the barrier between volumes i and j given that fire exists in volume i. Clearly, these probabilities need to be derived by professional assessors.

As an illustration consider the single-story building portrayed in Figure 1. It has ten volumes, two of which, numbers 1 and 6, are corridors. (The dotted lines in volume 6 are sites of possible barriers which will be referred to later.) The network representation of the building is shown in Figure 2.

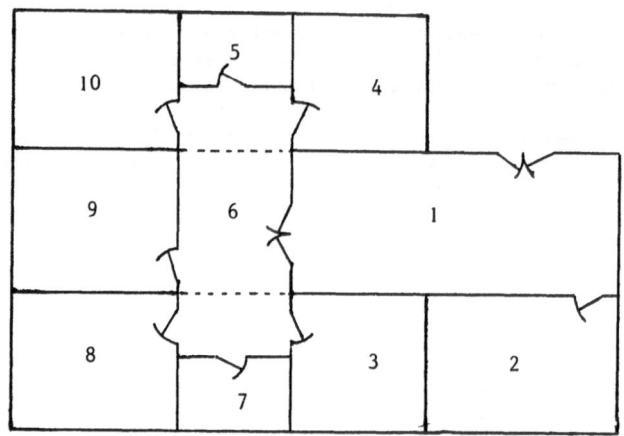

Figure 1.   Plan of a ten volume building.

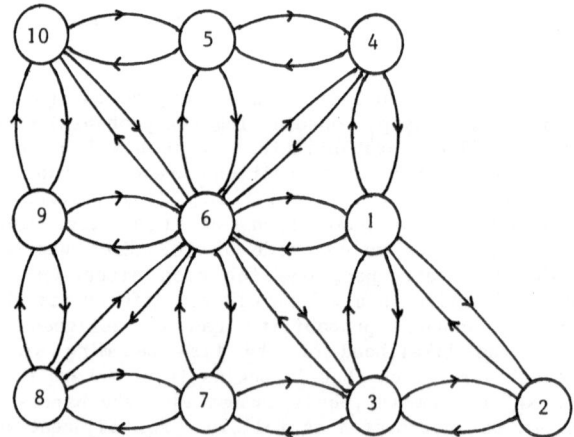

Figure 2.   Network representation of the ten volume building.

Typically, the sort of calculation required in the fire spread model is to determine the probability of fire spreading to a target volume, k say, when fire initially breaks out in volume i. One way of doing this would be to find all feasible distinct paths through the network from node i to node k and accumulate the probabilities associated with each path. The probability associated with a single path is the product of the breach probabilities on its route. Computationally this task becomes formidable for any practically useful number of volumes. To ease this computational burden a restriction is made to those fire paths having probabilities which exceed a specified 'cutoff' value. This cutoff is a user controlled parameter of the model, the default value being based on a highest probability path between components obtained using a modification of Dijkstra's algorithm (6). An efficient computational scheme has been devised which anticipates fire paths that will have probability less than the cutoff value; such paths are not developed further. The efficiency is enhanced by using an appropriate data structure for storing sparse networks (7). (In general, the term 'fire path' is not restricted to a single string of volumes but may refer to forked strings when several components are involved.) For each fire path the conditional probability thus obtained is then multiplied by the ignition probability p_i to give the probability of fire starting in volume i and spreading to volume k. The total risk to volume k is obtained by repeating this calculation for all volumes, i, in the structure, including the case i=k, and adding the contributions.

More generally, when a probabilistic risk assessment is being performed which involves a vital component passing through or housed in several volumes of a structure, some further considerations are required. If the component is considered to be failed when compromised by fire in any volume in which it resides then the above calculation needs an appropriate modification. Further extensions to include several vital components within the structure have been considered, in particular, the situation where duplicated safety systems are involved. A computer program has been written in FORTRAN 77 which implements the model and performs the necessary probability calculations.

A NUMERICAL EXAMPLE

Suppose that a safety system is duplicated in volumes 8 and 10 of the example structure and that ignition and breach probabilities as shown in Table 1 have been assigned.

The question of interest for this example is, what is the probability that both safety systems will be compromised by fire reaching their respective volumes? Running the computer code with the given information produces a probability of 4.121×10^{-5}.

The model used to make this assessment depends entirely on the specified ignition and breach probabilities and results obtained from it must be viewed with that consideration in mind. However, useful information, such as the identification of the greatest risk fire paths, can be obtained when breach probabilities are specified only in relative terms rather than absolute terms.

The model can be used in a design context to indicate the reduction in risk obtainable by introducing new barriers, such as corridor doors, or by up-rating the fire resistance of selected existing barriers. In the example, the effect of placing new barriers in the positions of the dotted lines in volume 6, with each-way breach probabilities of 0.02 and taking the ignition probabilities in each of the three volumes replacing volume 6 to be 0.02, is seen as follows. The probability that both safety systems will be compromised by fire is now 4.359×10^{-6}, which represents a significant reduction in risk.

Table 1. Ignition probabilities p_i are given on the diagonal, and breach probabilities p_{ij} off the diagonal, for the example building

i	j = 1	2	3	4	5	6	7	8	9	10
1	.05	.02	.01	.01	0	.02	0	0	0	0
2	.02	.05	.01	0	0	0	0	0	0	0
3	.01	.01	.05	0	0	.02	.01	0	0	0
4	.01	0	0	.05	.01	.02	0	0	0	0
5	0	0	0	.01	.05	.02	0	0	0	.01
6	.02	0	.02	.02	.02	.06	.02	.02	.02	.02
7	0	0	.01	0	0	.02	.05	.01	0	0
8	0	0	0	0	0	.02	.01	.01	.02	0
9	0	0	0	0	0	.02	0	.01	.01	.01
10	0	0	0	0	.01	.02	0	0	.01	.01

EXTENSIONS TO THE MODEL

In the form described above the model takes no account of time or of external fire-fighting measures which may be brought to bear. The two are related in the following sense: the interval from the time of outbreak of fire to the time when external fire-fighting has an impact, is essentially the period available for spreading to take place within the assumptions of the model. Attention is, therefore, confined to this period of combustion.

Let T_{ij} represent the time taken for a fire of significant intensity to burn through the barrier from volume i to volume j. After allowing for sources of controllable variability, such as the thickness of the barrier, amount and type of ducting running through it, the residual variation in T_{ij} can be represented by a random variable with a continuous probability distribution defined by a density function $f_{ij}(t;\theta)$. Here, t denotes time and θ is a (vector) parameter of the distribution which controls its shape. The probability of burning through the barrier within a time t_0 is, therefore

$$\Pr(T_{ij} \leq t_0) = \int_0^{t_0} f_{ij}(t;\theta)dt.$$

The time-independent model can be thought of as having breach probabilities that are all based on the same value of t_0, which can be regarded as the unit of time.

The probability associated with a particular single string fire path containing m breached barriers can, therefore, be considered to be based on m units of time. One way of incorporating the time factor into the risk assessment is to model the spread of fire in discrete time units as defined above. The probability that a component is compromised within any specified number of time units can then be obtained. Another way is to develop the model in continuous time and determine the probability distribution of the total time T_{i-k} for fire to spread from volume i along a prescribed path to volume k. Since

$$T_{i-k} = T_{ij} + T_{jl} + \ldots + T_{mk}$$

the problem involves the convolution of the contributing random variables. If each contributor has a distribution from a common parametric family, then straightforward solutions exist in some cases, for example, the exponential, gamma or normal families.

There is some evidence to support the Weibull family of models for time to burn through a barrier. The limited data on times to breach clay tile masonry walls in (5), for example, can be fitted by one of this family. Work is continuing in this area.

ACKNOWLEDGEMENT

The program referred to in this article was developed under contract with the Fire Safety Section of the UKAEA where it is known as ARSSUN

REFERENCES

(1) Chitty, R., In 'Computers model buildings to keep fire at bay'. New Scientist, October 1987.

(2) Lie, T.T. Fire and Buildings, Applied Science Publishers, London, 1972.

(3) Castino, G.T. and Harmathy, T.Z. (Eds), <u>Fire Risk Assessment</u>,
     ASTM, 1982.

(4) Thomas, P.H. Behaviour of fire in enclosures - some recent
     results. Fourteenth Symposium (International) on Combustion
     (1973).

(5) Ling, W-C.T., and Williamson, R.B., Modeling of fire spread
     through probabilistic networks, <u>Fire Safety Journal</u>, 1985, <u>9</u>.
     287-300.

(6) Boffey, T.B. <u>Graph Theory in Operations Research</u>, MacMillan,
     London, 1982.

(7) Boffey, T.B. and Yates, D.F. <u>Introduction to Computer Data
     Structures</u>, Chartwell-Pratt, 1988.

A TRIGGER-COUPLING MECHANISM MODEL
FOR
DEPENDENT FAILURES

A M Games

and

G M Ballard

National Centre of Systems Reliability,
UK Atomic Energy Authority,
Safety and Reliability Directorate,
Wigshaw Lane, Culcheth,
Warrington, UK, WA3 4NE

ABSTRACT

The recognition of the significance of dependent failures in system safety and reliability has led to increased international efforts to establish adequate procedures, models and data sources for their treatment.

This paper introduces the concept of coupling mechanisms in relation to the propagation of single to multiple failures. The resultant model provides a method for the incorporation of structured engineering judgement with available data in the recognition and quantification of dependent failures.

INTRODUCTION

Dependent failure events, which involve multiple failures of equipment and which breach redundancy, are now regarded as significant to system safety and reliability. For this reason, there has been much international effort to establish the procedures by which dependent failures should be accounted for in safety assessments.

Whilst there has been a notable step forward in agreement as to the general procedures to be employed, no agreement has been reached as to the models to be employed in the quantitative phases and there is still concern as to the lack of operational data to support those models which are available. Some of the models lack a clear engineering basis which could contribute to the understanding of dependent failure events and others require more information than the source data can provide.

Work is in progress to improve data availability by means of classification schemes for source data analysis and the development of databases for the storage of data in a suitable format; but, in parallel, a framework for incorporating both data and engineering judgement is required in the modelling of dependent failures.

There is now a clear recognition of the need to improve the understanding of the mechanisms underlying dependent failure events, together with an estimation of both the impact of such events on any specific plant design and the means of mitigating or defending against such events.

DEPENDENT FAILURE EVENTS

By definition, a dependent failure event is one in which there are failures, of multiple equipments or systems, which are not independent. As a consequence, the probability of occurrence of such an event is not accurately reflected by the normal multiplication of individual failure probabilities and may in practice be of significantly higher probability that would be indicated by such a calculation which assumes failure independence.

Analysis of operational events which have involved dependent, multiple failures suggests that the anatomy of such events may be described as shown in Figure 1 in terms of a trigger, coupling mechanisms and defences.

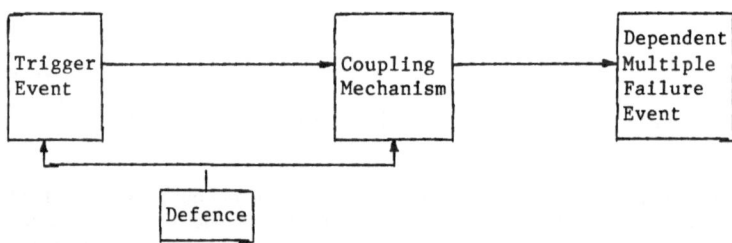

Figure 1 Essential anatomy of an event involving dependent failure of multiple components

Any multiple failure event must, obviously, involve at least one equipment failure. The root cause of that equipment failure is always a potential trigger for the failure of other items of equipment. Given that a cause for one equipment failure exists, there has to be a reason why other equipment should fail in a dependent manner. Thus the coupling mechanism is the link between one equipment failure and other equipment failures within the multiple failure event. In order for the trigger to lead to a dependent multiple failure event there has to be a coupling mechanism. If there is no coupling then a multiple failure event can involve only independent failures.

A coupling mechanism may be a deterministic or probabilistic effect and may propagate different trigger events to different degrees.

The coupling mechanism determines the conditional probability of multiple failures given the cause of one failure.

In order to reduce or mitigate the significance of multiple dependent failures there are various design or operational actions that can be taken. Clearly these actions may affect either the frequency of the trigger events or the likelihood of event escalation via the coupling mechanism and can thus be viewed as defences against dependent failure events.

When examining operational data it is important to realise that events are occurring in environments where some defences have been deployed on the plant in question. The occurrence of the event indicates which defences were not employed in relation to the equipment involved. Operational data does not usually indicate defences which were applied. Furthermore, the analyst using the operational data has to decide whether or not defences are present on his system which would mitigate against a similar event.

Another crucial factor is whether or not defences are applied to the triggers or the coupling mechanisms.

COUPLING MECHANISMS

From the above discussion it is clear that the driving force behind multiple dependent failure events is the existence of a coupling mechanism. So what is a coupling mechanism?

If there are two EXACTLY identical components being operated in EXACTLY identical environments (where "environment" signifies every aspect of a component's operation, operating conditions and maintenance) and if one component fails, the other component must also fail because the two are, in every way the same.

Of course, in practice, two components are never exactly identical and they are never operated in exactly identical environments. Thus there is some probability of the other component failing. This probability is dependent on the cause of failure and the extent to which the components and their environments really are identical. The cause of failure is the trigger and the measure of similarity between the components and their environments is expressed in the coupling mechanism.

The absolute defence against dependent failures is, therefore, DIVERSITY. The defences which affect the coupling mechanisms are those features which reduce the extent to which the components are the same.

To better explain the role of coupling mechanisms in failure events, the following examples are given:

Example 1

With the reactor at 20% power two reactor coolant pumps (RCP) were shut down for testing. the reactor protection system (RPS) failed to see either pump trip. The two RCPs were restarted and the other two RCPs were tripped. Again the RPS failed to see either pump trip.

Investigation of this incident, which involved the failure of 4 underpower monitoring circuits to detect pump trip, revealed that the relays had been adjusted two days earlier in accordance with the relay manufacturers specification. However, for this particular application, the relays had to be adjusted to operate at a specific pre-determined power level.

Analysis of this incident from a dependent failure point of view would indicate:-

• the design involved redundancy of relays which were of the same type and the same manufacture

• the relays were reset at the same time

• the relays were probably reset by the same staff using the same procedure in each case.

Consequently most aspects of the design and operation of these relays were the same. The only elements of diversity were those achieved by using multiple redundant components whose failures would have a degree of randomness associated with the manufacturing and operational history. Since inappropriate setting/calibration of relays is a recognised cause of relay failure (trigger event) the high degree of sameness between the relays (coupling mechanism) indicates a high susceptibility to multiple dependent failure of these relays.

Example 2

With the reactor at 84% power the high pressure coolant injection system (HPCI) was found to be inoperable. This situation required that the automatic depressurisation system (ADS) be checked. Due to improper use of the surveillance procedures all the ADS valves were inadvertently opened.

Analysis of this incident from a dependent failure perspective would indicate:-

• the valve surveillance was performed by the same staff member

• the valve surveillance was performed using the same procedure

• all the valves were surveilled at essentially the same time.

The multiple valve failures were due to an operator error (trigger event) propagated by a common timing, procedure and staffing of the valve surveillance process (coupling mechanism). In practice there was very limited diversity in this design but it is unlikely that diverse ADS valves would have materially influenced the incident because the sameness of the surveillance was the dominant mechanism for the incident.

In the above examples it can be seen that the dependent multiple failure events are being driven by coupling mechanisms which are related to the sameness between the redundant items or their operating environments.

The coupling mechanism provides the means for a trigger to propagate from one to many unavailabilities.

Since the triggers are the well-known root causes of component failure and are not necessarily related solely to dependent multiple failures, an analysis of events based on coupling mechanisms is a more meaningful approach.

Those models which look to the relationships between causes and defences without considering coupling mechanisms run into problems. Is a defence affecting only the dependent failure rate or both the independent and dependent failure rates? The use of coupling mechanisms enables the analyst to identify which defences attack root causes or triggers and which defences attack specifically the coupling mechanisms.

The implementation of diversity as a defence against coupling mechanisms could extend from the randomness of the manufacturing and operational history which exists for components in simple redundant systems to the almost complete diversity existing in a system with designed equipment, functional, location and environment differences.

A shortlist of coupling mechanism might be:

- same hardware
- same procedures: operation, test maintenance
- same staff: operation, test, maintenance
- same environment
- same location.

These categories could be sub-divided to add more discrimination. The list currently used by NCSR in its data analysis considers:

Same management and supervision
Same procedures
Same staff
Same equipment used
Same quality assurance

at each stage in the lifecycle of an engineering system, namely:

Specification
Design
Manufacture
Installation
Operation
Maintenance
Test
Calibration

in addition to:

same location
same environment
same hardware
and same timing.

308

RELATIONSHIP BETWEEN THE COUPLING MECHANISM
CONCEPT AND OTHER MODELS

The base β factor model (1) fits into the trigger-coupling mechanism framework shown in Figure 1 in the following way:-

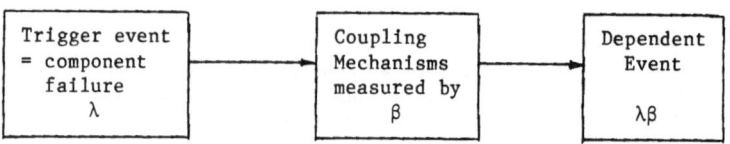

At this level of modelling all causes of the trigger events have been summed together in a component failure rate. Any multiple dependent failure must involve the failure of a component and the frequency of this from <u>all</u> causes is represented by λ. The reason why other component failures are linked to that failure is described in the coupling mechanisms. The conditional probability of a linkage being present is β.

The frequency of dependent failure events is related to component failure rates since such failure rates include all causes of component failure, including those that have a potential to propagate or couple to involve multiple failures. The conditional probability of individual component failures being part of multiple failure events will depend on many factors including the trigger event and the design defences involved to reduce potential coupling mechanisms.

The partial parameter models (2,3,4) have been introduced to reduce the degree of averaging of effects in terms of triggers and defences.

The partial β factor model represents an option of increasing the detail at the level of system defences. The model implicitly assumes that these defences affect the coupling mechanism. In practice some of the defences identified really apply primarily to the trigger event but they are assumed to affect β rather than λ. The defences are specified in terms of specific activities or design features and not directly related to specific coupling mechanisms.

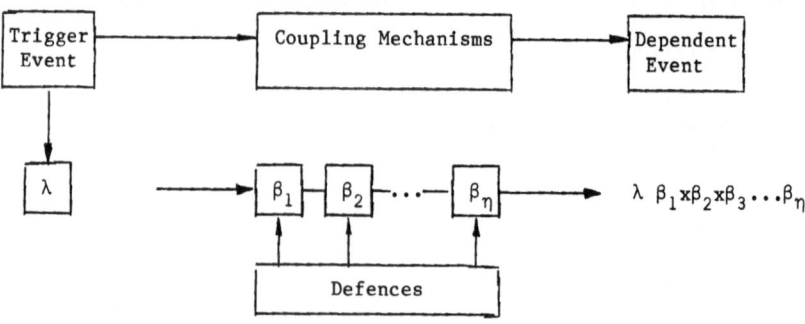

Since an overall limitation on all model development will be the availability of data to provide the basis for parameter estimation the extent of model detail should be viewed carefully. Even at the level of existing partial β factors the assignment is largely based on engineering assessment. Nonetheless, in the interests of clarity at least, it is worth stating the more detailed model which results from a division of the trigger events.

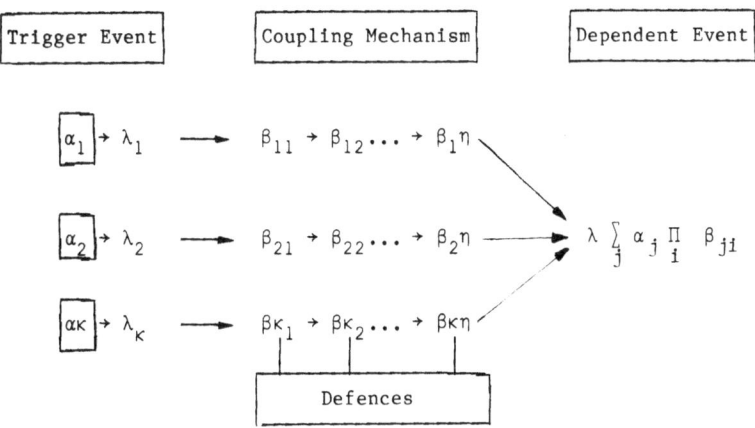

At first sight this may appear to be a comprehensive model which represents all the required detail. The only remaining problem being the assignment of the parameters of the model.

It should be noted that the model makes no reference to specific coupling mechanisms since the defences are typically couched in terms of activities or design features. Thus while the division of trigger events by cause of component failure (the categorisation currently suggested) recognises the initiating cause, there is no explicit recognition of the cause of propagation or coupling which leads to a multiple failure event. This issue is fundamental since it is at the level of coupling mechanism that the most effective action against dependent events can be taken. In addition it has proved difficult to relate dependent failure defences to the causes of component failure which provide the trigger events.

In existing models, therefore, the β factor is an expression of the concept of failure coupling but the coupling mechanisms are not explicitly identified and measured.

Another method which has recently been proposed for the quantification of dependent failures is the Distributed Failure Probability (DFP) method (5). It is a method which returns to the fundaments of the dependent failures problem in mathematical terms. The method recognises that

a   GIVEN a particular "environment", a component has a probability of failure, which can be expressed as a probability distribution function. $P(F/E)$.

b   The existence of any particular "environment" is probabilistic. $P(E)$.

The concept of "environment" embodies every aspect of a component's behaviour, operation, maintenance and ambient conditions.

The "environment" in which a single component exists is not described in terms of coupling mechanisms. The current "environment" in which a component operates can be described in terms of triggers or root causes (eg maintenance or operator error) and a general description of each component's continuous operating environment. The probability of an individual component's failure in a particular environment does not depend on the coupling mechanisms, as they are defined herein. That is to say, the coupling mechanisms do not describe an environment, but refer to similarities between the characteristics of "environments".

The advantage of explicitly recognising coupling mechanisms, however, is that it is easier to recognise potential defences against multiple failure. In removing a coupling mechanisms, one may produce a difference in component "environments". For example, introducing a fire barrier between two pumps creates separate "environments", or introducing separate maintenance teams for each redundant pump may create separate environments. The failure probabilities for the pumps are then altered and are different for each "environment". The concept of coupling mechanisms is therefore linked to the DFP model in terms of providing a systematic approach to determining whether components are operating in the same or different environments.

MODEL DEVELOPMENT

Returning to the basic model of dependent failure events the following points can be observed:-

i   Dependent failure events only matter when there is redundancy of equipment or function.

ii   If the redundancy involves truly identical parts in all aspects then theoretically the coupling probability should be one. In particular it is known that there is inadvertent difference which leads to some inherent randomness and thus a coupling probability less than 1. In addition some difference is introduced by design and this will also reduce the coupling probability.

iii   The identical nature of redundancy would not matter if there were no component failures, ie no trigger events. Thus reduction of the trigger events is also a defence against the frequency of dependent events. However, reduction of component failure rates is not specific to dependent failure issues and thus could be left to one side while remembering its potential significance.

The major issue therefore appears to be the degree of sameness between the redundancy elements which causes the coupling probability to vary. The coupling mechanisms are thus those factors in which redundancy elements can be considered to some degree to be the same. Thus, for example, an analysis of 100 reactor events involving dependent failures indicated that the relative contributions of the coupling mechanisms, irrespective of trigger event or type of component (within a largely mechanical orientation) were as listed in Table 1.

For this analysis coupling mechanisms which made a significant contribution to the occurrence of the event were assessed (not just a primary mechanism) and the relative frequency of occurrence of each coupling mechanism calculated. If the spectrum of trigger events was assumed fixed this table gives the relative significance, in dependent failure terms, of each coupling mechanism and thus measures the relative potential improvements in reliability that can be achieved by the use of various aspects of diversity.

Table 1. Relative frequencies of active coupling mechanisms in failure events

Same hardware (all aspects)		.47
Same staff	– operation	.08
	– test	.05
	– maintenance	.05
Same procedure	– operation	.06
	– test	.11
	– maintenance	.04
Same environment		.11
Same location		.03
		Σ 1.00

Further data analysis (using averaging across classes of components or engineering systems) should make significant progress towards the enumeration of a more detailed model.

Prior to the quantification of a dependent failure frequency/ probability the information from the qualitative reliability analysis should be formulated in a systematic way. The qualitative analysis will have generated minimum cut sets for system failure probabilities and event sequence frequencies. The cut sets will contain component failure probabilities. The task of the qualitative analysis is to identify those features of the components within individual cut sets that are the same. Thus each cut set is viewed as a mini-system in which we are concerned about potential dependent failures breaching the redundancy. In reference 6 a matrix formulation was used to display some of the features of the cut sets. More widely, however, it is necessary to be systematic in the identification of characteristics of the components that are the same including not only manufacture, type etc but also location, environment, test/maintenance procedures and timing, operational staff, etc.

The trigger event is some particular category of component failure causes which will depend on the data available. At the simplest level it may be the total failure rate of the component while, if more detailed data is available, it may be broken down into classes such as internal failure, human error in operations, test or maintenance, environment including all external influences.

The coupling mechanisms, which for practical purposes are independent are those features of sameness which may be expected to couple component

failures caused by the particular category of trigger events under
consideration. Thus, if λ_i is the total failure rate of the component, all
coupling mechanisms will need to be included. By contrast, if λ_i
represents component failures due to an internal fault, then the coupling
mechanism likely to be most effective is the same hardware.

The coupling mechanisms which are active in propagating a trigger
event depend upon the nature of the trigger. Three trigger event categories
are identified in Table 2 and these are linked to coupling mechanisms.
This list is not exhaustive.

Table 2. Suggested links between triggers and coupling mechanisms

Trigger Event	Coupling Mechanisms

Internal component failure	Same hardware
	  - manufacturer
	  - type
	  - design principle
	
	Same environment
	
	Same location

External caused component failure (including environment, energetic events etc) but not human error	Same hardware
	  - manufacturer
	  - type
	  - design principle
	
	Same environment
	
	Same location

Human error caused component failure (operations, test, maintenance etc)	Same hardware
	  - manufacturer
	  - type
	  - design principle
	
	Same staff
	  - operations
	  - test
	  - maintenance
	
	Same procedures
	  - operations
	  - test
	  - maintenance

If it is assumed that the nuclear power plants, covered by the event
data utilised, are typical of an industry average then the parameters
derived in Table 1 can be viewed as a base case reflecting past design
practice. The motivation for modelling is to be able to assess the
effectiveness of design and operational actions taken to reduce the impact
of dependent failure events on plant safety. The task is therefore to

modify the base case in a structured and systematic way to reflect the specific defensive actions taken. Since diversity has not been a widely used aspect of plant design in the past there is little data available and thus the assessment process will be largely based on engineering judgement guided by the structure and parameters of the coupling mechanism model.

In accord with the set of coupling mechanisms listed in Table 1 the list of defensive actions will be:-

- Diversity of hardware   - manufacture
                          - type
                          - design principle

- Diversity of staff   - operations/management
                          - test
                          - maintenance

- Diversity of procedures - operations
                          - test
                          - maintenance

- Diversity of environment

- Diversity of location (segregation)

- Diversity of timing

The potential effectiveness of these individual defensive actions is to eliminate entirely the contribution of a coupling mechanism whose industry average strength has been derived from available data. Within an event sequence cut set it should be possible to judge the extent of diversity, of the various types, using an engineering understanding of the potential failure modes of the components. On a scale of 1 (no diversity) to zero (total diversity) a multiplier on each coupling mechanism would thus be derived and would appropriately modify the strength of that coupling mechanism in establishing the conditional failure probability of a trigger event propagation. The application of engineering judgement is structured because the questions related to identified aspects of "sameness" between components in a cut set and not to an unspecific spectrum of potential dependent failure causes and defences.

Having derived the average strength of the coupling mechanisms the assessment of the degree of coupling between components in a minimal cut set would be assessed as follows:-

Coupling Mechanism	Average Strength	Degree of Diversity	Specific Strength
		$\gamma = (1 \to 0)$	
1	α_1	γ_1	$\alpha_1 \gamma_1$
2	α_2	γ_2	$\alpha_2 \gamma_2$
.	.	.	.
.	.	.	.
.	.	.	.
.	.	.	.
η	α_η	γ_η	$\alpha_\eta \gamma_\eta$
	1		≤ 1

where $\gamma = 1$ - industry average (as reflected in incidents)
      $\gamma = 0$ - high degree of diversity for coupling mechanism

The strength of coupling calculated in this way is relative to the industry average. Any parameters derived from the generic data, therefore, can be adjusted to allow for improvements introduced in new plant by way of increased diversity between environments.

Using the previous discussion the framework for performing a qualitative and quantitative evaluation of dependent failure probabilities for a specific design involves the following steps:-

i   Identify the event sequence or system cut sets of interest.

ii   Use available data to quantify the failure rate for appropriate trigger event categories and the conditional probabilities for coupling mechanisms within the trigger event categories. This is then the base case.

iii Identify the aspects of hardware, staff, procedures etc which apply to the components within each cut set.

iv   Within the coupling mechanisms (aspects of "sameness") assess the degree of coupling for the specific cut set and trigger event category. This assessment would be on a scale of 0 - 1 and reflect the extent of diversity achieved by specific intent.

v   Combine (ii) and (iv) to assess the dependent failure probabilities for each cut set.

Intentionally no specific quantification method has been discussed because the above framework can be used in association with many such methods including the β factor, Binomial Failure Rate, MGL etc.

The intent has been to provide an engineering framework to assess the parameters of these methods and to make them more specific to any particular application. This is in accord with the European Benchmark Exercise (7) which showed that the methods will give equivalent answers but major differences can be caused by parameter estimation. In that exercise SRD used a "partial β factor" approach which involved a combination of data and engineering judgement. The framework discussed here is an evolution from that approach to provide an improved basis for the incorporation of the more extensive data available and the essential engineering judgements.

CONCLUSIONS

It is widely agreed that events involving multiple, dependent failures are of major significance to the safety of complex systems. Such events occur because redundant elements within a system share common design and operational features and hence have common susceptibilities.

The coupling mechanism model provides the basis for combining actual data and structured engineering assessment in an analysis of dependent failure probabilities for any specific plant. The potential advantage of employing this concept as opposed to defence-orientated partial parameter models is that it is possible to clearly identify coupling mechanisms and assess their impact, whereas to assess the degree of "training", "awareness" or "design review", which are typical defences (not related to diversity) is very difficult. In the same way it is difficult to relate root causes directly to defences and to determine when applied defences are affecting the rate of independent failures or rate of dependent failures.

Coupling mechanisms bridge this gap between cause and defence and place diversity as the crucial and measurable factor in reducing dependent failure event probabilities.

REFERENCES

1   Fleming, K N, A Reliability Model for Common Mode Failures in Redundant Safety Systems. General Atomic Report, GA-A13284, 1974.

2   Edwards, G T, A Method of Assessment of Common-Mode Failures, Lecture Notes, National Centre of Systems Reliability, TTU/4.15.2, 1981.

3   Vesely, W E, Extending the Partial Beta Factor Approach to Cause Considerations, SAIC, Internal Report/Preprint, May 1986.

4   Paula, H M, A Restructured Approach to the Partial Beta Factor Method, JBF Associates Inc, Internal Report/Preprint, Oct 1986.

5   Hughes, R P, A New Approach to Common Cause Failure, CEGB Berkley Nuclear Laboratories, TPRD/B/0813/R86, May 1986.

6   Johnston, B D and Crackett, J, Common Cause Failure Reliability Benchmark Exercise, SRD Report R-383, UKAEA, 1985.

7   Poucet, A, Amendola, A, Cacciabue, P C, CCF-RBE Common Cause Failure Reliability Benchmark Exercise, EUR.11054 EN, JRC Ispra, April 1987.

BENEFITS OF RELIABILITY AND QUALITY ASSURANCE PROGRAMMES IN U.K. MANUFACTURING INDUSTRY

M.H. Abed
A.S. Sohal
A.Z. Keller

Postgraduate School of Technology and Management Science
University of Bradford
Bradford BD7 1DP
U.K.

ABSTRACT

This paper presents a preliminary analysis of the organisational and structural relationships between reliability and quality assurance programmes. The operation of these functions and the use of specific techniques in these areas and the barriers to their acceptance have also been discussed. The collection of reliability and quality costs data and its use by top management in decision-making relating to future improvements have also been covered.

The paper is based on a postal questionnaire survey carried out in the U.K. manufacturing industry. The preliminary analysis shows that the importance placed on reliability by top management is approximately one-third of that placed on quality.

INTRODUCTION

As well as price and dependable deliveries, high quality and reliable products and services are now the competitive weapons being used by many manufacturing and service industries. In order for companies to survive against international and other competition they need to achieve substantial cost reductions. The application of reliability and quality assurance programmes represent powerful instruments for major savings. To quantify the benefits of reliability and quality assurance programmes in industry the Postgraduate School of Studies in Technology and Management Science at the University of Bradford (U.K.) initiated a research project in October 1986. Another aim of this research project was to identify the organisational and structural relationships existing between reliability and quality assurance. Earlier research in this area has concentrated on the use of specific reliability and quality techniques and some work has also been carried out relating to costs. The present research focuses on the organisational and structural relationships and the benefits of using quality and reliability programmes in the manufacturing industry.

THE SURVEY

The use of postal questionnaires to gather representative amounts of data from a given population are now widely accepted. Many studies in the

United Kingdom investigating management systems and techniques have made use of the postal questionnaire and found it to be a successful method of data gathering. Although the membership of an appropriate body or association is a common and convenient population from which to draw a sample, it does lead to a bias towards respondents who should be somewhat advanced in thier knowledge and/or usage of reliability and quality assurance programmes. Keeping this in mind, the sampling frame used for this study was largely the membership of the British Quality Association affiliated to the Institute of Quality Assurance.

The questionnaire was designed to fulfil the following main objectives:

(i)   To establish the organisational and structural relationship between reliability and quality assurance.
(ii) To establish the organisation and operation of reliability and quality assurance functions.
(iii) To establish the nature, extent of usage and barriers to acceptance of specific techniques in these areas.
(iv) To establish cost and economic benefits of reliability and quality assurance programmes.
(v)   To establish the educational and professional qualifications of practitioners and their extent of communication with professional associations and institutions.

The present paper presents a preliminary analysis relating to (i), (ii), (iii) and (iv) above. Results relating to (v) will be presented elsewhere.

ANALYSIS OF THE RESPONDENTS

Within the time allowed (10 weeks) a total of 112 valid, completed questionnaires were received, giving a response rate of 26.6 percent.

The sample covered all sizes of manufacturing units and industrial sectors (standard industrial classification) except for Leather, Leather Goods and Furs, Clothing and Footwear and Timber and Furniture. Just over three quarters (76.9%) of the respondent companies were subsidiaries or divisions of a parent company. Almost one-fifth (18.8%) of the respondent companies supplied their products to government agencies and nearly two-thirds (64.3%) supplied to other manufacturing companies. Only 9.8% supplied directly to the public and 7.1% to own agencies.

To simplify the analysis the size of manufacturing unit was categorised into one of the following:

        small   - up to 199 employees
        medium - from 200 up to 999 employees
        large   - over 1,000 employees.

The rate of technological change in the production processes of the responding companies over three different time periods (10 years ago, 5 years ago and current) varied considerably. Over the past 10 years there had been moderate change in technology, however, the current rate of technological change was much faster, indicating that most of the manufacturers had now realised that one way of competing internationally was to adopt new or advanced manufacturing technologies.

An analysis of the importance given to competitive priorities which
manufacturers were setting for themselves showed some interesting results.
Table 1 shows the rank ordering of these priorities. Also shown in Table 1
are the competitive priorities rank ordering for Europe, North America and
Japan taken from the 1985 Global Manufacturing Futures Survey [1]. The
present survey showed that price was the most important competitive factor
in competing internationally which was in agreement with the Japanese but
this differed from the Americans and Europeans who placed high priority on
their ability to offer consistently high quality products. Quality,
reliability, reputation and dependable deliveries were considered to be the
important competitive priorities by all the major manufacturing nations.

Table 1:   Competitive Priorities

Ranking	Present Survey (U.K.)	Global Manufacturing Futures Survey 1985		
		Europe	North America	Japan
1	Price	Consistent Quality	Consistent Quality	Low Prices
2	Quality	High Performance Products	Dependable Deliveries	Rapid Design Changes
3	Reliability	Dependable Deliveries	High Performance Products	Consistent Quality
4	Reputation	Low Prices	Fast Deliveries	High Performance Products
5	Stated Delivery	Fast Deliveries	Low Prices	Dependable Deliveries
6	Design	Rapid Design Changes	After-Sales Service	Rapid Volume Changes
7	After-Sales Service	After-Sales Service	Rapid Design Changes	Fast Deliveries
8	Periodical Delivery	Rapid Volume Changes	Rapid Volume Changes	After-Sales Service

The style of management within an organisation is of paramount importance
to the efficient running of the company and in understanding the quality
and reliability functions. Nearly half of the respondents (44.2%) said
that the style of management within their organisations was based upon
encouraging autonomy within clearly defined limits and regular head-office
monitoring. Another 36% said that the style was to encourage autonomy
of operating companies and departments, but without head-office monitoring
and for 18% of the responding companies, head-office provided central
guidance and clearly specified operating procedures. The larger the
operating company the less the involvement of the parent company. These
results agreed with the findings of the research carried out by the Aston
Group [2].

ORGANISATION OF QUALITY AND RELIABILITY

Financial Budgets

More than three-quarters (76.8%) of the respondents claimed that their
organisations prepared a formal budget for the quality control department
whereas only one in six companies (16.2%) did so for reliability. The
proportion of companies having a formal financial budget increased with
an increase in the size of the manufacturing unit. For example 89% of the
larger companies had a formal quality budget compared with 54% of the

smaller companies. An analysis by industrial sectors showed that a higher proportion of the companies in the Food, Drink and Tobacco, Vehicle, Aerospace and Electrical Engineering industries prepared a quality budget than companies in other industrial sectors. A higher proportion of the companies in the Vehicle and Aerospace industry were found to have a budget for reliability than companies in other industries. This finding is perhaps not surprising because the products manufactured by this industry require high reliability.

Status and Structure of Quality and Reliability Functions

Over ninety percent (91.1%) of the respondents said that their companies had a separate quality control department whereas only a fifth (18.8%) claimed to have a separate reliability department. This reflected the fact that most of the companies considered reliability as part of the engineering function or as part of quality. The proportion of companies having a separate reliability department increased steadily as the size of the manufacturing unit increased. However, this is not the case for quality. 92.3% of the smaller size companies, 85.4% of the medium sized companies and 95.6% of the larger sized companies had a separate, independent quality department.

The person responsible for the quality control department had the title of Quality Assurance Manager (35.3%), Quality Manager (34.3%) or Quality Control Manager (14.7%) (Others 8.8%). Only 6.9% of the respondents responsible for the quality control department had the title of Quality Director, this reflects that few companies had board members directly responsible for the quality function. Titles given to the person responsible for reliability varied considerably, the most common being Technical Manager (22%) and Reliability Engineering Manager (10.1%). It should be noted that reliability was not directly represented at board level.

Senior management involvement in quality and reliability is an indication of the importance placed on these functions within an organisation. The higher the level of management involvement, the more likely it is that company policy is influenced by reliability and quality considerations.

Nearly three-quarters (74.4%) of the respondents indicated that the quality control function received adequate support and approval from top management. This finding was consistent with that reported in earlier literature - see for example Crosby [3] and Feigenbeum [4]. In comparison less than one-third (31.4%) of the respondents indicated that top management gave their full support to the reliability function.

The use of outside consultants on quality and reliability problems is also an indication of senior management commitment to these functions. Table 2 shows the proportion of companies using consultants on quality and reliability problems. Outside consultants are tended to be used on quality problems.

Table 2: Use of Outside Consultants on Quality and Reliability Problems
        (n = 112)

Frequency of Usage	%age of Companies Using Outside Consultants on:-	
	Quality Problems	Reliability Problems
Always	-	-
Often	1.8	1.8
Sometimes	25.9	9.8
Seldom	39.3	15.2
Not at all	33.0	36.6
Not Answered	-	36.6

The style of communication is another indicator of the importance of
quality and reliability within an organisation. It affects the efficiency
and progress of individual departments. Table 3 shows the respondents
responses to the level of formality and regularity and mode of communication.
It can be seen that communication relating to quality is mostly formal, in
writing and communicated on a regular basis. The larger the size of the
organisation the more formal and regular the style of communication. A
formal style of communication increased certainty to a greater degree than
an informal system [5]. This does not appear to be the situation regarding
reliability communication. The style is less formal and less regular than
that for quality.

Table 3: Management Style of Communication Related to Quality and
        Reliability

Style of Communication		% of Total Respondents	
		Quality	Reliability
Level of Formality	Very Formal	22.8	14.3
	Mostly Formal	64.5	16.2
	Somewhat Informal	11.2	18.1
	Very Informal	1.9	7.6
Regularity of Communication	Very Regular	38.8	16.4
	Mostly Regular	54.4	16.4
	Seldom Regular	3.9	11.8
	Completely Ad-hoc	2.9	13.6
Method of Communication	Always in Writing	29.1	18.5
	Mostly in Writing	68.0	26.9
	Seldom in Writing	2.9	9.3
	Never in Writing	-	2.8

Nearly three-quarters (71.5%) of the quality control department heads
report directly to senior management at 'director' level (38.4% reporting
directly to the Managing Director, 17.9% to the Technical Director and
15.2% to the General Director). The remaining 28.5% report to various
managerial levels, i.e. Manufacturing Site Manager, Works Manager and
Operations Manager. It is clear that most companies consider quality as
a separate function from production in terms of responsibility, however, it
is not adequately represented at the board level, i.e. by a Quality Director.

Over half (57.1%) of the heads of the reliability department reported
directly to senior management at 'director' level, most of them (47.6%) to
the Managing Director. The remaining (42.9%) of the reliability heads
reported to various managerial levels, i.e. Operations Manager, Technical
Manager and Engineering Manager. In terms of the size of manufacturing
unit, proportionately more of the reliability heads in the larger companies
report directly to the Managing Director than in the smaller companies.

Table 4 presents an analysis of the number of persons employed in
quality and reliability as a function of the size of the organisation. The
larger the company the larger the number of persons employed in the quality
department, however, no similar relationship can be deduced for the
reliability function.

Table 4: Number of Persons Employed in Quality and Reliability as a
Percentage of the Total Sample
(n = 112)

Number of Persons	SIZE OF THE COMPANY						TOTAL SAMPLE	
	SMALL		MEDIUM		LARGE			
	Q n=26	R* n=8	Q n=41	R* n=20	Q n=45	R* n=21	Q n=112	R* n=49
Less than 6	57.8	100.0	9.8	65.0	8.9	52.3	20.5	65.3
6 - 10	34.6	-	2.4	15.0	4.4	9.5	10.7	10.2
11 - 15	3.8	-	17.1	15.0	6.7	-	9.8	6.1
16 - 20	3.8	-	19.5	-	2.2	9.5	8.9	4.1
Over 20	-	-	51.2	5.0	77.8	28.7	50.0	14.3

Q = Quality
R = Reliability

*Note only 18.8% of the companies had reliability departments and
not everyone of these answered this question.

OPERATION OF QUALITY AND RELIABILITY DEPARTMENTS

Responsibilities

Tables 5 and 6 show the level of responsibility (total, partial or none) for a range of quality and reliability activities. More than three-quarters of the respondents claimed complete responsibility for quality audit or quality assurance (84.8%) and quality instruction activities (80.4%) in their companies, and another 9.8% and 17.0% respectively claimed partial responsibility. The proportion of respondents claiming total responsibility for vendor rating, scrap and rework and quality costs reporting, which could be considered as essential components of the job of a quality control manager, were considerably lower. Within the various industries the highest proportion of quality control managers having complete responsibility for quality audit, quality instruction manuals, vendor rating, quality control budget and quality costs reporting were those in the Electrical Engineering, Mechanical Engineering and Vehicle and Aerospace industries. The lowest proportion of quality control managers having complete responsibility for many of these functions were those in the Food, Drink and Tobacco industries. Only 37.5% of the managers in this industry claimed complete responsibility for quality audit and quality instruction manual compared with 85% and 78% respectively in the other industries.

Table 5: Level of Responsibility of Quality Control Managers for Different Activities

Activity	Level of Responsibility of Respondents (%) (n=112)			
	Complete	Partial	None	Not Applicable
Quality Audits or Quality Assurance	84.8	9.8	5.4	-
Quality Instruction Manuals	80.4	17.0	2.6	-
Vendor Rating	48.2	44.6	6.3	0.9
Customer Liaison	19.6	70.5	8.0	1.9
Scrap Rework Costing	17.9	48.2	33.0	0.9
Quality Control Budget	59.8	27.7	9.8	2.7
Quality Cost Report	37.5	33.9	25.0	3.6
Product Design	5.4	58.0	33.9	2.7
Other	-	2.7	92.0	5.4

Total Responsibility for quality audit and quality instruction manuals increased with an increase in the size of the manufacturing unit whilst total responsibility for product design and customer liaison decreased with increasing size. This is perhaps explained by the fact that these functions tend to be controlled by specialists in the larger companies. It can be concluded that quality control managers are now being involved with all activities within the manufacturing organisation, consistent with a gradual shift toward quality assurance. Although only 18.8% of the responding companies had separate reliability departments, many of the reliability activities were being undertaken by a number of other companies. Hence, the percentages given in Table 6 are based on the total sample size (n=112). A higher proportion of the respondents claimed complete responsibility for reliability instruction manuals, product reliability analysis and measurement, and part evaluation and failure analysis than any of the other activities listed in Table 6. A lower proportion had complete responsibility for reliability budget, maintainability and reliability assurance and statistical processing. This is surprising since these activities could be considered as essential components of the reliability manager's job. Within the various industries, a higher proportion of reliability managers from the Electrical Engineering and Vehicle and Aerospace industries have complete responsibility for product reliability analysis and measurement, and maintainability and reliability assurance than those in consumer product industries. The lowest proportion of reliability managers having complete responsibility for many of these activities were those from the Mechanical Engineering and Food, Drink and Tobacco industries. Low responsibility/involvement of the reliability managers in the various reliability activities could be explained by the lack of knowledge of senior managers of reliability and the poor encouragement given by them to the reliability managers.

Table 6: Level of Responsibility of Reliability Managers for Different Activities

Activity	Level of Responsibility of Respondents (%) (n=112)			
	Complete	Partial	None	Not Applicable
Reliability Instruction Manual	17.3	9.1	14.5	59.1
Annual Warranty Payment	3.6	5.5	29.1	61.8
Reliability Budget	8.2	12.7	19.1	60.0
Product Design	7.3	20.9	15.5	56.4
Product Reliability Analysis and Measurement	20.9	13.6	8.2	57.3
Maintainability and Reliability Assurance	11.8	20.0	9.1	59.1
New Product Development	6.4	21.8	9.1	59.1
Part Evaluation & Failure Analysis	14.5	17.3	10.0	58.2
Statistical Processing	8.2	20.0	13.6	58.2
Annual Repair and Maintenance Costs	0.9	10.0	28.2	60.2
Other	0.9	1.8	97.3	0.9

Inspection

In the present survey 68.7% of the respondents indicated that inspection was considered as part of the quality control department in their organisation which is in agreement with Kidwell [6] and Reddy [7]. Inspection must be considered as part of production since it is an integral part of the manufacturing process.

Usage of Specific Techniques and Barriers to their Acceptance

A number of studies have been carried out in the U.K. investigating the extent of usage of quality control and reliability techniques and the barriers to their acceptance (see References [8] and [9]). The present survey has shown that a slightly higher proportion of respondents make use of the various techniques of statistical quality control and reliability. One can draw two conclusions from this. One that the situation in the U.K. industry has improved or secondly, the higher usage of the techniques has resulted from the use of a biased sampling frame. However, the reasons given for non or low usage of various techniques are similar to those identified in previous studies. These are 'not applicable', 'benefits not identified', 'successful without using the techniques', '100 percent inspection' or 'small batches'.

Most companies do not understand the benefits that can be obtained from using relevant techniques in the quality and reliability areas and those companies understanding the benefits, often do not understand the application of the techniques.

Reliability Data Banks

Less than a quarter (23.2%) of the respondents said that they used a data bank relating to reliability activities, i.e. failure rates, reliability estimates, warranties, performance and market research. The larger the size of the organisation the more likely it was to establish its own data banks, however, there was considerable variation between the different industrial sectors. A higher proportion of the companies in the Electrical Engineering and Vehicle and Aerospace industries have established their own data bases.

EXPENDITURE ON QUALITY AND RELIABILITY

Almost two-thirds (63.4%) of the respondents claimed that their organisations collected data relating to quality costs compared with only 19.6% collecting data relating to reliability costs. The proportion of companies collecting quality cost data has increased significantly over the past few years indicating that manufacturers are now much more conscious of quality costs than in the 1970's. Roche [10] found that only 39% of the companies were collecting quality costs data, whilst Dale and Dunclaf [11] found that 32.7% of the respondents to their survey collected quality cost data. In this context it is interesting to note that, although companies were claiming to collect such costs, few had established systems for collecting the whole elements relating to quality costs. Previous researchers have shown that few companies operate a cost data collection system which allows any kind of meaningful analysis to be carried out with a view to future cost saving [12].

Table 7:   Quality and Reliability Costs as a Percentage of Turnover

Costs Percentage Range	%ages of Respondents (n=112)	
	Quality	Reliability
Unknown	7.1	9.8
Less than 3	12.5	8.0
3 - 5	17.9	5.4
6 - 9	10.7	-
10 - 19	11.6	-
Over 20	3.6	-
Not Answered	36.6	7.1
Not Applicable	-	69.7

The range of quality costs relative to turnover are shown in Table 7.   Not surprisingly a large proportion of the respondents did not give an answer to this question or admitted that they did not know.   An analysis by industrial sectors indicated a considerable variation in costs across the different industries.   This agrees with Crosby [3] that cost definitions and elements may differ within a single company, between accounting systems used and even among fully informed personnel [13].   This leads to the observation that no standard available can be regarded as definitive in this respect and cannot be considered as an indicator of the costs.   The proportion of respondents claiming that their organisations record or estimate the various cost sources is shown in Table 8, again indicating that all sources of costs are not identified, estimated or measured.

Table 8:   Quality and Reliability Failure Elements

Cost Source	%age of Respondents (n=112)
Scrap or Waste	79.5
Repair and Rework	70.5
Troubleshooting	16.2
Scrap and Rework Fault of Vendor	52.7
Returned Product Downtime and Reworking	62.5
Warranty Replacement	50.0
Complaints	48.2
Warranty Repair Cost	43.8
Spares, Production and Inventory Costs	25.9
Maintenance Costs	27.7

More than half (56.2%) of the respondents claimed that they used quality costs data in decision-making relating to quality improvements although 63.2% claimed that they collected quality cost data.   In comparison, 24.1% claimed that they used reliability cost data in decision-making relating to reliability improvements although only 19.6% had indicated that they

collected reliability cost data. Because of the ambiguity of these
results further study is required to resolve it.

A number of the respondents indicated the need to modify their existing
accounting system in order to provide the necessary information relating
to quality and reliability costs. 57% of the respondents said that their
existing accounting system was not adequate for providing information on
quality costs and 48% said the same regarding reliability costs.

The method of measuring the performance of quality and reliability
programmes is of fundamental importance in assessing their successful use.
Table 9 shows the various methods used by the respondents. A number of
companies used more than one method. The most commonly used methods are
'cost-benefit analysis' and 'rate of return'. The least used method of
assessing performance is turnover. Various other methods of measuring
the performance of reliability programmes were indicated by the respondents.
These included size of warranty, reliability measurement, mean time between
failures and percentage of failures.

Table 9: Methods of Measuring the Performance of Quality and
          Reliability

Method	% of respondents (n=112)	
	Quality	Reliability
Cost-Benefit Analysis	33.6	15.6
Rate of Return	18.7	14.7
Turnover	8.4	1.8
Profit on Turnover	14.0	3.7
Profitability	19.0	4.6
Other	20.0	6.4

CONCLUSIONS

The preliminary analysis of the survey shows that the importance placed on
reliability by senior management is approximately one-third of that placed
on quality. Just under a fifth (18.8%) of the respondents claimed that
their organisations have a separate reliability department and only 16.2%
said that their organisations prepared a formal financial budget for
reliability. In comparison, more than three-quarters (76.8%) claimed that
their organisations prepared a formal financial budget for quality with
91.1% having a separate quality control department. In none of the
responding companies was reliability directly represented at the board
level whereas 6.9% of the companies had board members directly responsible
for the quality function having the title of Quality Director. Senior
management's style of communication regarding reliability is less formal
and less regular than that for the quality function. The larger the
organisation, the larger the number of persons employed in the quality
function, however, no similar relationship could be deduced for the
reliability function.

The usage being made of specific reliability and quality techniques is
low and most companies do not understand the benefits that can be derived

from their usage. Companies that understand the benefits, often do not understand the application of the techniques.

Only one in five companies said that they collected data relating to reliability costs whereas two-thirds collected quality costs data. Around half of the respondents said that their existing accounting system was not adequate for providing information on reliability and quality costs and indicated the need to modify their existing accounting systems. There was some ambiguity regarding the use of reliability and quality cost data in decision-making relating to future improvement. A higher proportion claimed to use reliability cost data in decision-making than actually claimed to collect it. Further study is required to resolve this ambiguity.

Further in-depth analysis of the questionnaire survey is underway and this will be followed by in-company work which will look at a number of reliability and quality assurance programmes. The results of this work will be reported in due course.

REFERENCES

1. Ferdows, K., Miller, J.G., Nakane, J. and Vollmann, E.E., "Evolving Global Manufacturing Strategies: Projections into the 1990's", Int. J. Operations and Production Management, 1986, 6, 4, pp 6-16.
2. Pugh, D.S. and Hickson, D.J., Organisational Structure in its Context, The Aston Programme I, Saxon House, Great Britain, 1976, (pp 43-76).
3. Crosby, P.B., Quality is Free, McGraw-Hill, New York, 1983, pp 132-9.
4. Feigenbaum, A.V., "Total Quality Control", McGraw-Hill, New York, 1983, pp 823-9.
5. Price, J.L., Organisational Effectiveness: An Inventory Proposition, Richard D. Irwin, INC., 1971, pp 163-183.
6. Kidwell, J.L., Inspection Perspective and Prospectives, Quality Progress, 1976, 9, 8, pp 28-30.
7. Reddy, J., Incorporating Quality in Competitive Strategies, Solan Management Review, 1980, 21, 3, pp 53-60.
8. Lockyer, K.G., Oakland, J.S. and Duprey, C.H., Quality Control in the U.K. Chemical Manufacturing Industry - a Study, Part I, Int. J. Production Research, 1981, 19, pp 317-325.
9. Oakland, J.S. and Sohal, A., Production Management Techniques in U.K. Manufacturing Industry: Usage and Barriers to Acceptance, Int. J. Operations and Production Management, 1986, 7, 1, pp 8-37.
10. Roche, J.G., National Survey of Quality Control in Manufacturing Industry, National Board for Science and Technology, Ireland, 1981, pp 3-45.
11. Dale, B.G. and Duncalf, A.G., Quality - the Linchpin to British Manufacturing Success, Management Decision, 1984, 22, 1, pp 30-39.
12. Dale, B.G. and Plunkent, J.J., "A Study of Audits, Inspections and Quality Costs in the Pressure Vessel Fabrication Sector of the Process Plant Industry, Proceedings of the Institution of Mechanical Engineers, 1984, Part B, 198.
13. British Standard Institution, "The Determination and Case of Quality Related Costs", BS 6143, 1981, HMSO, London, pp 1-14.

328

THE TAGUCHI APPROACH TO DESIGNING FOR RELIABILITY

John Disney and Professor Tony Bendell
Department of Mathematics, Statistics and Operational Research
Trent Polytechnic
Burton Street Nottingham NG1 4BU

ABSTRACT

Taguchi methods of optimum product and process design are now well established in the United States as very efficient methods for ensuring reliable and robust products. Corporations such as Ford, Bell, ITT and Xerox are using them extensively. They have become embedded in the whole of the American car and electronics industries.

In the UK and Europe, however, the developments of the last 5-10 years in the United States have largely gone unheeded. A few companies, notably those with American links, have begun to apply these methodologies with extremely successful results. These companies such as Lucas, Rank Eeros and Eaton Transmissions are now well established. The potential interest in the methodologies, however, is enormous and recently more than 150 people attended a one-day conference in London, specifically on Taguchi methods, organised by the Institute of Statisticians Working Group on Taguchi Methodology. Under the Chairmanship of Professor Tony Bendell, this Group is attempting to establish the potential for the application of Taguchi methods within British manufacturing industry via the recent formation of the UK's Taguchi Club.

Taguchi methods are concerned with the routine optimisation of the product and process prior to manufacture, rather than emphasising the achievement of quality through inspection. Instead concepts of quality and reliability are pushed back to the design stage where they really belong. The method provides a highly efficient technique to design product tests prior to entering the manufacturing phase. It is essentially a prototyping methodology which makes use of efficient statistical methodology for carrying out the experimental design.

In this paper the authors review the nature of Taguchi methodology and indicate the current state of application within British industry.

HISTORICAL BACKGROUND

Dr Genichi Taguchi was assigned to the Electrical Communications Laboratory of the Nippon Telephone and Telegraph Company in 1949 with the task of improving Japan's poor post-war telephone communication system. He soon discovered that excessive time and money was spent by Research and Development departments in conducting experiments and introduced the idea of experimental design to improve efficiency. Taguchi also discovered the limitation for industrial experimentation of both R A Fisher's Analysis of Variance techniques and Dr Motosaburo Masuyama's orthogonal Latin squares, developed during the war for medical research. (Taguchi, [1]).

An experiment consists of testing combinations of different values, termed levels, of factors thought likely to influence the characteristic of interest. Each combination or experimental unit is known as a trial. If the factors are not independent, their interaction may also be considered. To dampen the effect of systematic changes, the trials should ideally be conducted in a random order, known as randomization, and replicated.

Industrial experiments involve a large number of factors, levels and interactions between factors. The basic traditional one-at-a-time method measures the effect of varying one factor whilst holding the remaining factors constant, requiring a large number of experiments for even a small number of factors and levels. A full factorial experiment combines the levels of each factor with each of the levels of all the remaining factors. With a smaller number of experimental trials required, randomization of the order of experiments becomes easier and replication becomes feasible. The full factorial design has the advantage of being able to estimate interactions between factors.

However, full factorial designs become very large as the number of factors increases: seven two-level factors require $2^7 = 128$ experiments in a full factorial design; twenty three-level factors require over three billion experiments to be conducted. Performing the latter design at the rate of one trial every second would take 110 years to complete. Fractional factorial designs assume that higher-order interactions are negligible and run a fraction of the full factorial, typically a $(\frac{1}{2})^p$ fraction of a two-level factorial design or a $(\frac{1}{3})^p$ fraction of a three-level factorial design. Fractional factorial designs are used in screening experiments to identify factors which have a large effect for further, more intensive analysis. However, they still have the disadvantage of requiring a large number of experiments to be performed.

Against this background of large cumbersome experiments with associated complicated analysis, Taguchi has developed orthogonal arrays, to aid efficient conduct of experiments, and simple graphical analyses.

QUALITY AND THE LOSS FUNCTION

In contrast to Western definitions, Taguchi defines the quality of a product as "the (minimum) loss imparted by the product to society from the time the product is shipped". (Taguchi [1], [2]). This loss includes not only the loss to the company through costs of reworking or scrapping, maintenance costs, downtime due to equipment failure and warranty claims, but also costs to the customer through poor product performance and reliability, leading to further losses to the manufacturer as his market share falls. Taking a target value for the quality characteristic under consideration as the best possible value of this characteristic, Taguchi associates a simple quadratic loss function with deviations from this target. This loss function shows that a reduction in variability about the target leads to a decrease in loss and a subsequent increase in quality (see Fig 1).

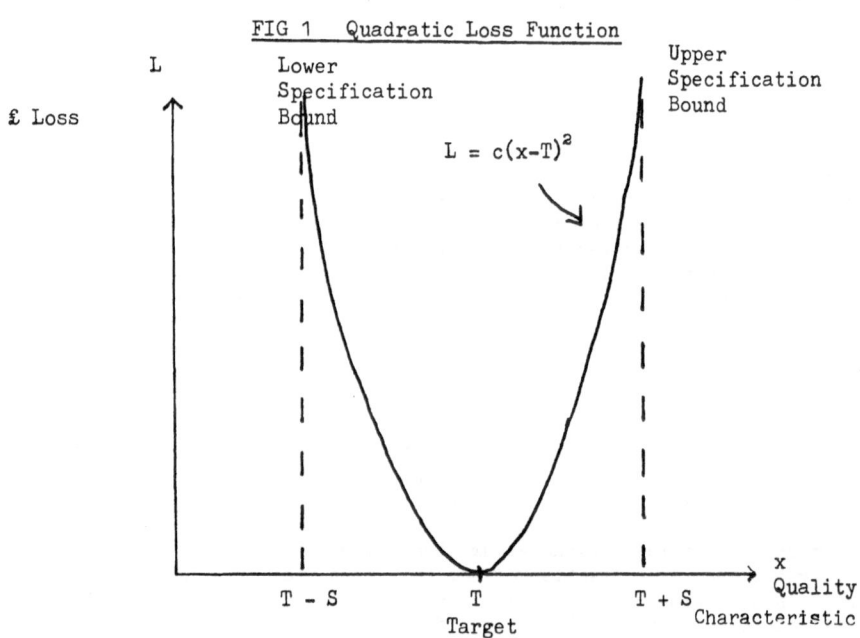

FIG 1   Quadratic Loss Function

$$L = c(x-T)^2$$

L = £ Loss
c = cost co-efficient (constant)
x = quality characteristic under consideration
T = target value
S = tolerance

Notice that with this conception a loss will occur even when the product is within the tolerance allowed, but is minimal when the product is on target. If the quality characteristic or response is required to be maximised (e g strength) or minimised (e g shrinkage) then the loss function becomes a half-parabola (see Fig 2).

FIG 2   Quadratic Loss Functions for Minimised
        and Maximised Responses

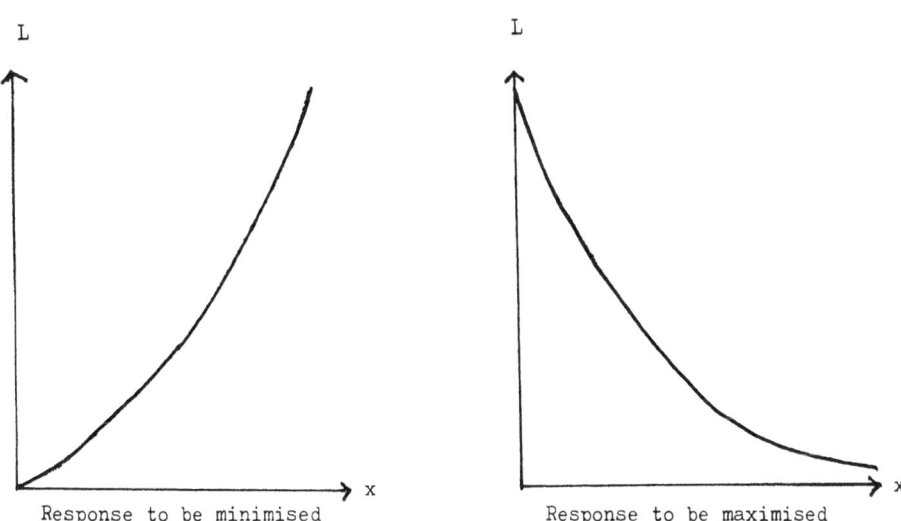

Response to be minimised   Response to be maximised

The loss function may be used to evaluate design decisions on a financial
basis to decide whether additional costs in production will actually prove
to be worthwhile in the market place.

OFF-LINE QUALITY CONTROL

Taguchi methodology may be applied as off-line quality control in the
design stage or, less commonly, as on-line quality control during
production.   In the firm belief that if quality is designed into a
product, on-line quality control becomes much less important, emphasis
will be concentrated on off-line quality control.   For a full discussion
on the use of Taguchi methods in on-line quality control see Taguchi [3].

Taguchi breaks down off-line quality control into three stages
(Fig 3)

(1) System design

(2) Parameter design

(3) Tolerance design

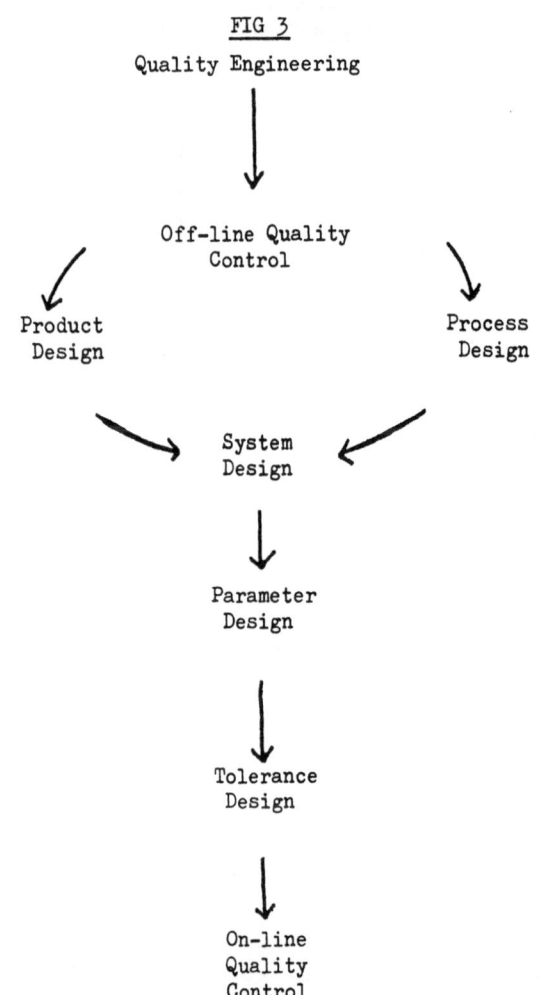

FIG 3
Quality Engineering

Off-line Quality
Control

Product
Design

Process
Design

System
Design

Parameter
Design

Tolerance
Design

On-line
Quality
Control

In the system design stage, parts and materials are selected and possible product parameter levels determined for product design, whilst for process design decisions are taken on the selection of equipment and possible levels for process factors. System design is best achieved by a "brainstorming" session involving engineers and designers.

Parameter design is the crucial step - this is where the Japanese excel at achieving high quality levels without an increase in cost. The nominal factor levels selected by system design are tested and the combination of product parameter levels or process operating levels least sensitive to changes in environmental conditions and other uncontrollable (noise) factors is determined. There are several unique aspects to this part of Taguch's work, notably the emphasis on the signal-to-noise ratio, the use of efficient orthogonal arrays to reduce the number of experimental trials necessary (Table 1) and the evolutionary "zooming in" by sequential experimentation.

TABLE 1

Common Orthogonal Arrays

Orthogonal Array	Factors and Levels	Number of trials in a full factorial experiment
(Suffix indicates number of trials)		
L$_4$	3 x 2 levels	8
L$_8$	7 x 2 levels	128
L$_9$	4 x 3 levels	81
L$_{12}$	11 x 2 levels	2 048
L$_{16}$	15 x 2 levels	32 768
L$_{16}$	5 x 4 levels	1 024
L$_{18}$	1 x 2 levels 7 x 3 levels	4 374
L$_{36}$	3 x 2 levels 13 x 3 levels	12 754 584

The two L$_{16}$ arrays are different but are labelled the same because they both involve 16 trials.

Finally, tolerance design is employed to reduce variation further if required, by tightening the tolerances on those factors shown to have a large impact on variation. This is the stage at which, by utilising the loss function, more money is spent if necessary buying better materials or equipment, emphasising the Japanese philosophy of "invest last", not "invest first".

CURRENT STATE OF APPLICATIONS IN THE UK

The success of Taguchi methods in recent years in the United States has led to some awareness of their crucial importance within British Industry. The Institute of Statisticians has founded a Taguchi Methodology Working Group and a Taguchi Club to promote Taguchi methods within Britain and several companies are now aware of the opportunities to reduce inspection and reworking costs, by designing quality into new products or existing processes using Taguchi methodology.

Within Britain, Taguchi methods are being applied perhaps most extensively by Lucas Industries and Rank Xerox. Lucas embarked on an ambitious internal training programme in May 1986 with a seminar given by Dr Genichi Taguchi and one year later were on target to having all one thousand engineers within Lucas trained as Taguchi practitioners together with twenty Taguchi specialists by December 1988. Within twelve months, twenty Taguchi projects have been completed with thirty more underway in all aspects of product and process design.

Rank Xerox, under the guidance of Taguchi specialist Ray Greenall, have also embarked upon a large training programme and have completed thirty experiments in nine months on design activities and manufacturing processes. A typical example of the success of these methods is given by studying an experiment involving an injection-moulded photocopier housing supplied by an outside trade moulder with a 20-30% rejection rate. To have conducted a full factorial experiment involving all possible factors would have been an immense task, but following a "brainstorming" session, seven control factors were identified and an L_{18} orthogonal array was

used to find a completely different optimum combination of factor levels which, on verification with a production run of 3000, reduced the rejection rate to virtually zero. This has eliminated the need for the process to be supervised, thus freeing manpower and reducing unit cost, and a much better photocopier housing is now being produced. Rank Xerox are so delighted with such successes that Ray Greenall is now planning to use Taguchi methods to aid the design of a new photocopier, starting with the fifteen major sub-assemblies and aiming to achieve a robust design as quickly as possible so that design changes during development can be kept to an absolute minimum.

References

[1]   Taguchi, G and Wu, Y-I (1979), Introduction to Off-line Quality Control, Central Japan Quality Control Association

[2]   Taguchi, G (1986), Introduction to Quality Engineering, Asian Productivity Organisation

[3]   Taguchi, G (1981), On-line Quality Control During Production, Japanese Standards Association

A full bibliography is obtainable on request from the authors

OVERHAUL COST ESCALATION DUE
TO MECHANICAL WEAR-OUT

R.F. de la Mare
University of Bradford Management Centre, England

ABSTRACT

Over the past 15 years, the Technological Management Unit of the
University of Bradford has investigated the breakdowns and repairs of plant
and machinery working under different operating and maintenance conditions.
A common conclusion of this research is that mechanical equipment
invariably deteriorates, in a reliability sense, despite maintenance
attention to prevent this from happening. This conclusion has been found
so widespread as to suggest the need for an underlying theory to explain
the reasons for the deterioration of repaired mechanical equipment.

The case study presented here was developed to test whether this
conclusion applies to mining equipment which receives the most
comprehensive overhauling upon breakdown. The results confirm that these
machines exhibit a similar deterioration, which is statistically
significant, and that the resulting escalation in their overhaul costs
is of such a magnitude as to warrant their systematic replacement. This
deterioration results from the wear-out of the components, parts and
bodies of these machines which cannot be rectified completely by overhaul.

INTRODUCTION

Previous research by several authors such as Berg (1977), Davies (1952)
and Kugel (1965) has shown that mechanical equipment is seldom repaired to
a standard which is 'as good as new' in a reliability sense. They concluded
that a deterioration in reliability performance, made manifest by declining
Mean Times to Failure with successive repair, inevitably occurs. Berg(1977)
examined this matter in detail and confirmed these findings whilst
researching the breakdowns and repairs of large populations of centrifugal
pumps, valves and compressors. He discovered that their deterioration was
so significant as to point the way to substantial economies through their
systematic replacement by brand new machines. He suggested that this
deterioration in reliability resulted from inappropriate maintenance
policies, which fostered an ad hoc approach to the repair and replacement
of parts, which had failed or shown signs of incipient failing and the

use of maintenance procedures which lacked proper measurements and standards.

The purpose of the work reported here was to test the hypothesis that standard proprietary mechanical equipment, which does not undergo further development, exhibits a similar decline in reliability performance even when the repairs conform to the most exacting standards which are industrially feasible and commercially viable. This project involved a detailed study, lasting four man-years, of the breakdowns, repairs and costs of coal-mining machinery which receives the most comprehensive overhauling on failure.

A DESCRIPTION OF THE EQUIPMENT

This project involved Powerloader equipment which is used to shear mineral coal from the coal faces of underground mines and loads it onto armoured flexible conveyors for removal to the pit-head. These machines are electro-pneumatically operated and they comprise four distinct units:

1. The gear-head which is the cutting unit and can be elevated to contact the coal face.

2. The mechanical haulage unit, which propels the machine along the mine corridor into the coal face.

3. The electric motor, which provides the automotive power for the complete machine and

4. Reduction gear-boxes.

Due to limits imposed on the size of this paper, subsequent results deal only with the reliability performance and costs associated with one type of Gear-head. These Powerloaders extend to 10 metres in length and cost in excess of £100,000. They are used in a variety of mineral environments containing seams of coal ranging from sandstone to siltstone and mudstone, which exhibit different degrees of hardness and toughness. Furthermore the environmental conditions of coal seams can vary considerably due to differences in termperature, humidity and corrosivity.

In the event of any major breakdown which would render a Powerloader inoperable, it is removed to a work-shop for overhaul.

POWERLOADER MAINTENANCE POLICY

To prolong the economic lives of these machines in a cost effective manner, routing maintenance schedules are formulated for all the Powerloader assemblies at every mine. Each schedule comprises a detailed checklist itemising the nature of the jobs to be done. The frequency of the examination and subsequent repairs is chosen to take account of the conditions under which the machine is working but, in any event, each Powerloader receives at least one detailed inspection and subsequent repair per week, which is undertaken by skilled tradesmen. It will be appreciated that the very nature of the mining environment precludes major repair work underground, so these repairs involve the replacement of broken and worn external parts, the repair of electrical connections, fixing and tightening loose parts and lubrication.

A machine is sent to the workshop for three reasons:

1. Due to a breakdown failure which cannot be remedied within the mine.

2. When a coal face is exhausted and the machine cannot be deployed immediately elsewhere within that mine and

3. Due to wear beyond acceptable limits or due to the detection, during periodic inspection at the mine, of other symptoms which would cause the machine to fail if left in operation.

It is policy for all machines sent to the workshop to undergo a major overhaul if their repair is justified. This results from the very high costs of transporting these machines between the workshop and the coal face, coupled with the extremely high opportunity costs, due to their failure, making the substantial cost of their complete overhaul economically justifiable.

Upon receipt at the workshop, each machine is first inspected externally for missing parts, which are logged and then it is cleaned and again inspected for damage. It is then stripped and examined in detail by highly trained inspectors who determine whether an overhaul is justified. Each inspection involves the measurement of parts and tolerances and their comparison against standards to determine the work content of the overhaul. In the event of an overhaul being justified the inspectors determine the overhaul strategy for the tradesmen to follow initially and later they decide what exactly is to be repaired or replaced and how the overhaul should proceed. As the machine is reassembled so the inspectors test the proficiency of each maintenance action against standard measures and other technical criteria. In effect therefore the overhaul of these machines is most comprehensive and proficient.

THE RELIABILITY ANALYSIS OF GEAR-HEAD FAILURES

For the purpose of this research, a failure of a gear-head is defined as any event which necessitated its removal from the coal face to the workshop for overhaul, due to its breakdown.

Two reliability attributes were monitored; the operating time and the coal production between successive overhaul. The former measured the time during which the machine was positioned at the coal face whereas the latter measured the linear metres of coal actually cut, since the machine was initially installed or reinstalled at the coal face until it subsequently failed.

Cu-sum graphs which plotted the cumulative numbers of failures against the cumulative operating time and the cumulative production indicated that there was a deterioration in the performance of these machines. This was confirmed using the Reverse Attribute Test for trend which is credited to Mann (1945) and the method ascribed to Prochan and Pike (1967).

338

To enable a much more detailed investigation of this phenomenon to be
undertaken it was decided to conduct a reliability analysis using order
statistics. In this research a Breakdown Order Number is the serial
number of the failure of a machine as it occurs such that the first
breakdown has an order number of unity whereas the next breakdown of
the same machine has a breakdown order number of 2. By these means it
was possible to differentiate between samples of gear-heads which had
different repair histories.

Due to the relatively large number of sample statistics involved
with these gear-head failures (69 gear-heads sustained in excess of
300 failures during the test period) it is not possible to provide the
full listing of their failure attributes here. Instead, only processed
data are provided. Table 1 illustrates the changes in the Mean Time
to Failure (MTTF) and the Mean Production to Failure (MPTF) with each
successive breakdown. These values allow for the fact that some of
these machines had not failed by the time the tests were complete.

TABLE 1

Gearhead Average Time and Production to
Failure for each Successive Breakdown

Breakdown Order Number	Sample Size	MTTF (days)	MPTF (1000 metres)
1	69	279	132
2	69	251	123
3	69	272	115
4	69	294	130
5	52	289	128
6	35	216	79
7	21	237	54
8	15	207	73

To gain a better understanding of the reliability attributes of
these machines, their times to failure and production to failure for
each breakdown order number were analysed to discern those statistical
distributions and their parameters which could model, most accurately,
their failure behaviour. A suite of computer programs which use
randomisation tests, Maximum Likelihood estimators, the Kolmorgorov-
Smirnov D-statistic and Maximum Likelihood Ratios was used for this
purpose. The most relevant results of these analyses are given in
Table 2 where it will be noticed that the 2-parameter Weibull
Distribution was the best model for the gear-heads' times to failure.
Very similar results were obtained for the corresponding 'production

to failure' statistics. The corresponding parametric values for the Weibull Distribution and each breakdown order number are given in Table 3, where it will be noticed that the best estimates of the shape parameter 'B' exceed unity thereby inferring that those machines incurred increasing hazard rates. This conclusion was confirmed using other statistical tests.

TABLE 2

Distributional Tests for Successive Times to Failure for Gearheads

Breakdown Order Number	D-max					Best Distribution	
	Weibull	Exponential	Normal	Log-Normal	Gamma	First	Second
1	0.0803	0.2171	0.1056	0.1443	0.0925	W	G
2	0.0579	0.2516	0.0905	0.0877	0.0728	W	G
3	0.0777	0.2021	0.0784	0.1308	0.1062	W	N
4	0.0821	0.1895	0.1686	0.1115	0.0932	W	G
5	0.135	0.191	0.245	0.091	0.1363	LN	W
6	0.170	0.246	0.283	0.282	0.203	W	G
7	0.144	0.228	0.254	0.149	0.189	W	LN
8	0.291	0.349	0.297	0.309	0.313	W	N

G = Gamma
LN = Log-Normal
N = Normal
W = Weibull

A Pareto Analysis of the repairs showed that many parts and similar remedial actions were taken during every overhaul. By way of an example, Table 4 shows the major parts which were repaired or replaced on average. However, it was discovered that the average number of parts repaired or replaced per overhaul increased, at successive overhaul, as shown by Table 5, thereby confirming that some form of deterioration was taking place.

A COST ANALYSIS OF THE GEARHEAD OVERHAULS

An analysis of the actual costs of each overhaul showed how they escalated with successive repair. This conclusion applied to each gearhead taken singly and machines taken in concert which had the same breakdown order number. Table 6 illustrates the extent of the resultant cost escalation. However, it must be realized that these machines are of different vintages in that some were originally installed in 1974 whereas others have only recently been installed. In effect therefore

TABLE 3

Weibull Parameters for Successive
Times to Failure for Gearheads

Breakdown Order Number (i)	Maximum Likelihood Estimators		
	$\hat{\beta}$	$\hat{\eta}$	$\hat{\sigma}_{\hat{\beta}}$
1	1.558	310	0.020
2	2.062	283	0.036
3	1.887	306	0.034
4	1.573	323	0.027
5	1.37	312	0.025
6	1.46	234	0.043
7	1.29	254	0.049
8	2.325	209	0.541

the costs which they incurred cannot be compared directly because they
involve different purchasing power due to the effects of inflation. To
remedy this deficiency, the costs of every overhaul were deflated,
using the appropriate factors, so that they all correspond to money values
in 1974-terms. These 'real' costs are also shown in Table 6. Although
they do not feature the tremendous excalation witnessed by the actual
costs nevertheless these real costs exhibit approximately a two fold
increase between the first and eighth overhauls.

A detailed correlation analysis which examined the relationship
between the numbers of parts repaired and the 'real' cost for every
overhaul showed that over 90% of the variation of the latter could be
explained by variations in the fomer, thereby confirming that the
escalation in real costs was due to the increase in the number of parts
replaced or repaired per overhaul.

DISCUSSION OF RESULTS

Table 1 indicates that these machines deteriorated substantially over

TABLE 4

Pareto Analysis of the Major Parts Repaired or
Replaced during Gearhead Overhauls

		No of Times Repaired	Percentage of total Repair
	Total For all the 69 Gearheads;	322	100%
	MAJOR PARTS		
A	Bearing Housing Assembly	302	94%
B	Casting Rewelded	290	90%
C	Gearbox Pipes for Haulage End	270	84%
D	Green and Bingham Wet Cutting	261	81%
E	Gearbox Tell Tale Sub-assembly	258	80%
F	Gearbox Du plex and Drive Sub-assembly	174	54%
G	Gearbox David Brown Coupling Assembly	171	53%
H	Gearbox Pipes for LH. and RH. Motor Ends	155	48%
I	Gearbox Emergency Stop Switch Material	148	46%
J	Headshaft and Planet Gear Assembly	132	41%
K	Gearbox Variable Material	122	38%
L	Common Parts	113	35%
M	Gear Box Common Material	97	30%
N	Gearbox Lubricating Pump Assembly	90	28%
O	Boom Trunnion and Cowl Ring Material	86	25%

TABLE 5

Average Number of Major Parts Repaired or
Replaced per Gearhead Overhaul

Breakdown Order Number	1	2	3	4	5	6	7	8
Average Number of Parts	7.57	7.52	8.14	8.53	8.27	8.25	8.42	8.7

TABLE 6

Gearhead Cost Escalation with Successive Overhauls

Overhaul	1	2	3	4	5	6	7	8
Average actual Overhaul Cost	4849	6223	7814	8622	9417	10054	11642	12059
Average real Overhaul Cost in 1974-terms	2299	2620	2990	3244	3361	3521	3909	3933

(Units: £ per overhaul)

time. Albeit that both their mean times to failure and production to failure altered only modestly during the first five breakdowns nevertheless their deterioration was substantial thereafter. The levels of confidence one has that this deterioration was statistically significant are given in Table 7. This result was inferred by the cusum graphs and later confirmed using more precise statistical tests for trend. Furthermore, this result applied equally to tests using 'production' instead of 'time'.

A detailed analysis of the effects of different mine environments produced results very similar to those presented here. In effect therefore there exists a considerable body of evidence to substantiate the conclusion that these machines deteriorated despite efforts to prevent this from happening by the application of comprehensive overhaul policies, procedures and practices.

The fact that the average numbers of parts repaired and/or replaced per overhaul also increased with successive overhaul lends weight to the conclusion that some form of deterioration was taking place. Furthermore the escalation in the real costs of overhauls corroborates this conclusion.

TABLE 7

Confidence Levels that Successive MTTF Deteriorate

Breakdown Order Number (i)	MTTF (i) Days	$\dfrac{\text{MTTF}(i)}{\text{MTTF}(i+1)}$	$\dfrac{\text{MTTF}(i+1)}{\text{MTTF}(i)}$	Confidence Level % $\text{MTTF}(i) > \text{MTTF}(i+1)$ or $\text{MTTF}(i+1) > \text{MTTF}(i)$
1	279	1.112		> 96
2	251		1.084	> 93*
3	272		1.081	> 90*
4	294	1.017		> 70
5	289	1.33		> 90
6	216		1.097	> 65*
7	237	1.145		> 80
8	207			

*denotes that $\text{MTTF}(i+1) > \text{MTTF}(i)$

The key to the reason why this deterioration was occurring is given by
the values of the shape parameters in Table 3. These values indicate
that these machines were failing due to a fairly strong 'wear-out process'
which in turn necessitated the replacement and repair of so many parts
thereby incurring high overhaul costs. What is interesting however is
that despite the comprehensive efforts to maintain these machines to very
high standards nevertheless their reliability performance deteriorated with
successive overhaul. One might surmise that the true reason for this
deterioration was due to a subtle change in maintenance policy or practice
with time. We have investigated this matter but have discovered no
evidence to support this indictment. We have noticed however that the
frames and bodies of those machines do require a lot of maintenance
attention during each overhaul and we believe that frame distortion
and creep may be the fundamental reason why the maintenance fitters
are unable to retain the reliability of these machines at their previous
levels. We have certain evidence to support this belief but it is not
conclusive.

The extent of this deterioration in reliability performance coupled with the escalation in overhaul costs is such that optimum replacement strategies exist for those machines which would dictate their systematic replacement. Unfortunately our research has shown that many of these machines should have been replaced already and that the company is forgoing substantial economic benefits by retaining its machines for so long.

The overall results which have been promulgated here are very similar to those offered by Berg (1977) except that, in his case, the maintenance of machines was nothing like so comprehensive and proficient as the case reported here. It would appear therefore that there are grounds for believing that a deterioration in the life performance of standard proprietary machines is inevitable despite maintenance efforts to the contrary.

CONCLUSIONS

This case study has demonstrated that these machines deteriorate in life performance, as measured by their mean time and production between failure, despite the detailed, proficient and expensive overhauls which they receive. Furthermore it has shown that, in an attempt to stem what appears to be an inevitable deterioration, increasing quantities of resources have been consumed which have led to a significant increase in real costs with successive overhaul. The extent of the deterioration in life performance and the escalation in overhaul costs has been such as to warrant replacing these machines by brand new machines in order to make their ownership more cost effective

REFERENCES

1. Berg, Ø., The terotechnological implications of capital plant breakdowns, unpublished PhD thesis, University of Bradford, 1977.

2. Davies, D.J., An analysis of some failure data, Journal of the American Statistical Association, June 1952, 258, 47. 113-149.

3. Kugel, R.W., Equipment reliability, Russian Engineering Journal, III, 6 3-9

4. Mann, H., Non-parametric test against trend, Econometrica, 1945, 13, 245-259.

5. Proschan, F., and Pyke R., Tests for monotone failure rate, Proceedings of 5th Berkely Symposium, Maths, Statistics and Probability, 1967, III, 293-312 ·

CALCULATION OF RELIABILITY CHARACTERISTICS IN THE CASE OF CORROSION USING THE RCP-BASED APPROACH

J. KNEZEVIC
University of Exeter,
Department of Engineering Science,
Exeter EX4 4QF,
U.K.

ABSTRACT

The classical approach to the calculation of reliability characteristics is based on the probability distribution of the recorded times
when the transition from a state of functioning to a state of failure
occur. The system/component under consideration is accepted as a "black
box" which performs the required function until it fails. The RCP-based
approach to the calculation of reliability characteristics presents an
engineering approach which provides the same results based on information
about "what is going on inside the box". Present study shows how the
reliability characteristics can be determined in cases where uniform
corrosion is the main cause of deterioration using the RCP-based approach.
Two examples illustrate the benefit of this study for designers and
maintenance engineers.

INTRODUCTION

Uniform attack is the most common form of corrosion (1). It is
normally characterised by a chemical or electrochemical reaction which
proceeds uniformly over the entire exposed surface, or over a large area,
as a result of which the metal becomes thinner and eventually fails.
This type of corrosion in engineering practice is monitored and controlled
by corrosion rate expressed in some unit (percent weight loss, milligrams
per square centimeters per day, miles per year, etc.). This could be a
good indicator of the deterioration of material caused by corrosion but
we must not forget that this is only an average value for a whole set of
similar systems/components. Corrosion rates vary between individual
systems due to differences in temperature, concentration and other environmental features. Hence, deterioration of the material caused by uniform
corrosion for similar systems/components is a random process whose fuller
determination requires some other characteristics than corrosion rate.
For example, any make of car corrodes at a different rate in Europe and in
Africa; corrosion rate differs between European countries and within
individual countries as well. Therefore it is meaningless to say that
corrosion rate for one particular make is such and such. Characteristics

which better describe corrosion as a random process are reliability characteristics (R(t), h(t), B-life, etc.).

The classical approach to the calculation of reliability characteristics is based on the probability distribution of the recorded times when the transition from a state of functioning to a state of failure occur. The system/component under consideration is accepted as a "black box" which performs the required function until it fails.

The RCP-based approach (2) to the calculation of reliability characteristics presents an engineering approach which provides the same results based on information about "what is going on inside the box"

The purpose of the present study is to show how the reliability characteristics could be determined in cases where uniform corrosion is the main cause of deterioration using the RCP-based approach.

RUDIMENTS OF RCP-BASED APPROACH TO RELIABILITY

According to this approach reliability characteristics have been determined using relevant condition parameter, RCP. This is a parameter which fully quantifies the condition of a system at every instant of operating time.

As the studies of processes of change of relevant condition parameters show that they are random processes they can only be described through random variable and its probability distribution.

For a system/component to be capable of functioning its relevant condition parameter must lie between initial value, RCPin, and limiting value, RCPlim. Hence, in the case considered, the probability of the value of relevant condition parameter being within the tolerance range is also the probability of the reliable operation of the whole system, thus:

$$R(t) = P(RCP, t < RCPlim) = \int_{RCPin}^{RCPlim} f(RCP,t)dRCP$$

where f(RCP,t) is the probability density function of the relevant condition parameter at the instant of operating time t (see figure 1, where t=tk).

This presents novelty in Reliability Theory because it is based on the real condition of the system/component at some instants of operating time rather than the classical approach which is based only on the instants of an operating time when transitions to a state of failure occur.

The proposed approach is, in general, applicable to all engineering systems, but it is most likely to apply to those which are subjected to the gradual deterioration of material.

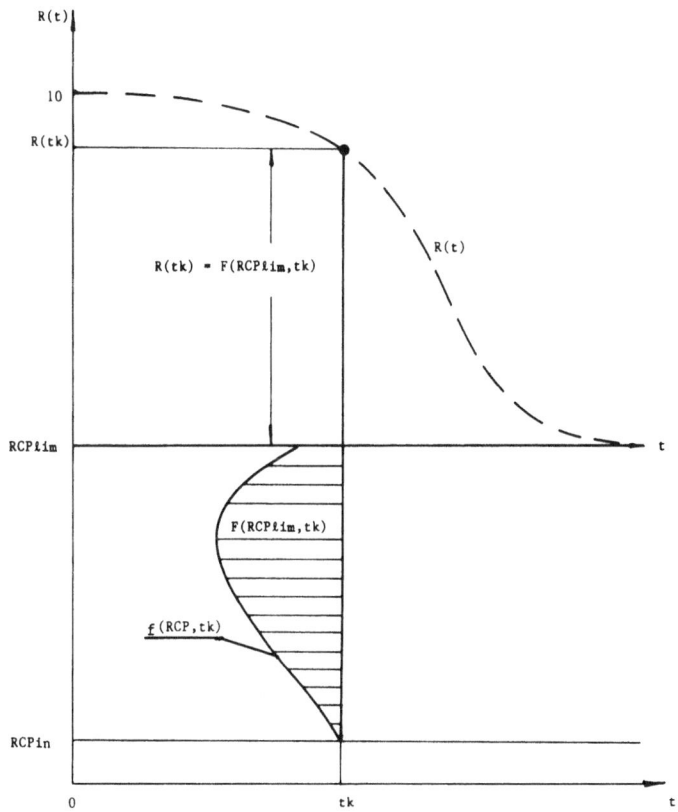

Figure 1.   Relation between reliability function and probability
distribution of relevant condition parameter at instant tk.

APPLICATION OF RCP-BASED APPROACH TO UNIFORM CORROSION

Deterioration of material caused by corrosion is a gradual process
of change in condition of the system/components on which RCP-based
approach can be successfully applied. Since corrosion is a penetrating
action it is not difficult to determine the parameter which describes
the condition of the system at every instant of operation. Thus the
relevant condition parameter in case of uniform corrosion is material
thickness expressed in some units.

For a practical application of RCP-based approach to reliability a
computer program has been developed by the author. In order to use it
a condition matrix must be provided. This represents the set of numerical
values of relevant condition parameter of several identical systems
obtained at several different instants of operating time. The program
calculates reliability characteristics of the system/component based on
measured values of relevant condition parameter. The application of RCP-

based approach to reliability will be demonstrated with two illustrative examples.

EXAMPLE 1   Mild steel high pressure gas transmission pipelines with a diameter of 900mm and wall thickness of 18mm will be considered.   The condition parameter which fully quantifies the condition of the pipelines at every instant of operating time is the reduction of the wall thickness of the pipe expressed in millimetres.

According to the diameter, material and the design pressure of the chosen pipeline the limit value of relevant condition parameter in this example was adopted as 8mm.

Let us assume that the data given in table 1 presents the condition matrix of a hypothetical network obtained during routine annual inspections.

Table 1. The maximum wall thickness reduction (mm)

pipeline	1	2	3	4	5	6	7
1 Year	0.22	0.53	0.68	0.46	0.77	0.62	0.57
2 Year	1.24	1.39	1.35	1.59	1.48	1.44	1.51
3 Year	2.30	2.69	2.59	2.37	2.53	2.42	2.49
4 Year	3.39	3.50	3.73	3.92	3.56	3.59	3.81
5 Year	5.29	4.61	5.16	4.97	4.94	4.88	4.78

The above results were used by the computer program to derive the reliability function which in this example will have the following form

$$R(t) = P\ (RCP, t < 8) = \theta l \theta\ [<8 - 0.5467t^{1,37}\ )\ /\ (0.05t)]$$

where $\theta l \theta$ is the standard Laplace function.

Condition matrix and calculated reliability function are presented in Figure 2. A three-dimensional graphical representation of the process of change in the wall thickness due to uniform corrosion expressed through probability density function of RCP is presented in Figure 3 where f(t) is the probability density function of instants of operating time when the reduction of wall thickness exceeds the limit value of 8mm.

Reliability function obtained can be used for the control of maintenance procedure as shown in (3).

This example clearly illustrates the advantage of the approach presented here over the classical one, i.e. reliability function has been calculated according to the real condition of the system during operating life whereas in the case of the classical approach it is necessary to wait until systems/components fail.

EXAMPLE 2   This example is based on data published by Nissan Motor Company regarding their experience of vehicle corrosion caused by de-icing salts where a decrease in the sheet thickness was used as relevant condition

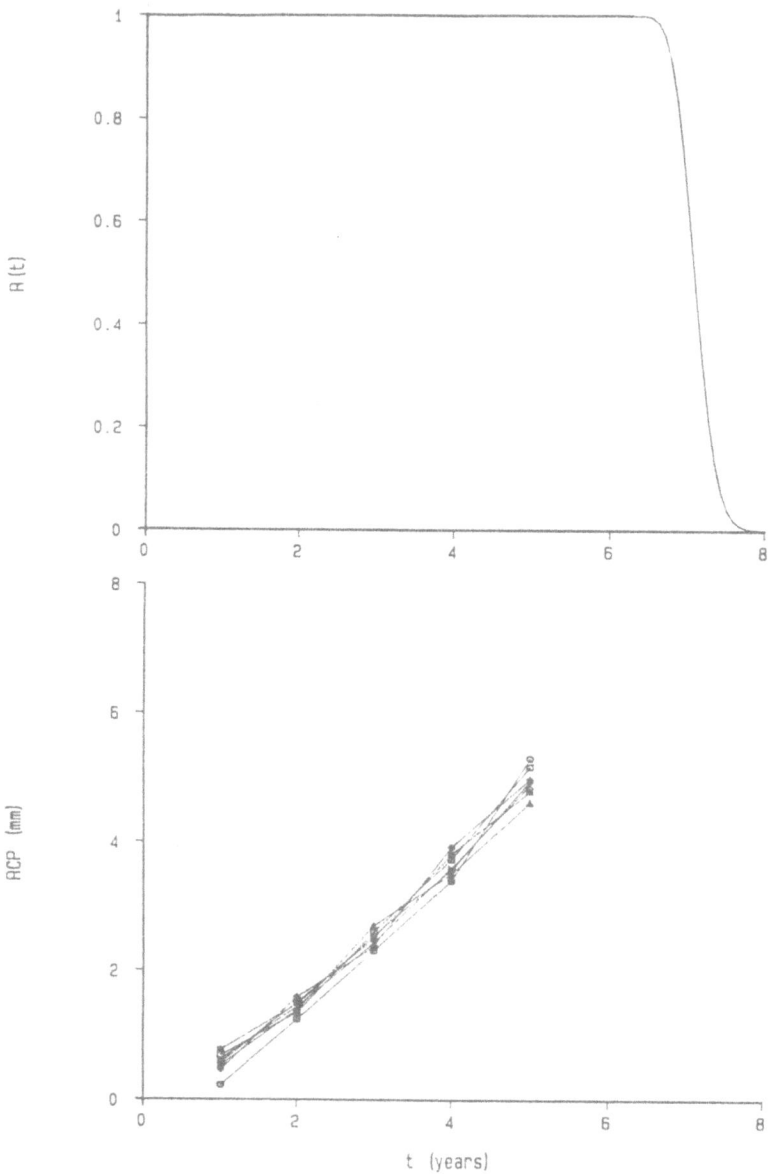

Figure 2.   Graphical illustration of example 1.

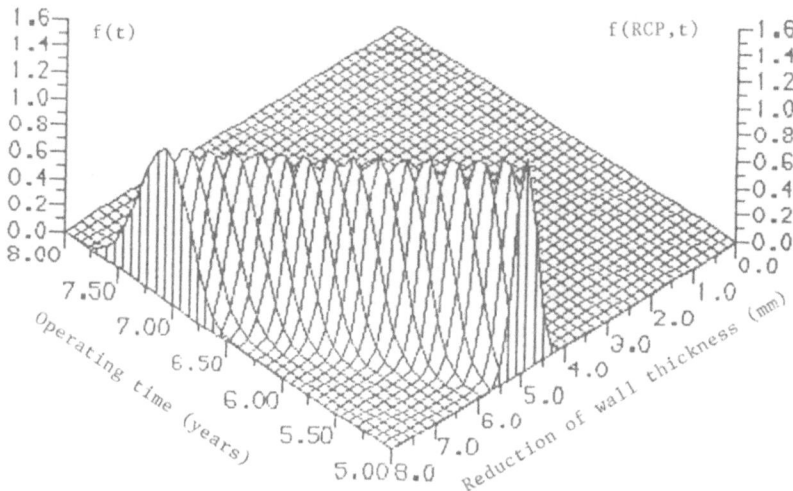

Figure 3. Graphical representation of the reduction of the
wall thickness

parameter (4). The main aim of this example is to show how the RCP-based
approach can be used as an accelerated method for determination of
reliability characteristics. The accelerated laboratory test and proving-
ground test have shown that there is a linear relationship between a
decrease in sheet thickness expressed in mm and the length of a specially
designed corrosion test expressed in hours. According to (4) it is
necessary to perform a test for 1000 hours in order to achieve a 0.6mm
(approx.) decrease in sheet thickness.

By applying the RCP-based approach reliability characteristics can
be predicted much sooner. As a linear relationship exists between the
relevant condition parameter and operating time f(RCP,t) has to be
determined at only two instants of time in order to predict reliability
characteristics. Assuming that the measured values of sheets during
the test, say after 100 and 200 hours obey Weibull distribution with
parameter B1=3,9/A1=0,0075 and B2=3,/A2=0,015 respectively, the predicted
reliability function determined by RCP-approach is presented in figure
4 whose analytical expression has the following form:

$$R(t) = P{<}RCP,t < 0.6) = 1 - \exp\left(-\left(\frac{0.6}{0.00075t}\right)^{3.9}\right)$$

The above example clearly illustrates the suitability of the new
approach for the prediction of reliability characteristics at an early
stage in the life of the system/component in cases where it is known
that a linear relationship exists between reduction of material thickness
and operating time.

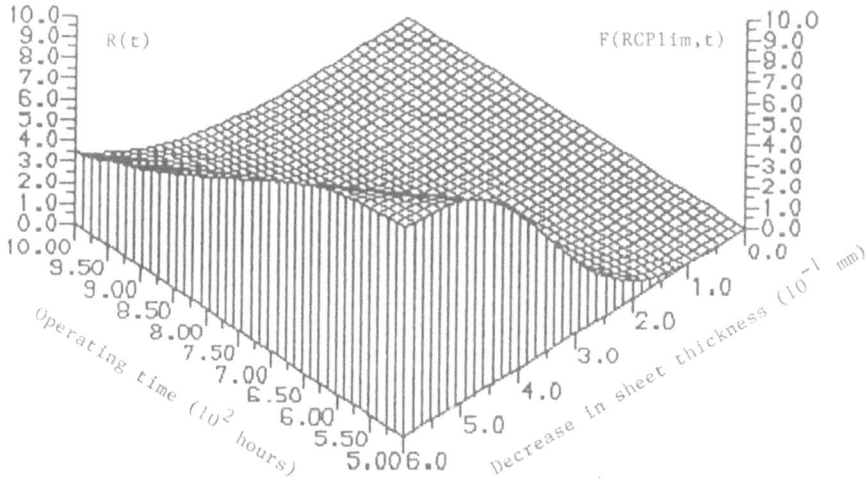

Figure 4. Graphical illustration of example 2.

CONCLUSION

The examples shown clearly demonstrate the potential usefulness of the RCP-based approach to the calculation and prediction of reliability characteristics of engineering systems/components during operating life as well as during accelerated tests in cases where uniform corrosion is the deteriorating mechanism. Hence, it can be used as a tool for designers and maintenance engineers in the "battle" against corrosion.

The main difficulties in the practical application of this approach are the measurement of relevant parameter and the determination of the mechanism of change in condition during operating time. These could limit the application of the approach presented. The former should be solved by organised monitoring of material deterioration and the latter by applying available computer programs or even by experience.

REFERENCES

1. Fontana, M.G., Corrosion Engineering, Third edition, McGraw-Hill Book Company, New York, 1986, pp.556.

2. Knezevic, J., Condition Parameter Based Approach to Calculation of Reliability Characteristics. J. Reliability Engineering, 1987, 19, pp.29-39.

3. Bland, R.J. and Knezevic, J., A Practical Application of a New Method for Condition-Based Maintenance. J. Maintenance Management Inter-

national, 1987, $\underline{7}$, pp.31-35.

4. Kawamoto, I. and Nishii, K., The Development of Evaluation Test Procedures for Vehicle Corrosion Caused by De-icing Salts. In proceedings of Design Conference, Impact of Vehicle Design on Whole Life Costing, I.Mech. E, 1980, pp.93-99.

353

A MODIFIED GO METHODOLOGY FOR SYSTEM AVAILABILITY ASSESSMENT

R. Billinton   M. Patwardhan
Power Systems Research Group
University of Saskatchewan
Saskatoon, Canada S7N 0W0

ABSTRACT

The analysis of systems consisting of series components normally assumes element independence when calculating the system availability. The utilization of this assumption for series systems in which no further component failures can occur once a system failure has occurred provides an underestimation of system availability. The GO methodology which is one of the popular techniques for analyzing system availability, assumes independence of series elements in the determination of system availability. This paper presents a modification of the basic calculation procedure used in the GO methodology which utilizes equivalent components in place of the individual series elements. The effect on the calculated system availability due to the recognition of series component dependence is clearly illustrated by application of both the basic and modified GO methodology to a practical system example.

INTRODUCTION

The GO methodology [2,3,4] is a success-oriented probabilistic system performance analysis technique which can be used to quantify system availability. This technique originally developed by the Kalman Sciences Corporation for the defence industry, has been further developed and its capabilities extended for application in the power industry. The approach employs a straightforward inductive logic to analyse system performance and can be used to evaluate system availability and reliability. This is accomplished by developing a representation known as a GO model or a GO chart which then becomes an integral element in the analysis.

The failure of any component in a series system leads to the failure of the entire system. The simple product rule [1] used to combine series elements assumes statistical independence of the individual components. In some systems, the likelihood of component failure is reduced considerably during those times when the system is not operating and in certain cases, system failure eliminates any possibility of further component failures occurring. The calculated availability for such a system obtained assuming independence of the individual components, is an

354

underestimation and can be considerably in error.

The basic GO methodology combines series elements using the product rule during the process of calculation. The basic GO method, therefore, must be modified in order to recognize series element dependence while analysing a system.

Figure 1 shows a simple example system.

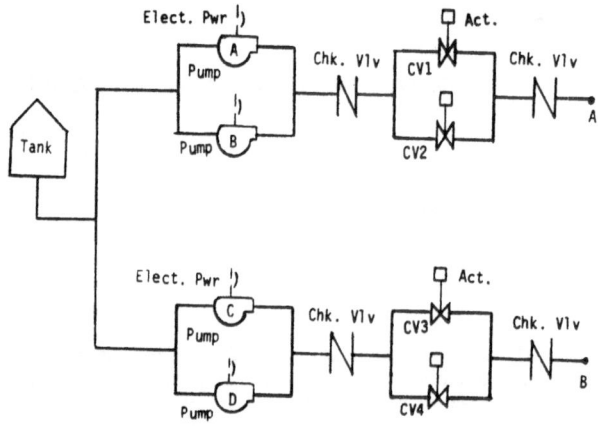

Figure 1.   Example system.

The complete GO chart for the system in Figure 1 is given in Figure 2.   GO operators are the building blocks in the GO method and are used to

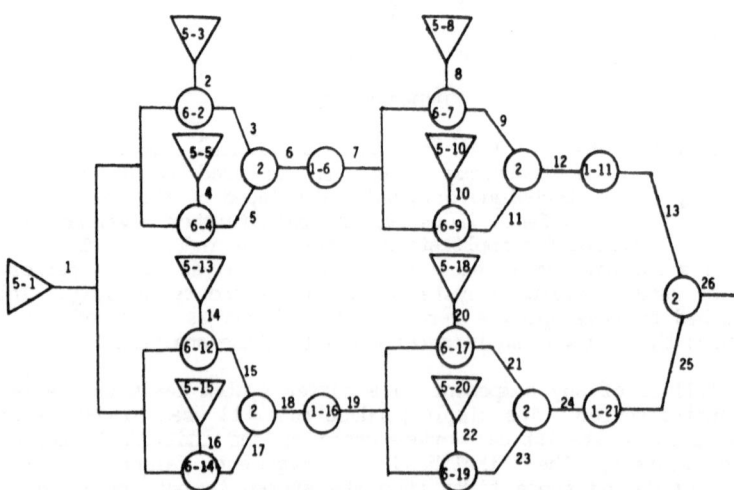

Figure 2.   GO chart for the example system.

represent the function of the system components. The operators consist of symbols which were developed to represent, in mathematical terms, the way that physical components operate, or to represent logic functions for combining the relationship between components. There are seventeen GO operators of which there are three basic types, independent, dependent and logic. The seventeen operators in use are numbered 1-17. The most commonly used are 1, 2, 5, 6, 10 and 11 as most systems can be modeled using these operators alone. These operators are briefly described in the Appendix. More detail is provided in References 2-4. The simple system shown in Figure 1 only requires operator types 1, 2, 5 and 6.

The GO method proceeds as follows. The calculation process is initiated with the main input to the system and the availability of each point in the GO chart is calculated. The availability at point 1 is the availability of the component represented by the operator 5-1. The next operator encountered along the signal path is 6-2. The availability at point 1 is multipled by that of the component represented by 5-3 and that of the component 6-2 itself. The value thus obtained is the availability of signal 3 in the GO chart. The availability at signal point 5 is similarly obtained. These two signals must be combined logically and this is done by the logical operator 2. The operator encountered next is 1-6 and the availability of the component represented by this operator is multipled by the value at signal point 6. Values at signal points 9 and 11 are similarly obtained and are combined logically to obtain a value at signal point 12 which is then combined with the operator 1-11 giving the value at point 13. Signal point availabilities along the other leg are obtained in a similar manner and the final two values at signal points 13 and 25 are combined to give the final value. The value at signal point 26 is the availability index of the system.

The calculation routine follows the signal paths and uses a combinatorial process to link the probabilities of the components encountered in that path before logically combining the outputs according to the specified success requirement. The components encountered along a path are considered to be in series and the combination process assumes independence of the components encountered in the path followed by the signals. This concept is discussed in the following section.

INDEPENDENCE IN SERIES SYSTEMS

Many engineering systems are made up of a combination of series and parallel configurations. Simple reduction techniques exist [1] which can be used to sequentially reduce the overall configuration to a simple block. Series elements are combined using the product rule [1] which states that the reliability value for the individual components making up the series chain are multiplied together to obtain the system reliability. The basic assumption is that the individual components making up the series system are statistically independent. This assumption can lead to an overestimation of the system unavailability.

Two points can be made regarding a series system. The first one is that the failure of any component in the system will result in system failure. The second point is that once the system has failed, further component failures may or may not be possible.

The assumption of statistical independence normally made implies that further component failures are possible after the system has failed. This can be illustrated using a simple two component series system and a Markov model of the system.

Consider the series system shown in Figure 3 made up of two components, 1 and 2, with the following transition rates:

$\lambda_1 = 0.05$ f/hr $\qquad\qquad \mu_1 = 0.4$ r/hr

$\lambda_2 = 0.04$ f/hr $\qquad\qquad \mu_2 = 0.5$ r/hr

Figure 3. Simple two component series system.

The state space diagram for this system is shown in Figure 4.

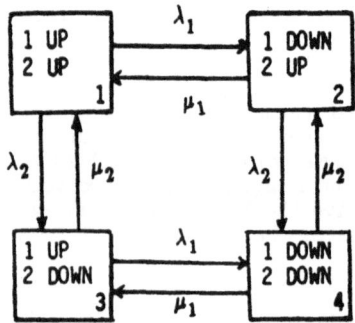

Figure 4. State space diagram for a two unit series system.

The state probabilities can be found using the stochastic probability matrix and the procedure described in Reference 1.

In the model shown in Figure 4,

$P_{up} = P_1 =$ System Availability,

and

$P_{dn} = P_1 + P_2 + P_3 =$ System Unavailability

Using the transition rates given earlier,

$P_1 = 0.82304530$

$P_2 = 0.10288070$

$P_3 = 0.06584360$

$P_4 = 0.00823045$.

The availability of the system is $A = 0.82304530$.

The state space diagram shown in Figure 4 assumes component independence and therefore the system availability could have been obtained quite directly using the product rule.

System Availability = $A_1 * A_2$ (1)

where

A_1 = Availability of Component 1 = 0.4/(0.4 + 0.05)

= 0.88888890

A_2 = Availability of Component 2 = 0.5/(0.5 + 0.04)

= 0.92592590

System Availability = $A_1 * A_2$ = (0.88888890)*(0.92592590)

= 0.82304530 .

The state space diagram shown in Figure 4 contains all the possible states for the system and the transitions between these states. In this simple case, State 1 is the 'up' or the success state and the remaining three states constitute the system 'down' or the failed state.

Failure of any component in a series system results in failure of the system. If in this case, no further component failures can occur once the system has failed, then State 4 in the state space diagram shown in Figure 4 cannot exist. State 1 is the state in which both the components are operable. Failure of Component 1 will send the system into State 2. In the case of the failure of Component 2, the system will transit to State 3. In both cases, the system moves into a failed state. Since the system is no longer operational having entered either one of the two failed states, there is no possibility of further failures occurring and the system cannot travel into State 4 from either State 3 or State 2. Repair of the failed component will put the system back into State 1 and in effect State 4 does not exist. The state space diagram of the two component series system in this case is shown in Figure 5.

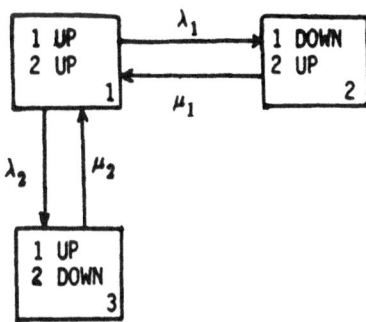

Figure 5.   State space diagram for a two unit system in which when one failure occurs no further failures can occur until the system is restored to service.

Using the transition rates given earlier, the state probabilities are:

P_1 = 0.8298755

P_2 = 0.10373440

P_3 = 0.06639000

The availability of the system P_1, is:

A = 0.82987550 .

This value is higher than the one obtained previously for the case where all the four states were considered. It can be seen that physical attributes of the system such as the dependence considered in this example must be considered in an availability analysis and it may not be correct to simply multiply the individual availabilities to obtain the overall system availability.

THE CONCEPT OF EQUIVALENCE

Billinton and Hosain [5] examined the concept of reliability equivalents and suggested their applicability in a general reliability analysis. A reliability equivalent is a reduced model which retains the pertinent system parameters required for a particular study. A general method for calculating equivalent rates is described in Reference [5].

Figure 4 shows the state space diagram for the simple two component series system described in Figure 3. This state space diagram assumes that all the states shown in the figure can exist.

The two components can be replaced by an equivalent component defined by an equivalent failure rate and an equivalent repair rate, which represents the original components.

The availability of this equivalent component is:

$$A_s = \frac{\mu_s}{(\lambda_s + \mu_s)} \qquad (2)$$

Equation 1 gives the availability of a two independent component system and should be equal to Equation 2 for the equivalent component to represent the two component system.

$$\frac{\mu_1 \mu_2}{(\lambda_1 + \mu_1)(\lambda_2 + \mu_2)} = \frac{\mu_s}{(\lambda_s + \mu_s)} \qquad (3)$$

The transition rate from the system up state for the two equivalent component system is λ_s, and for the two component system is $\lambda_1 + \lambda_2$. Therefore,

$$\lambda_s = \lambda_1 + \lambda_2 \qquad (4)$$

Substituting Equation 4 into Equation 3 and replacing the repair rates, μ_i, by their reciprocals (the average repair times r_i) gives

$$r_s = \frac{1}{\mu_s} = \frac{\lambda_1 r_1 + \lambda_2 r_2 + \lambda_1 \lambda_2 r_1 r_2}{\lambda_s} \ . \tag{5}$$

In some systems, the product $\lambda_i r_i$ is very small and therefore, $\lambda_1 \lambda_2 r_1 r_2 \lll \lambda_1 r_1$. In which case, Equation 5 reduces to

$$r_s = \frac{1}{\mu_s} = \frac{\lambda_1 r_1 + \lambda_2 r_2}{\lambda_s} \ . \tag{6}$$

Equation 6 is an approximate equation in the case of two independent series components but is an exact equation for a system described by the state space diagram shown in Figure 5. The equivalent transition rates therefore recognise that when one component failure occurs, no further failures can occur until the system is restored to service.

The failure and repair rates of a general n-component series system can be determined in a similar manner and are as follows:

$$\lambda_s = \sum_{i=1}^{n} \lambda_i \tag{7}$$

$$r_s = \frac{\sum_{i=1}^{n} \lambda_1 r_i}{\lambda_s} \tag{8}$$

$$\mu_s = \frac{1}{r_s} \ . \tag{9}$$

MODIFICATION OF THE GO METHODOLOGY TO RECOGNIZE DEPENDENCE

The series dependence concept can be extended to the GO methodology to include dependence in the calculation process. Individual components are combined by forming equivalent components to represent the individual components in a series path. This section describes how the basic GO can be modified. Figure 6 shows the GO chart of a simple system which can be used to illustrate the formation of equivalent components.

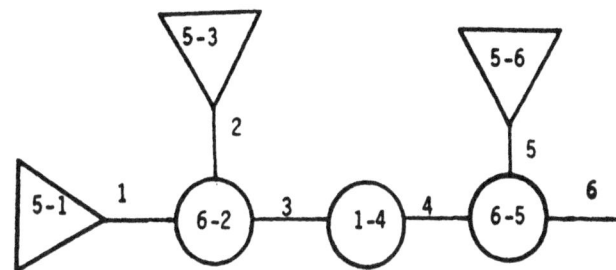

Figure 6. GO chart of a single system to illustrate the formation of equivalent components.

The components along the path can be combined successively using equivalent components to represent the individual components. Starting with component 5-1, the first components encountered are 6-2 and 5-3. Component 6-2 which is a type 6 operator, requires two inputs to produce an output, as described in the Appendix. Therefore, components 5-1, 6-2 and 5-3 are essentially in series. These three components can be combined to form an equivalent component A as shown in Figure 7.

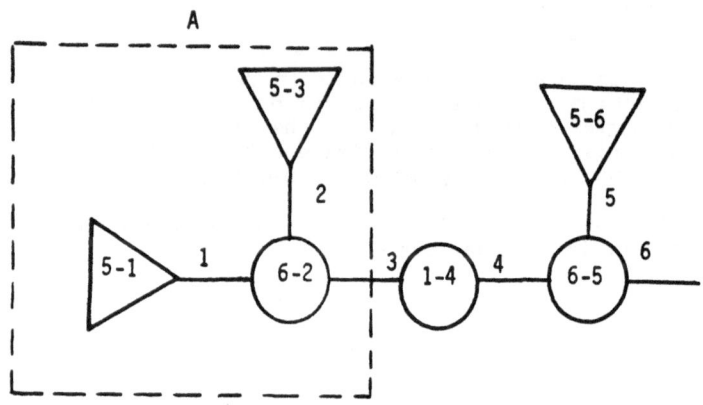

Figure 7. Formation of equivalent component A.

The equivalent failure rate of component A, λ_A can be obtained using Equation 7 and the equivalent repair rate, μ_A using Equation 8. The equivalent availability A_{eq1} at point 3 can be calculated using λ_A and μ_A. The next step involves combining component A with component 1-4. This results in an equivalent component B as shown in Figure 8.

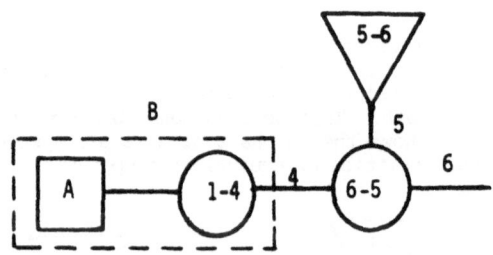

Figure 8. Formation of equivalent component B.

The failure rate λ_A of equivalent component A and the repair rate μ_A of the equivalent component A are combined with the failure and repair rates of component 1-4 using Equation 7 and Equation 8 to produce an equivalent failure rate λ_B and equivalent repair rate μ_B which can be used to calculate the availability A_{eq2} of component. This is the index at signal point 4. Figure 9 shows how the remaining components in the chain can be combined to produce the final equivalent component C.

The failure and repair rates, λ_B and μ_B, of equivalent component B can now be combined with the failure and repair rates of components 5-6 and

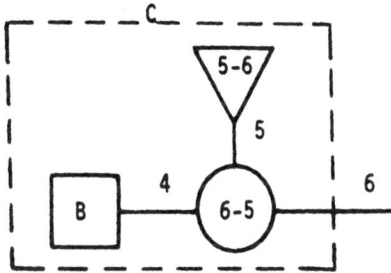

Figure 9. Formation of equivalent component C.

6-5 to produce the equivalent failure rate λ_C and the equivalent repair rate μ_C of equivalent component C. The availability of this component and the availability at signal point 6, A_{eq3} can be calculated using λ_C and μ_C.

Table 1 shows the successive formation of equivalent components which leads to the availability at signal point 6 in the example shown in Figure 6. In this example, operator type 5 elements have failure and repair rates of 3 f/yr and 62 r/yr respectively while operator type 6 elements have values of 2 f/yr and 63 r/yr. Operator type 1 has values of 2 f/yr and 58 r/yr.

Table 1. Formation of equivalent components

Signal Point	Components Involved	Equivalent Component	Eq. Rates f/yr	Eq. Rates r/yr	Availability
1	5-1	–	3	62	0.953846
3	5-1,5-3,6-2	A_{eq}	8	62.247	0.886116
4	A_{eq},1-4	B_{eq}	10	61.3485	0.859843
6	eq,5-6,6-4	C_{eq}	15	61.6937	0.804417

Table 2 shows the difference in the signal point availabilities using the original and the modified method.

Table 2. Comparison of signal point availabilities

	Availability	
Signal Point	Original Method	Modified Method
1	0.95384616	0.95384616
3	0.88182790	0.88611620
4	0.85243300	0.85984300
6	0.78807241	0.80441700

APPLICATIONS AND COMPARISON OF THE TWO METHODS

This section illustrates the application of the two methods in a more practical situation. The impact on the calculated availability using the

modified method is a function of the number of components in the series
system. This situation can be illustrated using a simple series example.

The number of components in a series system affects the calculated
availability index. The two methods were applied to a simple series
system in which the number of components in series was steadily increased.
Identical components with $\lambda = 0.002$ f/hr and $\mu = 0.2$ r/hr were considered.
The availability of a single component is 0.990099 in this case. The
variation in the difference between the calculated values as a function of
the number of series components is shown in Table 3.

Table 3. Difference in calculated availability with the increase in
system size.

No. of Components	Original Method	Modified Method	Difference
2	0.98029608	0.98039222	0.00009614
3	0.97059017	0.97087377	0.00028360
4	0.96098036	0.96153843	0.00055808
5	0.95146573	0.95238096	0.00091523
6	0.94204527	0.94339627	0.00135100
7	0.93271810	0.93457943	0.00186133
8	0.92348325	0.92592591	0.00244266
9	0.91433984	0.91743118	0.00309134
10	0.90528697	0.90909094	0.00380397
11	0.89632374	0.90090090	0.00457716
12	0.88744926	0.89285713	0.00540787
13	0.87866265	0.88495570	0.00629306
14	0.86996299	0.87719297	0.00722998
15	0.86134952	0.86956525	0.00821573
16	0.85282129	0.86206895	0.00924766

The calculated availability of the system is obviously very dependent
on the availability of the individual components making up the system.
The difference between the original and modified method for calculating
system reliability is also very dependent on the individual component
availability. In order to illustrate this, consider a system consisting
of a number of identical components. Table 4 shows the variation in the
availability of a series system containing sixteen components when the
availability of the individual components is varied.

Table 4. Variation of system availability with component availability

| Component Availability | System Availability | | |
	Original	Modified	Difference
0.990099	0.852821	0.862069	0.009248
0.980392	0.728446	0.757575	0.029129
0.970873	0.623167	0.675675	0.052508
0.961538	0.533908	0.609756	0.075848
0.952381	0.458112	0.555561	0.097444

The two methods were applied to a practical system. Figure 10 shows
the component level block diagram of a propulsion subsystem in a ship [6].
The system component data and a description of the system components are

363

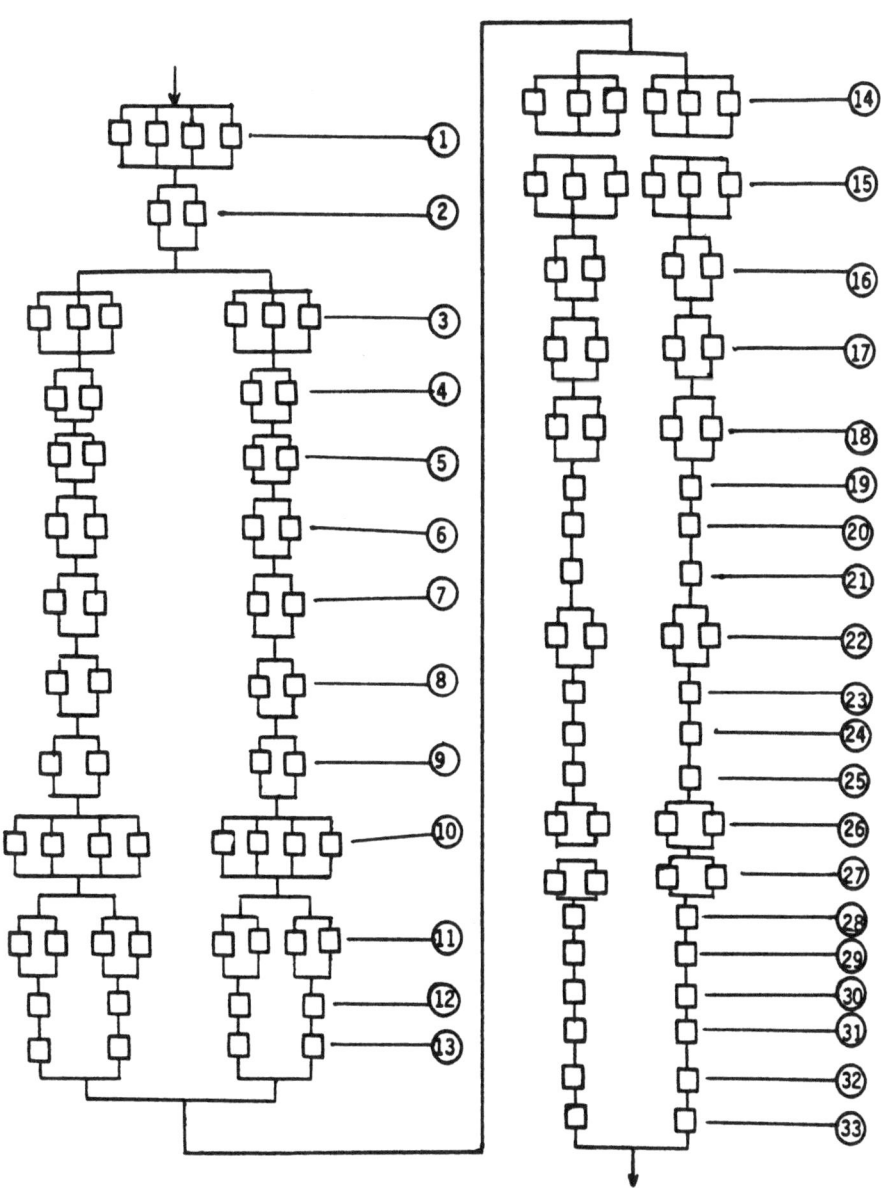

Figure 10.   Propulsion subsystem block diagram.

given in Table 5.

Table 5. Component data of system

Block No.	Item	Failure Rate f/hr	Repair Rate r/hr
1	Drain Pump	0.000038	0.04
2	Deaerating Pump	0.000044	0.05
3	Main Feed Booster Pump	0.000352	0.10
4	Main Feed Pump	0.000239	0.29
5	Main Feed Control	0.000048	0.31
6	Main Feed Lines and Fittings	0.000019	0.36
7	F.O. Service Tanks	0.000057	0.45
8	F.O. Service Pumps and Strainer	0.000045	0.24
9	F.O. Lines and Fittings	0.000019	0.36
10	F.O. Heaters	0.000013	0.17
11	Forced Draft Blower	0.000057	0.25
12	Main Boiler	0.000285	0.06
13	Steam Piping and Fittings	0.000127	0.22
14	L.O. Pumps	0.000041	0.36
15	L.O. Storage Tanks	0.000045	0.32
16	L.O. Setting Tanks and Heaters	0.000006	1.00
17	L.O. Cooler	0.000009	1.00
18	L.O. Strainer	0.000012	0.50
19	L.O. Heater, Pump and Coales	0.000016	0.17
20	L.O. Purifier	0.000066	0.19
21	L.O. Piping and Valves	0.000018	0.22
22	Condensate Pumps	0.000050	0.22
23	Turbine Sot and Controls	0.000072	0.21
24	Gland Fan	0.000044	0.20
25	Main Condenser	0.000062	0.27
26	Circulating Water Lines and Fittings	0.000111	0.36
27	Circulator Pump	0.000121	0.35
28	Main Air Ejector	0.000010	0.26
29	Ejector Condenser	0.000009	0.11
30	Reduction Gear	0.000060	0.45
31	Thrust Block and Bearings	0.000008	0.59
32	Shafts and Bearing Assemblies	0.000061	0.37
33	Propeller	0.000061	0.37

The system was analyzed using both the original and the modified GO methods. Table 6 presents the availability values obtained using the two methods. The first set of readings were obtained using the original data. The unavailabilities of the components were then increased by the factors shown in Table 6.

The results presented in Table 6 show the improvement in the calculated index when the modified method is used. The values obtained when the original data are used are quite comparable as the individual component availabilities are very high. The difference between the values gradually increases as the individual component availability values decrease. Table 6 clearly shows the variation in the system availability with individual component availability. The number of series blocks or components is a major factor affecting system availability. The relatively high system availability in the example shown is due to the high degree of redundancy in this system.

Table 6. Comparison of system availability using the two methods

Factor	Availability		
	Original Method	Modified Method	Difference
1	0.99999458	0.99999869	0.00000411
2	0.99992520	0.99998295	0.00005775
3	0.99962997	0.99992132	0.00029134
4	0.99882048	0.99972624	0.00090575
5	0.99702096	0.99917716	0.00215619
6	0.99345732	0.99779779	0.00434047
7	0.98693055	0.99474436	0.00781381
8	0.97573227	0.98877537	0.01304310
9	0.95769006	0.97835517	0.02066511
10	0.93039894	0.96188325	0.03148431

CONCLUSION

This paper presents a basic but simple modification to the GO methodology. The basic GO approach assumes independence of series elements which can lead to an severe underestimation of the system availability. The modification uses equivalent components to replace the individual components in a series system and incorporates the concept of dependence into the calculation process. The application of the modified package is illustrated using a practical example and a comparative study is presented using the original program and the modified program. The results of the analysis clearly show that the calculated availability is higher when the modified approach is used. The difference in the calculated values is illustrated for a range of individual component availabilities.

The concepts described in this paper and the modified GO methodology are useful tools for analysing maintainable systems. Series systems and subsystems in which no further failures can occur once a failure has occurred can be easily evaluated without resorting to detailed state space modeling and the modified method should be used in those systems for which it is applicable.

REFERENCES

1. Billinton, R. and Allan, R.N., "Reliability Evaluation of Engineering Systems: Concepts and Techniques", Longman, London (England)/Plenum Press, New York, 1983.

2. "GO Methodology", Volume 1: Overview Manual NP-3123-CCM, Volume 1, June 1983.

3. "GO Methodology", Volume 2: Application and Comparison of the GO Methodology and Fault Tree Analysis NP-3123-CCM, Volume 2, June 1983.

4. "GO Methodology", Volume 3: GO Modeling Manual NP-3123-CCM, Volume 3, November 1983.

5. Billinton, R. and Hosain, K.L., "Reliability Equivalents – Basic Concepts", Reliability Engineering, Vol. 5, No. 4, 1983, pp 223-237.

6. Absher Jr, George W., "Reliability Approach to Ship Systems", Proceedings, Symposium on Reliability, Washington D.C., January 1971.

APPENDIX

Commonly Used GO Operators

Type 1 — This operator is defined by two states, success or failure. This is perhaps the most common operator type.

Type 2 — This operator acts like an OR gate and is a logic operator. In a two state success orientated model, the OR function provides a success state output when one or more success states are present at the input.

Type 5 — This operator is used to generate a sequence of operations. Triangular symbols are used to identify the initiator. Generators, batteries and water supplies are typical examples of this type of operator.

Type 6 — This operator is used to represent components that do not produce an output signal unless they have two input signals, a primary signal and a secondary (actuation) signal.

Type 10 — This is a logic operator and is associated with the AND function which implies success when all inputs are successful.

Type 11 — This operator is used to simplify the modeling of the logical combination of m out of n signals.

GEOMETRIC TRANSFORM METHOD FOR NUMERICAL EVALUATION OF INSTANTANEOUS AVAILABILITY

N. Nemat-bakhsh
A.Z. Keller

University of Bradford

ABSTRACT

Evaluation of performance of equipments or entire complexes of components assembled into systems has become increasingly important as a result of rapid technological progress and the need to evaluate reliability of such systems.

In the present paper a method for calculation of availability based on renewal theory is proposed. The availability is expressed by an integral equation and numerical solutions are obtained with the use of the geometric transform. Results are compared with those obtained using a simulation techique. It is concluded that the geometric transform method is of particular value when the failure and repair distributions differ from exponential and one is interested in initial behaviour of availability after either initial start up or major overhaul.

INTRODUCTION

Only the asymptotic value of the availability, A_∞ is largely used in practice [Hamadani, 1980]; however, it is necessary sometimes to know the detailed behaviour of the availability $A(t)$ of a device or a system over a shorter period of time such as immediately after start up or major overhaul. Exponential breakdown and repair distributions are often used to calculate instantaneous availability [9] because of mathematical simplicity. There are several definitions of availability in the literature, the one adopted in this paper is that of reference [1]; developed by ARINC (1964). Here availability is defined as "the probability that the system is operating satisfactorily at any point in time under a stated condition".

AVAILABILITY MODELLING USING RENEWAL PROCESSES

The renewal process has been studied by several authors [Smith 1958, Cox 1960]. Investigators who have used renewal processes for modelling and calculating availability include [Tilman et al 1983, Ikebe and Inagaki 1985]. In the theoretical treatment of availability it is convenient to consider the system under study as being in either a working state S_1 or a failed state S_2. It is further assumed that it is in S_1 and S_2 for periods of times t_1 and t_2 respectively and that t_1 and t_2 are random variables. It is further usually assumed that initially the system is in an operative condition and that on restoration to S_1 from a failed condition, the condition of the equipment is identical with that as at start up. The cycle of being operative and inoperative for random periods is successively repeated and it is assumed that the successive failure and repair times are all independent.

GENERAL FORMULATION

Following [Parzen, 1962] it is convenient to define a probability density function h(t) which is the convolution of the failure p.d.f. f(t), and the repair p.d.f. g(t) by:

$$h(t) = \int_0^t f(t-t') \, g(t') \, dt' \tag{1}$$

Since the availability function is the probability that a unit will be functioning at time t, it can be shown [Parzen, 1962] that the availability function A(t) can be written in the form of the renewal equation:

$$A(t) = R(t) + \int_0^t A(t-t') \, h(t') \, dt' \tag{2}$$

where R(t) is the reliability function given by:

$$R(t) = \int_t^\infty f(t) \, dt$$

It can be further shown by the renewal theorem [Parzen, 1962] that the availability function A(t) tends to a stationary value as t tends to infinity.

Availability Evaluation using Laplace Transforms

Let:
$$\begin{aligned} \bar{A}(s) &= \mathcal{L}A(t) \\ \bar{h}(s) &= \mathcal{L}h(t) \\ \bar{R}(s) &= \mathcal{L}R(t) \end{aligned} \tag{3}$$

where $\bar{A}(s)$, $\bar{h}(s)$ and $\bar{R}(s)$ are the transforms of A(t), h(t) and R(t) respectively. The definition of the Laplace and the corresponding geometric transforms together with their convolution properties required for the present study are given in the Appendix. Following [Parzen, 1962] the transform of equation (2) yields the availability function which can now be written as:

$$\bar{A}(s) = \frac{\bar{R}(s)}{1 - \bar{h}(s)} \tag{4}$$

The above equation is general and applies to all failure and repair distributions that possess Laplace transforms. Unfortunately, the above solution is generally of little practical value as inversion of $\bar{A}(s)$ except for very special cases of limited interest is generally intractable.

In the following section an analogue of (4) using geometric transforms is derived and this unlike (4) can be numerically inverted with little difficulty.

Availability Evaluation Using Geometric Transforms

The geometric transform method provides solutions analogous to those given by using Laplace transforms. With this method only times at the following discrete values are considered [Giblin, 1984], these are the values T_i = id, where i assumes only integer values and d is a suitable time interval. By choosing d sufficiently small any desirable degree of accuracy can be achieved. Defining probabilities of failures occurring in the time interval $[T_{i-1}, T_i]$ as

$$P(i) = R(T_{i-1}) - R(T_i) \tag{5}$$

A similar set of probabilities can be defined for the restoration process by:

$$Q(i) = G(T_{i-1}) - G(T_i) \tag{6}$$

where $G(t) = \int_t^\infty g(t) \, dt$

Furthermore the probability h(i) of a complete cycle being completed in the time interval [0, id] is given by:

$$h(i) = \sum_{j=0}^{i} P(j) \, Q(i-j) \tag{7}$$

which corresponds to (1) for the continuous case. The convolution can be readily evaluated numerically using the definition of the geometric transform given in the Appendix; the following transforms corresponding to (3) can now be defined:

$$
\begin{aligned}
A(z) &= \mathcal{L}_g A(i) \\
h(z) &= \mathcal{L}_g h(i) \\
R(z) &= \mathcal{L}_g R(i)
\end{aligned}
\tag{8}
$$

where $\mathcal{L}_g$ denotes geometric transform and A(i) is the value of A(t) at time T_i. With these definitions a solution corresponding to (4),

$$\bar{A}(z) = \frac{\bar{R}(z)}{1 - \bar{h}(z)} \tag{9}$$

is obtained.

For convenience put:

$$\bar{W}(z) = 1 - \bar{h}(z) \tag{10}$$

(9) can be rewritten as:

$$\bar{A}(z) \, \bar{W}(z) = \bar{R}(z) \tag{11}$$

Using the convolution properties given in Appendix (A.2), (11) can now be
inverted to give:

$$\sum_{j=0}^{i} A(j) \; W(i-j) = R(i)$$

This can be rewritten as:

$$A(i) \; W(0) + \sum_{j=0}^{i-1} A(j) \; W(i-j) = R(i)$$

Since all the $R(i)$ and $W(i)$ can readily be evaluated then starting with
$A(0) = 1$, all values of A_i can successively be evaluated by recursion. A
computer program "AVAIL" was written in Fortran 77 and implemented on the
University of Bradford CDC 180/830 computer to calculate $A(t)$. As input
any of the following distributions can be used for either $f(t)$ or $g(t)$:

Exponential
Gamma
Weibull
Truncated Normal
Bi-modal Weibull
Lognormal
Inverse Weibull

ACCURACY OF METHOD

In order to establish the accuracy of the above method, "AVAIL" was first
tested against known results, where failure and repair times are both
exponentially distributed. The results of the comparison are given in
Table 2. For purposes of comparison the distributions listed in Table 1
were used.

It is seen that the "AVAIL" and exact results agree with an accuracy of
.005%. To establish the validity of the method for other distributions,
values calculated with the method were compared with values of $A(t)$
obtained using simulation. The results for a Weibull shape parameter 2,
scale parameter 10 for failure and a lognormal ($\mu=2$ and variance=.5) for
repair (see reference [2]), are given in Table 3 and Figure 1. 50,000
trials were used to generate the simulated values of $A(t)$ with an accuracy
of .01.

Values of $A(t)$ for three different combinations of failure and repair
distributions were calculated using "AVAIL". The parameters in all cases
were chosen so that the mean failure and repair times and A_∞ were the same.
The results are shown in Figure 2.

A typical "AVAIL" run required a CPU time of 10 seconds. This compares
favourably with simulation runs of comparable accuracy requiring three
minutes.

TABLE 1

DISTRIBUTIONS USED

NAME	PROCESS	FORM	PARAMETER VALUES
EXP.	FAILURE	$\lambda \exp(-\lambda t)$	$\lambda = .7$ day
EXP.	REPAIR	$\mu \exp(-\mu t)$	$\mu = .3$ day
WEIBULL	FAILURE	$\beta \dfrac{t^{\beta-1}}{\alpha^{\beta}} \exp[-(t/\alpha)]^{\beta}$	$\beta = 2 \quad \alpha = 10$
WEIBULL	REPAIR	$\beta \dfrac{t^{\beta-1}}{\alpha^{\beta}} \exp[-(t/\alpha)]^{\beta}$	$\beta = 3 \quad \alpha = 4$
LOGNORMAL	REPAIR	$\dfrac{1}{t\sigma\sqrt{2\Pi}} \exp\left[-\dfrac{(\ln t - \mu)^2}{2\sigma^2}\right]$	$\sigma^2 = .5 \quad \mu = 2$

FOR ALL DISTRIBUTIONS
MTBF = 3.323 days
MTR = 1.427 days

TABLE 2

A(T) COMPUTATION AVAILABILITY COMPARISON

TIME (DAYS)	"AVAIL" RESULTS	EXACT VALUES
.00	1.0000	1.0000
.50	.8818	.8819
1.00	.8101	.8103
1.50	.7667	.7669
2.00	.7403	.7406
2.50	.7244	.7246
3.00	.7147	.7149
3.50	.7088	.7090
4.00	.7053	.7054
4.50	.7031	.7033
5.00	.7018	.7020

$A_{\infty} = .7$
Step Length d = .005 days
(Exponential Failure and Repair Distributions as Table 1)

372

TABLE 3

COMPARISON BETWEEN "AVAIL" AND SIMULATED A(T)
(Failure Weibull, Repairs Lognormal)

TIME (DAYS)	SIMULATED A(T)	"AVAIL" A(T)
.00	1.0000	1.0000
.25	.9860	.9801
.50	.9353	.9483
.75	.9102	.9142
1.00	.8851	.8813
1.25	.8608	.8511
1.50	.8285	.8241
1.75	.8014	.8005
2.00	.7856	.7802
2.25	.7728	.7629
2.50	.7580	.7483
2.75	.7402	.7361
3.00	.7329	.7260
3.25	.7239	.7178
3.50	.7173	.7110
3.75	.7120	.7055
4.00	.7078	.7011
4.25	.7038	.6976
4.50	.7005	.6948
4.75	.6989	.6926
5.00	.6954	.6908

Step Length d = .005 days
$A_\infty = .7$

TABLE 4

VARIATION OF A(T) WITH CHOICE OF DISTRIBUTION

TIME (DAYS)	F: EXP. R: EXP.	F: WEIBULL R: WEIBULL	F: WEIBULL R: LOGNORMAL
.00	1.0000	1.0000	1.0000
.25	.9335	.9826	.9801
.50	.8818	.9517	.9483
.75	.8415	.9140	.9142
1.00	.8101	.8735	.8813
1.25	.7857	.8336	.8511
1.50	.7667	.7975	.8241
1.75	.7519	.7673	.8005
2.00	.7403	.7439	.7802
2.25	.7314	.7269	.7629
2.50	.7244	.7154	.7483
2.75	.7189	.7078	.7361
3.00	.7147	.7030	.7260
3.25	.7114	.7001	.7178
3.50	.7088	.6984	.7110
3.75	.7068	.6975	.7055
4.00	.7053	.6970	.7011
4.25	.7041	.6969	.6976
4.50	.7031	.6970	.6948
4.75	.7024	.6972	.6926
5.00	.7018	.6974	.6908

Step Length d = .005 days

A_∞ = .7

F: FAILURE
R: REPAIR $\Big\}$ Type Distribution

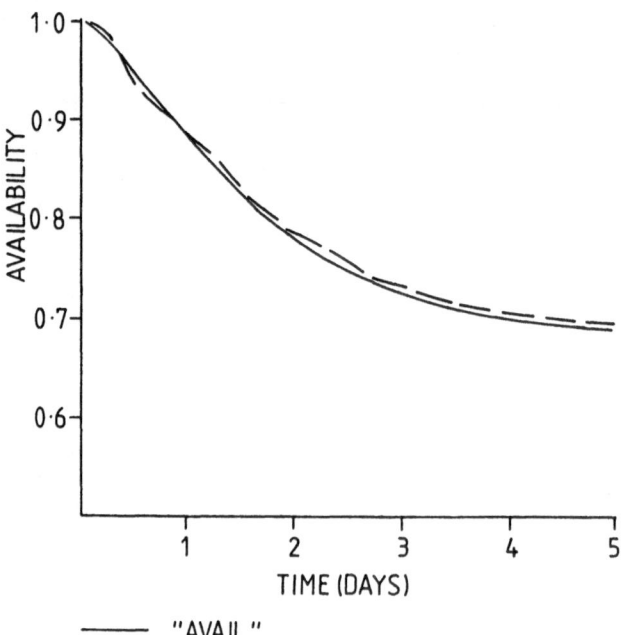

FIG. 1

Comparison of "AVAIL" and simulation values

——— "AVAIL"

– – simulation values

Failure distribution Weibull

Repair distribution Lognormal

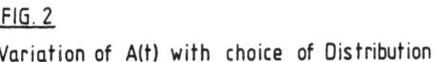

FIG. 2

Variation of A(t) with choice of Distribution

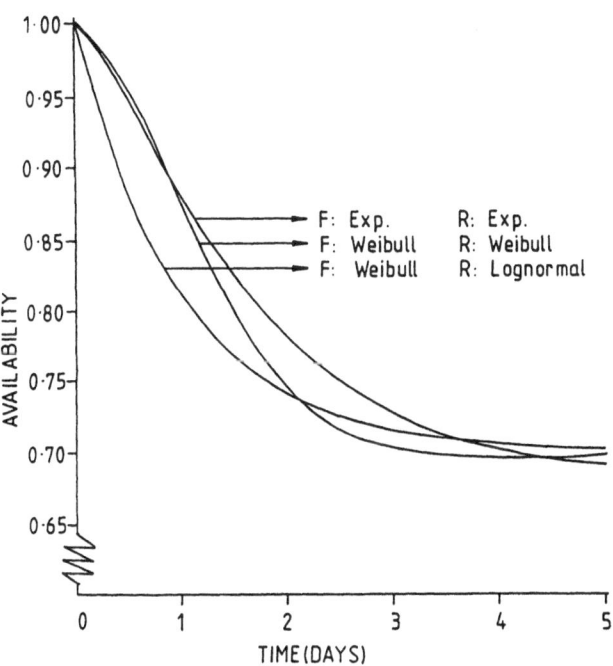

CONCLUSIONS

1. Discretisation using geometric transforms provides an accurate and rapid method for instantaneous availability calculations.

2. The method can be applied generally for all failure and repair distributions.

3. There are significant differences in values of instantaneous availability, dependent upon choice of distribution.

4. Comparison with simulated values suggest that in order to achieve comparable accuracy as obtained with other distributions a smaller step length is required when exponential distributions are used in "AVAIL".

REFERENCES

1. ARINC Research Corp., (1964), "Reliability Engineering", Englewood Cliff.

2. Hamadani, A., (1980), "Availability and Reliability Modelling", PhD Thesis, University of Bradford.

3. Cox, D.R., (1962), "Renewal Theory", Methuen & Co. Ltd.

4. Smith, W.L., (1958), "Renewal Theory and its Ramifications", J. of the Royal Statistical Society, Series B, No.2, pp 243-284.

5. Tilman, F.A., Kuo, W., Nassar, R.F. and Hwang, C.L., (1983), "Numerical Evaluation of Instantaneous Availability", IEEE Trans. on Reliability, Vol.R-32, No.1, pp 119-123.

6. Ikebe, Y. and Inagaki, T., (1985), "Numerical Solution of Integral Equations for Reliability Quantification", IEEE Trans. on Rel., Vol.R-34, No.1, pp 73-77.

7. Parzen, E., (1962), "Stochastic Processes", Holden-Day.

8. Giblin, M.T., (1984), "Derivation of Renewal Functions Using Discretisation", 8th Advances in Rel. Technology Symposium, 25th-27th April, University of Bradford, pp 1-10.

9. De Souza, E., Silva, E. and Richard Gail, H., (1986), "Calculating Cumulative Operational Time Distributions of Repairable Computer Systems", IEEE Trans. on Computers, Vol.C-35, No.4.

APPENDIX

A.1 Laplace Transform

The Laplace transform of f(t) is defined by:

$$\mathcal{L} f(t) = \int_0^\infty e^{-st} f(t) \, dt = \bar{f}(s)$$

Given two functions r(t) and g(t) it can be shown that:

$$\mathcal{L} \int_0^t r(t-t) \, g(t) \, dt = \mathcal{L} r(t) \, \mathcal{L} g(t)$$

$$= \bar{r}(s) \, \bar{g}(s)$$

A.2 Geometric Transform

Given a function h which can only adopt discrete value h(i) (i=1,2,..., ∞) the geometric transform h(z) is defined as:

$$\bar{h}(z) = \sum_{i=0}^\infty h(i) z^i$$

Given two functions h(i) and g(i). It can then be shown that:

$$\sum_{j=0}^i h(i-j) \, g(i) = \sum_{j=0}^i h(j) \, g(j-i)$$

and that the transform of either of these quantities is $\bar{h}(z) \, \bar{g}(z)$.

Hazard Rates and Generalized Beta Distributions

James B. McDonald
   Brigham Young University, Provo
Dale O. Richards
   Brigham Young University, Provo

Key Words -- Generalized beta distributions, Generalized gamma distribution, Distribution family tree; Bathtub shape, Inverted bathtub shape, Decreasing hazard rate, Constant hazard rate, Increasing hazard rate.

Reader Aids --
Purpose: Widen the state of the art
Special math needed for explanations: Probability, statistics
Special math needed to use results: None
Results useful to : Applied statisticians and reliability engineers

Abstract -- This paper considers the behavior of the hazard rates of the Generalized gamma, and beta of the first and second kind distributions. The hazard functions include strictly decreasing, constant, strictly increasing, $\cup$ or $\cap$ shaped hazard rates. By considering the generalized distributions a unified development for such distributions as beta type 1, beta type 2, Burr types 3 and 12, power, Weibull, gamma, Lomax, Fisk, uniform, Rayleigh, and exponential are included as special cases. The results are conveniently summarized in three figures.

The generalized distributions considered in this paper are seen to provide models for all of the different shaped hazard rates mentioned above. This flexibility permits the data to determine the nature of the hazard function without its being inadvertently imposed through the selection of an improper model. For example, the selection of a Weibull distribution permits a decreasing, constant, or increasing hazard rate, but not a $\cup$ or $\cap$ shaped one. The use of the generalized gamma or either of the generalized beta functions considered in section II does permit realization of these additional shapes for the hazard rate.

I. INTRODUCTION

The determination of a preferred distribution for a set of empirical data is not simply a matter of choosing the distribution having the most desirable properties as defined by some criteria for *best*, such as maximum likelihood. This selected distribution must also have certain properties. For instance, as a rule, many distributions used in reliability have the property of increasing failure rate (the failure rate increases with age) at least for larger values of age.

Earlier work of McDonald & Richards [2, 3] presents a rather general procedure for selecting the *best fit* distribution from a family of generalized beta functions. These distributions contain most of the distributions used in statistical data analysis as limiting or special cases; the interrelationships are conveniently depicted with a distribution family tree. Maximum likelihood was the basis of all fitting and comparison in [2, 3]; however, the hazard rates of the distributions in this family tree were not considered. This paper summarizes the characteristics of the hazard rates for this set of distributions. The Distribution Family Tree (figure 1), Key to Notation used in Distribution Tree, table I (pdf's, Cdf's, Moments) and table II (hazard rates), provide continuity, notational consistency and a base from which to work. Both pdf and Cdf are given because the Cdf is not always an easily recognized closed form result. In the density functions involving lal, a minus values merely indicates the random variable to be the reciprocal of the original (ie. 1/y). For further notational clarification, see the appendix. The log t, log cauchy, half-normal, and half-*t* distributions are part of the Distribution Family Tree of [2] but are not included here because they are not serious contenders in reliability studies. By analyzing the hazard rates for the generalized distributions (depicted by GB1, GB2, and GG) in figure 1 the behavior of the hazard rates of the remaining cases can readily be determined.

TABLE I
Density Functions, Distribution Functions and Moments

Model	pdf	Cdf	Moments $E(y^h)$
GB1(y;a,b,p,q)	$\dfrac{\|a\|y^{ap-1}(q\beta^a - y^a)^{q-1}}{(q\beta^a)^{p+q-1}B(p,q)}$	$\dfrac{y^{ap}}{p\beta^{ap}q^p B(p,q)}\,{}_2F_1\left[p, 1-q; p+1; \dfrac{y^a}{q\beta^a}\right]$	$\dfrac{\beta^h q^{h/a} B(p+q, h/a)}{B(p,h/a)}$
GB2(y;a,b,p,q)	$\dfrac{\|a\|y^{ap-1}q^q \beta^{aq}}{B(p,q)(q\beta^a + y^a)^{p+q}}$	$\dfrac{y^{ap}}{(q\beta^a + y^a)^p}\,{}_2F_1\left[p, 1-q; p+1; \dfrac{y^a}{q\beta^a + y^a}\right]$	$\dfrac{\beta^h q^{h/a} B(p+h/a, q-h/a)}{B(p,q)}$
B1(y;b,p,q)	$\dfrac{y^{p-1}(q\beta - y)^{q-1}}{(q\beta)^{p+q-1}B(p,q)}$	$\dfrac{y^p}{p\beta^p q^p B(p,q)}\,{}_2F_1[p, 1-q; p+1; y/q\beta]$	$\dfrac{\beta^h q^h B(p+q, h)}{B(p,h)}$
GG(y;a,β,p)	$\dfrac{\|a\|y^{ap-1}\exp[-(y/\beta)^a]}{\beta^{ap}\Gamma(p)}$	$\dfrac{(y/\beta)^{ap}\exp[-(y/\beta)^a]}{\Gamma(p+1)}\,{}_1F_1\left[1,p+1;(y/\beta)^a\right]$	$\dfrac{b^h\Gamma(p+h/a)}{\Gamma(p)}$
B2(y;p,q)	$\dfrac{y^{p-1}(q\beta)^q}{B(p,q)(q\beta+y)^{p+q}}$	$\dfrac{y^p}{(q\beta + y)^p}\,{}_2F_1\left[p,1-q;p+1;\dfrac{y}{q\beta+y}\right]$	$\dfrac{(q\beta)^h B(p+h, q-h)}{B(p,q)}$
BR12(y;a,b,q) Beta-p	$\dfrac{\|a\|y^{a-1}q^q\beta^{aq}}{B(1,q)(q\beta^a + y^a)^{q+1}}$	$1-[1+(y^a/q\beta^a)]^{-q}$	$\dfrac{\beta^h q^{h/a} B(1+h/a,q-h/a)}{B(1,q)}$
BR3(y;a,b,p) Beta-K, K-3	$\dfrac{\|a\|\beta^a y^{ap-1}}{B(p,1)(\beta^a + y^a)^{p+1}}$	$\left(\dfrac{y^a}{\beta^a + y^a}\right)^p$	$\dfrac{\beta^h B(p+h/a,1-h/a)}{p,1)}$
P(y;b,p)	py^{p-1}/β^p	$(y/\beta)^p$	$\beta^h p/(p+h)$
LN(y;μ,σ^2)	$\dfrac{\exp[-(\ln y-\mu)^2/2\sigma^2]}{y\sigma\sqrt{2\pi}}$	$\dfrac{1}{2}+\left(\dfrac{\ln y-\mu}{\sigma\sqrt{\pi}}\right){}_1F_1\left(\dfrac{1}{2};\dfrac{3}{2};\dfrac{-(\ln y-\mu)^2}{\sigma^2}\right)$	$\exp[\mu h+(h\sigma)^2/2]$
WEI(y;a,β)	$\dfrac{\|a\|y^{a-1}\exp[-(y/\beta)^a]}{\beta^a}$	$1-\exp[-(y/\beta)^a]$	$\beta^h\Gamma(1+h/a)$
GAM(y;β,p)	$\dfrac{y^{p-1}\exp[-(y/\beta)]}{\beta^p\Gamma(p)}$	$\dfrac{(y/\beta)^p\exp[-(y/\beta)]}{\Gamma(p+1)}\,{}_1F_1\left[1,p+1;y/\beta\right]$	$\dfrac{\beta^h\Gamma(p+h)}{\Gamma(p)}$
FIS(y;v,u)	$\dfrac{\Gamma[(v+\mu)/2]\,(v/\mu)^{v/2}}{\Gamma(v/2)\,\Gamma(\mu/2)}\dfrac{y^{v/2-1}}{(1+vy/\mu)^{(v+\mu)/2}}$	$\left(\dfrac{vy}{\mu+vy}\right)^{v/2}\dfrac{{}_2F_1\left(\dfrac{v}{2},1-\dfrac{\mu}{2};\dfrac{v}{2}+1;\dfrac{vy}{v+\mu y}\right)}{(v/2)B(v/2,\mu/2)}$	$\dfrac{\left(\dfrac{v}{\mu}\right)^h B\left(\dfrac{v}{2}+h,\dfrac{\mu}{2}-h\right)}{B\left(\dfrac{v}{2},\dfrac{\mu}{2}\right)}$
L(y;b,q)	$\dfrac{q(q\beta)^q}{(q\beta+y)^{q+1}}$	$1-\left(\dfrac{q\beta}{q\beta+y}\right)^q$	$\dfrac{(q\beta)^h B(1+h, q-h)}{B(1,q)}$
FISK(y;a,b)	$\dfrac{\|a\|\beta^a y^{a-1}}{(\beta^a + y^a)^2}$	$\dfrac{y^a}{\beta^a + y^a}$	$b^h B(p+h/a, 1-h/a)$
U(y;b)	$1/\beta$	y/β	$\beta^h/(h+1)$
R(y;β)	$\{2y[\exp(-(y/\beta)^2)]\}/\beta^2$	$1-\exp[-(y/\beta)^2]$	$\beta^h\Gamma(1+h/2)$
Exp(y;β)	$\{\exp[-(y/\beta)]\}/\beta$	$1-\exp[-(y/\beta)]$	$\beta^h\Gamma(1+h)$
CSQ (y;p)	$\dfrac{y^{p-1}\exp[-(y/2)]}{2^p\Gamma(p)}$	$\dfrac{(y/2)^p\exp[-(y/2)]}{\Gamma(p+1)}\,{}_1F_1[1;p+1;y/2]$	$\dfrac{2^h\Gamma(p+h)}{\Gamma(p)}$

TABLE II
Hazard Rates

Model	Hazard Rates
GB1(y;a,b,p,q)	$$\dfrac{\text{lalpy}^{\text{ap-1}}(q\beta^{a}-y^{a})^{q-1}}{\beta^{a(q-1)}q^{q-1}\left[p(q\beta)^{p}\,d(p,q)-y^{\text{ap}}\,{}_{2}F_{1}\left(p,1-q;\,p+1;\,(y^{a}/q\beta^{a})\right)\right]}$$
GB2(y;a,b,p,q)	$$\dfrac{\text{lalp}(q\beta)^{q}y^{\text{ap-1}}}{(q\beta^{a}+y^{a})^{q}\left[pB(p,q)(q\beta^{a}+y^{a})^{p}-y^{\text{ap}}\,{}_{2}F_{1}\left(p,1-q;\,p+1;\,y^{a}/(q\beta^{a}+y^{a})\right)\right]}$$
B1(y;b,p,q)	$$\dfrac{y^{p-1}[1-(y/\beta q^{1/a})]^{q-1}}{(\beta q^{1/a})^{p}\left[B(p,q)-(y/\beta q^{1/a})^{p}\,{}_{2}F_{1}\left(p,1-q;\,p+1;\,(y/\beta q^{1/a})\right)\right]}$$
GG(y;a,β,p)	$$\dfrac{\text{lalpy}^{\text{ap-1}}\exp[-(y/\beta)^{a}]}{\beta^{\text{ap}}\left[\Gamma(p+1)-(y/\beta)^{\text{ap}}\exp[-(y/\beta)^{a}]\,{}_{1}F_{1}\left(1;\,p+1;\,(y/\beta)^{a}\right)\right]}$$
B2(y;p,q)	$$\dfrac{p(\beta q^{1/a})^{q}y^{p-1}}{(\beta q^{1/a}+y)^{q}\left[p(\beta q^{1/a}+y)^{p}B(p,q)-y^{p}\,{}_{2}F_{1}\left(p;1-q;\,p+1;\,y/(\beta q^{1/a}+y)\right)\right]}$$
BR12(y;a,b,q) Beta-p	$$\dfrac{\text{lalqy}^{a-1}}{\beta^{a}[1+(y^{a}/q\beta^{a})]}$$
BR3(y;a,b,p) Beta-K, K-3	$$\dfrac{\text{lal}\beta^{a}py^{\text{ap-1}}}{(\beta^{a}+y^{a})\left[(\beta^{a}+y^{a})^{p}-y^{\text{ap}}\right]}$$
P(y;b,p)	$$\dfrac{py^{p-1}}{\beta^{p}-y^{p}}$$
LN(y;μ,σ²)	$$\dfrac{\exp[-(\ln y-\mu)^{2}/2\sigma^{2}]}{y\sigma\sqrt{2a}\,(1-\left[\dfrac{1}{2}+\left(\dfrac{\ln y-\mu}{\sigma\sqrt{\pi}}\right){}_{1}F_{1}\left(\dfrac{1}{2};\dfrac{3}{2};-\dfrac{(\ln y-\mu)^{2}}{\sigma^{2}}\right)\right]}$$
WEI(y;a,β)	$$\dfrac{\text{laly}^{a-1}}{\beta^{a}}$$
GAM(y;β,p)	$$\dfrac{py^{p-1}\exp[-(y/\beta)]}{\beta^{p}\left[\Gamma(p+1)-(y/\beta)^{p}\exp[-(y/\beta)]\,{}_{1}F_{1}\left(1;\,p+1;\,y/\beta\right)\right]}$$
FIS	$$\dfrac{2\left(\dfrac{y}{\mu}\right)^{\frac{v}{2}}y^{\frac{v}{2}-1}}{v\left(1+\dfrac{vy}{\mu}\right)^{\frac{\mu}{2}}\left[\left(\dfrac{y}{2}\right)B\left(\dfrac{v}{2},\dfrac{\mu}{2}\right)(\mu+vy)^{\frac{v}{2}}-(vy)^{\frac{v}{2}}\,{}_{2}F_{1}\left(\dfrac{v}{2},1-\dfrac{\mu}{2};\dfrac{v}{2}+1;\dfrac{vy}{\mu+vy}\right)\right]}$$
L(y;b,q)	$$\dfrac{q}{\beta q+y}$$
FISK(y;a,b)	$$\dfrac{\text{laly}^{a-1}}{\beta^{a}+y^{a}}$$
U(y;b)	$$\dfrac{1}{\beta-y}$$
R(y;β)	$2y/\beta^{2}$
Exp(y;β)	$1/\beta$
CSQ	$$\dfrac{py^{p-1}\exp[-(y/\beta)]}{2^{p}\left[\Gamma(p+1)-(y/2)^{2}\exp[-(y/2)]\,{}_{1}F_{1}(1;\,p+1;\,y/2)\right]}$$

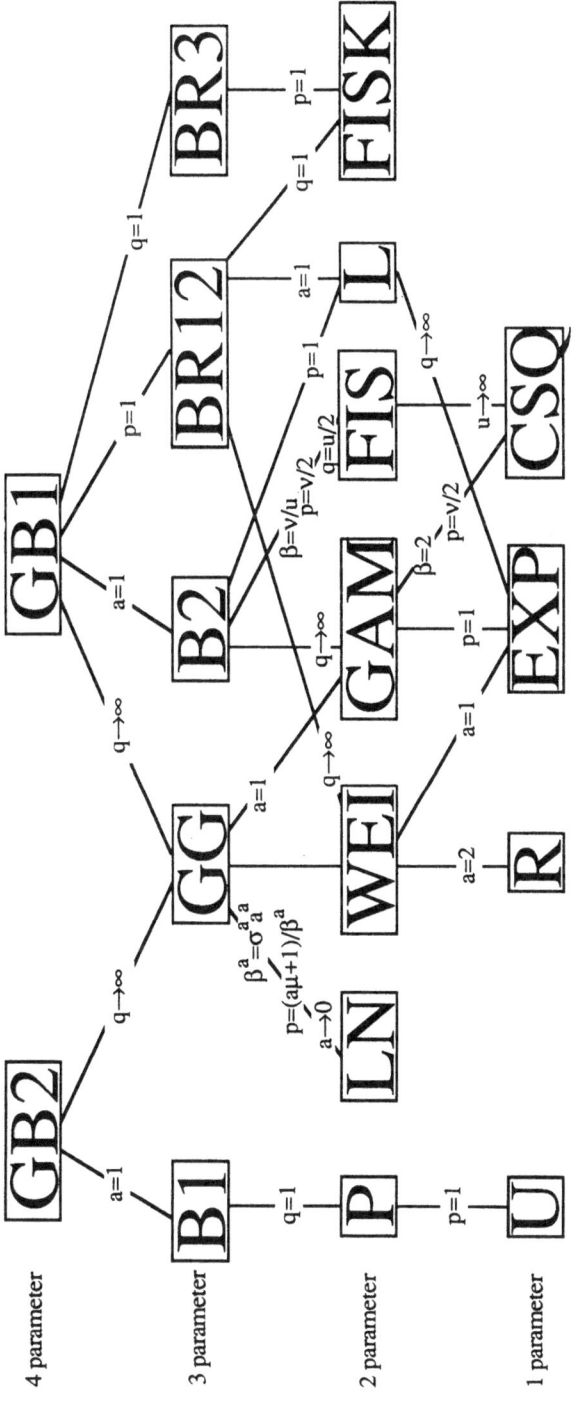

FIGURE 1. Distribution Family Tree

Nomenclature

Symbol	Notation	Distribution
GB1	$GB1(y; a, \beta, p, q)$	Generalized beta type I
GB2	$GB2(y; a, \beta, p, q)$	Generalized beta type II
B1	$B1(y; \beta, p, q)$	Beta of first kind
GG	$GG(y; a, \beta, p)$	Generalized gamma
B2	$B2(y; \beta, p, q)$	Beta of second kind
BR12	$BR12(y; a, \beta, q)$	Burr type 12, Beta-P, or Singh-Maddala
BR3	$BR3(y; a, \beta, p)$	Kappa (k - 3), Beta-K or Burr type 3
P	$P(y; \beta, p)$	Power
LN	$LN(y; \mu, \sigma^2)$	Lognormal
WEI	$WEI(y; a, \beta)$	Wiebull, Rosin-Rammler
GAM	$GAM(y; \beta, p)$	Gamma
FIS	$FIS(y; v, u)$	Fisher-Snedecor F
L	$L(y; \beta, p)$	Lomax
Fisk	$FISK(y; a, \beta)$	Fisk
U	$U(y; \beta)$	Uniform
R	$R(y; \beta)$	Rayleigh
Exp	$Exp(y; \beta)$	Exponential
CSQ	$CSQ(y; v)$	Chi-square

Hazard rate: Ratio of probability density function (pdf) to the survivor function (Sf), both evaluated at a particular value of time, y, and denoted by $hr(y)$.

Notation

$hr(y) = pdf(y)/Sf(y)$

$pdf^2(y) = (pdf(y))^2$

$pdf'(y) = d(pdf(y))/dy$

$\eta(y) \equiv -pdf'(y)/pdf(y)$

$\eta'(y) = d(\eta(y))/dy$

$$\eta_{GG}(y) = \lim_{q \to \infty} \{\eta_{GB2}(y)\}$$

$$= \lim_{q \to \infty} \{\eta_{GB1}(y)\}$$

II. HAZARD RATE CHARACTERISTICS

Table II gives the mathematical expressions for the hazard rates for each of the distributions in the Family Tree (figure 1). Although determining whether the hazard rate is decreasing, constant, or increasing is obvious for a few of the distributions, this is certainly not true for most of them. A few rather general properties can be helpful, however. Consider the following conditions:

A. Since $hr(y) \equiv pdf(y)/Sf(y)$,

$$\frac{d\,hr(y)}{dy} = [Sf(y)pdf'(y) + pdf^2(y)]/Sf^2(y)$$

$$= \frac{pdf'(y)}{Sf(y)} + \frac{pdf^2(y)}{Sf^2(y)} = \frac{pdf'(y)}{Sf(y)} + hr^2(y)$$

Thus for a distribution to have a decreasing, constant, or increasing hazard rate the quantity, $-$ pdf $'(y)$, must

be greater than, equal to, or less than $\left[\dfrac{pdf^2(y)}{Sf(y)} \right]$ respectively. The pdf's having a positive slope is a sufficient

but not a necessary condition for the hazard rate to be increasing. Hence the hazard rate is decreasing, constant, or increasing iff hr(y) is less than, equal to, or greater than:

$$- \frac{dln(pdf(y))}{dy} = \frac{-pdf'(y)}{pdf(y)}$$

B. The Ronald E. Glaser work [1] provides a clearer picture of how the hazard rate is affected by parameter value combinations. He considers parameter restrictions which insure that the hazard rate is strictly decreasing, D, strictly increasing, I, bathtub shaped, $\cup$, and upside down bathtub shaped, $\cap$. He derives the following results:

1. If $\eta'(y) > 0$ for all $y > 0$, then I.   (1)
2. If $\eta'(y) < 0$ for all $y > 0$, then D.   (2)
3. Suppose there exists $y_0 > 0$ such that --
   $\eta'(y) < 0$ for all $y \in (0, y_0)$, $\eta'(y_0) = 0$,
   $\eta'(y) > 0$ for all $y > y_0$ and

   a. If $\lim_{y \to 0} pdf(y) = 0$, then I.

           (3a)

   b. If $\lim_{y \to 0} pdf(y) \to \infty$, then $\cup$.

           (3b)

4. Suppose there exists $y_0 > 0$ such that --
   $\eta'(y) > 0$ for all $y \in (0, y_0)$, $\eta'(y_0) = 0$,
   $\eta'(y) < 0$ for all $y > y_0$ and

   a. If $\lim_{y \to 0} \{pdf(y)\} = 0$, then $\cap$.

           (4a)

   b. If $\lim_{y \to 0} pdf(y) \to \infty$, then D.

           (4b)

The Glaser results provide a useful approach to investigating the behavior of the hazard rates for the GG, GB2, and GB1 distributions. Table III includes expressions for $\eta(y)$ and $\eta'(y)$ for these distributions. The $\eta'(y)$ is linear in y^a for the generalized gamma, and is quadratic in y^a for both GB2 and GB1. The coefficients of the quadratic term in $\eta'(y)$ for the GB1 and GB2 approach zero as q increases without bound.

The properties of $\eta(y)$, eg, signs of the coefficients and the zeros of $\eta'(y)$, can be used to infer patterns of the hazard rate. This consideration is made for the generalized gamma, [1]. The analysis is summarized in table IV. The signs of the coefficients determine whether $\eta(y)$ is $I, D, \cup$, or $\cap$. The corresponding pattern for the hazard rate is a direct corollary of the Glaser result.

The coefficient of the linear term is positive for $a > 1$, and the constant term (ap - 1) is positive above the line $ap = 1$. The two lines in figure 2 divide the nonnegative quadrant into four areas, each corresponding to a case in table IV. Given parameter values for (a, β, p), the shape of the corresponding hazard rate is apparent.

The analysis for the GB2 and GB1 proceeds in the same manner as for the GG but are somewhat more involved, because $\eta'(y)$ involves quadratic terms and the results depend on the number of positive zeros for $\eta'(y)$.

The cases of interest for the GB2 can be enumerated as for the GG distribution. These results are conveniently summarized in figure 3. For positive a, the coefficient of the quadratic term is negative. The constant term in $\eta'(y)$ are positive (negative) for (a, p), which lie above (below) the $ap = 1$ line. Similarly, the coefficient of the linear term is positive or negative for (a, p) which are above or below the line

TABLE III
η(y) and η'(y) for GG, GB1, and GB2 Distributions

	$\eta(y)$	$\eta'(y)$
GG	$\dfrac{1-ap}{y} + \dfrac{ay^{a-1}}{\beta^a}$	$\left\{ y^a\left[a(a-1)\right] + (ap-1)\beta^a \right\} / y^2\beta^a$
GB1	$\dfrac{1-ap}{y} + \dfrac{a(q-1)y^{a-1}}{q\beta^a - y^a}$	$\left\{ y^{2a}[(ap-1) + a(q-1)] \right.$ $+ y^a q\beta^a[2(1-ap) + a(a-1)(q-1)]$ $\left. + (ap-1)q^2\beta^{2a} \right\} / y^2(q\beta^a - y^a)^2$
GB2	$\dfrac{1-ap}{y} + \dfrac{a(p+q)y^{a-1}}{q\beta^a + y^a}$	$\left\{ -y^{2a}(1+aq) + y^a q\beta^a[2(ap-1) + a(a-1)(p+q)] \right.$ $\left. + (ap-1)\beta^{2a}q^2 \right\} / (q\beta^a + y^a)^2 y^2$

TABLE IV
GG Analysis
(This same information is conveniently summarized in figure 2.)

Case	Sign of Linear Term $a(a-1)$	Sign of Constant $(ap-1)$	$\eta(y)$ Shape	Hazard Function
1.	−	−	D	D
2.	−	+	∩	∩
3.	+	+	I	I
4.	+	−	∪	∪

$p = \dfrac{2}{a(a+1)} - \left(\dfrac{a-1}{a+1}\right) q$. This line rotates clockwise through the point $(a, p) = (1, 1)$ for increases in q.

The limiting cases corresponding to a large q can be readily obtained from figure 2 or table IV, because of the limiting relationship between the GG and GB2.

The analysis for GB1 is more involved than for either the GG or GB2 in that the coefficient for the quadratic term depends on the relative magnitude of the parameters. The cases can be listed in a table similar to table IV. Each of the solid lines in figure 4 corresponds to one of the terms in the numerator of $\eta'(y)$. The "lines" corresponding to the linear term --

$p = \dfrac{1}{a} + \dfrac{(a-1)(q-1)}{2}$

rotates clockwise through the point (a, p) = (1, 1) with increases in q. The line corresponding to the quadratic term --

$p = \dfrac{1}{a} + 1 - q$

lies strictly below the line $ap = 1$ for $q > 1$ and move toward the horizontal axis with increases in q.

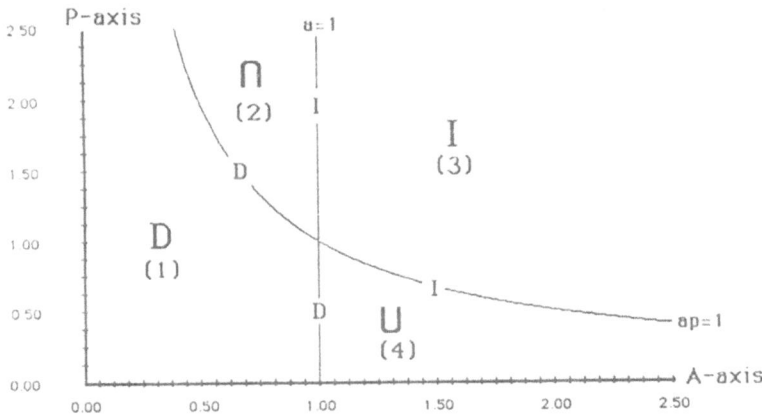

FIG. 2. GG Hazard Rates

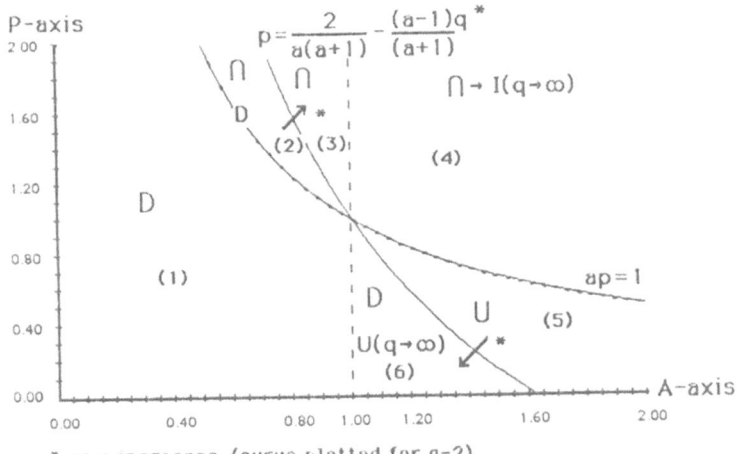

* as q increases (curve plotted for q=2)

FIGURE 3: GB2 Hazard rates

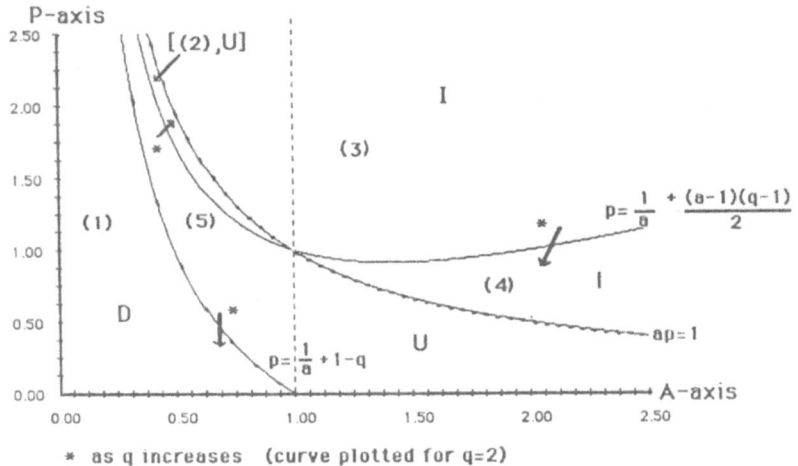

* as q increases (curve plotted for q=2)

FIGURE 4: GB1 Hazard rates

ACKNOWLEDGMENT

We appreciate the comments and suggestions made by the *Editor* and referees of IEEE Transactions on Reliability, and the research assistance of Steven White, Partial research support from the National Science Foundation SES-8509761 is gratefully acknowledged. This paper was originally published in IEEE Transactions on Reliability, Vol. R-36, No. 4 Oct. 1987.

REFERENCES

[1] Ronald E. Glaser, "Bathtub and related failure rate characterizations", *J. American Statistical Association*, vol 75, 1980, pp 667-672.
[2] James B. McDonald, Dale O. Richards, "Model selection: Some generalized distributions", *Communications in Statistics*, vol A16, No 4, 1987, pp 1049-1074.
[3] Dale O. Richards, James B. McDonald, "Some generalized models with application to reliability", *J. Statistical Planning and Inference*, vol 16, No 4, 1987, pp 365-376.

APPENDIX

$$_pF_q\,[a_1,...,\,a_p;\,b_1,...,b_q;\,x] = \sum_{i=0}^{\infty} \frac{(a_1)_i...(a_p)_i}{(b_1)_i...(b_q)_i}\,\frac{x^i}{i!}$$

where $(a_1)_i = (a)(a+1)...(a+i-1)$ for $i > 0$

$\qquad\quad = 1$ for $i = 0$